THE MASTER HANDBOOK OF IC CIRCUITS

BY THOMAS R. POWERS

TAB BOOKS Inc.
Blue Ridge Summit, PA 17294

FIRST EDITION

NINTH PRINTING

Printed in the United States of America

Reproduction or publication of the content in any manner, without express permission of the publisher, is prohibited. No liability is assumed with respect to the use of the information herein.

Copyright © 1982 by TAB BOOKS Inc.

Library of Congress Cataloging in Publication Data

Powers, Thomas R.
　The master handbook of IC circuits.

　Includes index.
　1. Integrated circuits—Handbooks, manuals, etc.
I. Title.
TK7874.P68　　621.381'73　　81-9280
ISBN 0-8306-0028-9　　　　　AACR2
ISBN 0-8306-1370-6 (pbk.)

Questions regarding the content of this book
should be addressed to:

　Reader Inquiry Branch
　Editorial Department
　TAB BOOKS Inc.
　Blue Ridge Summit, PA 17294

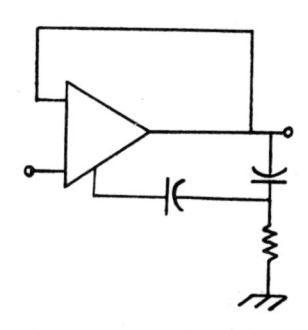

Contents

Part I Linear Integrated Circuits — 11

101	12
101A	25
102	39
106	45
107	48
108	51
110	58
111	64
118	72
119	74
121	75
122	77
124	78
139	88
143	94
144	95
146	97
148	99
155	100
157	104
158	106
170	115
193	117
198	124
311	129
318	131
324	131
339	135
347	139
348	142
349	143

353	144
356	149
357	154
377	154
378	157
379	159
380	162
381	167
381A	171
382	175
384	176
386	178
387	181
387A	186
388	187
389	190
390	193
398	194
531	197
540	200
555	201
706	207
709	208
715	209
725	211
739	214
741	215
747	219
748	222
749	225
760	226
775	227
776	229
777	231
791	235
798	236
799	240
0001	244
0003	245
0004	246
0005	247
0021	248
0022	249
0024	250
0032	252
0033	254
0042	255
0044	257
0062	258
1303	260
1306	261
1422	262
1436	264
1438	265
1439	272
1458	275

1494	276
1495	279
1555	284
1558	286
1776	289
1877	291
1900	293
3301	303
3302	308
3303	310
3380	315
3401	316
3403	319
3405	322
3458	323
3556	326
3900	329
4131	330
4132	332
4136	333
4250	342
4739	346
7514	347
8015	348
8341	350
52107	351
52702	351
72301	352
72709	353
72733	356
72741	357
72770	360
72810	361

Part II Voltage Regulators   362

100	363
104	367
105	368
109	370
117	374
120	380
123	381
145	382
340	385
342	389
0070	391
1468	392
1469	393
1723	394
4194	397
4195	400
7905	402
72309	404
78L05	405

Part III CMOS Integrated Circuits — 406

- 4001 407
- 4011 408
- 4012 413
- 4017 413
- 4023 415
- 4027 416
- 4028 419
- 4046 420
- 4049 420
- 4051 421
- 4066 422
- 4070 423

Part IV TTL/LS Integrated Circuits — 425

- 7400 426
- 7402 432
- 7408 435
- 7474 436
- 7476 439
- 7490 440
- 7492 443
- 74154 446
- 74192 447
- 74193 448
- 74LS13 448
- 74LS30 449
- 74LS85 451
- 74LS132 452
- 74LS151 454
- 74LS161 455
- 74LS175 456
- 74LS196 459
- 74LS373 461
- 74LS374 461

Part V Radio & Television Integrated Circuits — 462

- 172 463
- 175 464
- 703 466
- 720 468
- 732 469
- 1307 470
- 1310 471
- 1349 472
- 1350 473
- 1351 474
- 1352 475
- 1355 476
- 1357 477
- 1358 479
- 1364 480
- 1391 481
- 1496 482
- 1590 487

1733	491
1800	492
1808	493
1889	494
2111	495
3310	496
7511	498

Part VI Special Purpose Devices — 499

567	500
1913	501
2688	502
2907	502
3340	503
3909	504
3911	509
3914	513
5369	516
5600	517
5837	520
8281	521
9400	522
76477	523
76488	526

Index — 530

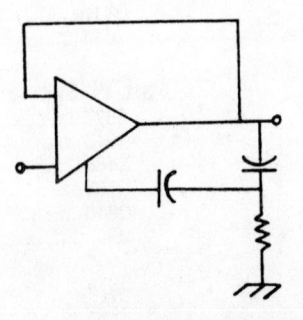

Introduction

No development in electronics has ever had the impact of the integrated circuit. That's a strong statement, considering such prior developments as the vacuum tube and transistor. But you'll probably be convinced of its truth as you scan through the pages of this book!

You're already familiar with such IC advantages as small size and low power requirements. What you may be unaware of is how ICs greatly simplify circuit design. Many ICs have been designed with specific purposes in mind, in contrast to many discrete components. We've gathered together over 900 applications circuits using over 200 popular ICs. The result is a smorgasbord of ideas and designs. All you have to do is connect the appropriate components to each IC and you're in business.

We've used a generic numbering system to refer to the ICs in this book. Various manufacturers of devices will add their company prefix (such as LM for National, SN for Texas Instruments, CA for RCA, etc.) to some of the devices in this book. Operationally, a device identified by a certain number will be identical from manufacturer to manufacturer.

Some of the circuits in this book may have a resistor and/or capacitor or two with no value indicated. The values of such parts should be determined experimentally for best circuit operation. Other circuits may have unlabeled transistors. In such circuits, the transistor can be any general purpose PNP or NPN transistor, as indicated.

With many ICs now available on the surplus market at amazingly low prices, you may have wanted to use more of them in your electronic projects and experiments but have been stymied by the lack of application information. That's the void this book hopes to fill. So go scrounging around in your junk box—odds are you'll find a couple of ICs you thought were useless, but there are some circuits using them in this book!

Part I

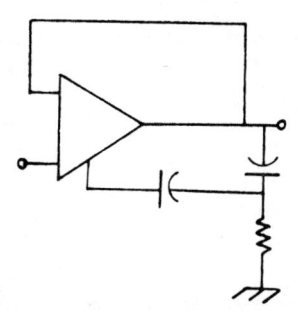

Linear Integrated Circuits

When most of us think of linear integrated circuits, we think of op amps. Certainly, operational amplifiers are the most common linear integrated devices, but they're far from the only ones. Take a look at this section and see for yourself.

101 — GENERAL PURPOSE OPERATIONAL AMPLIFIER

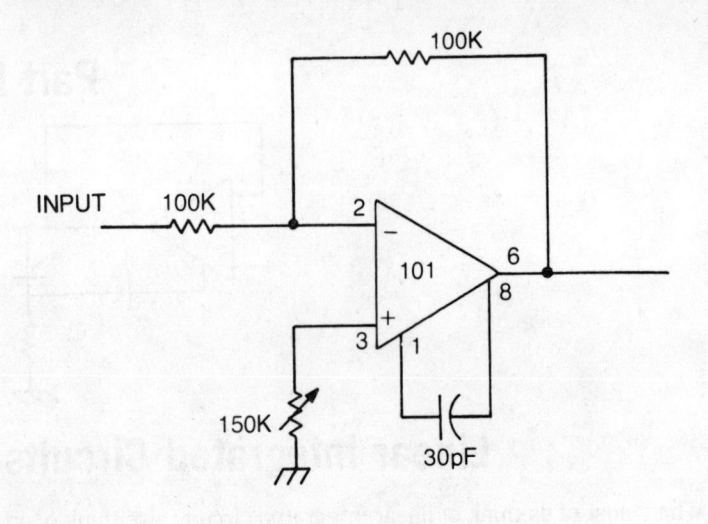

Summing amplifier with bias-current compensation

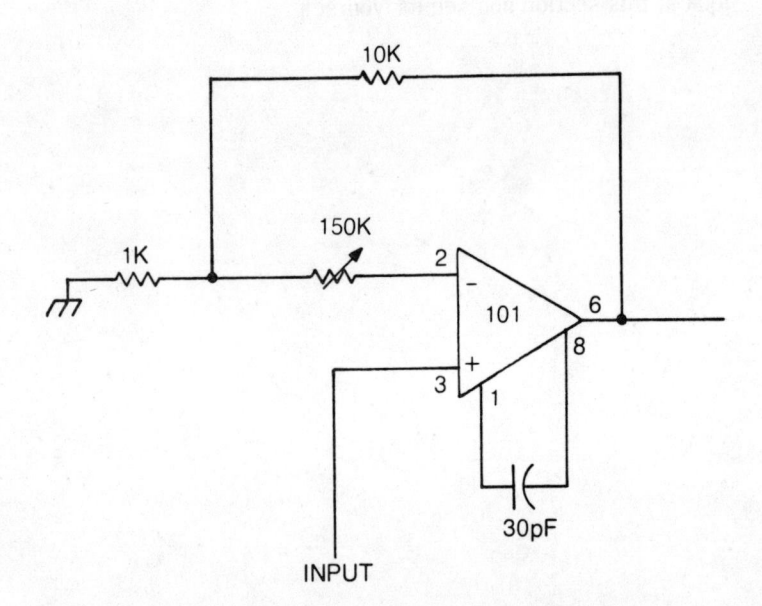

Noninverting amplifier with bias-current compensation

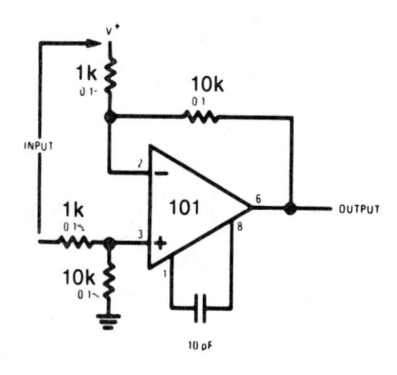

Standard differential amplifier

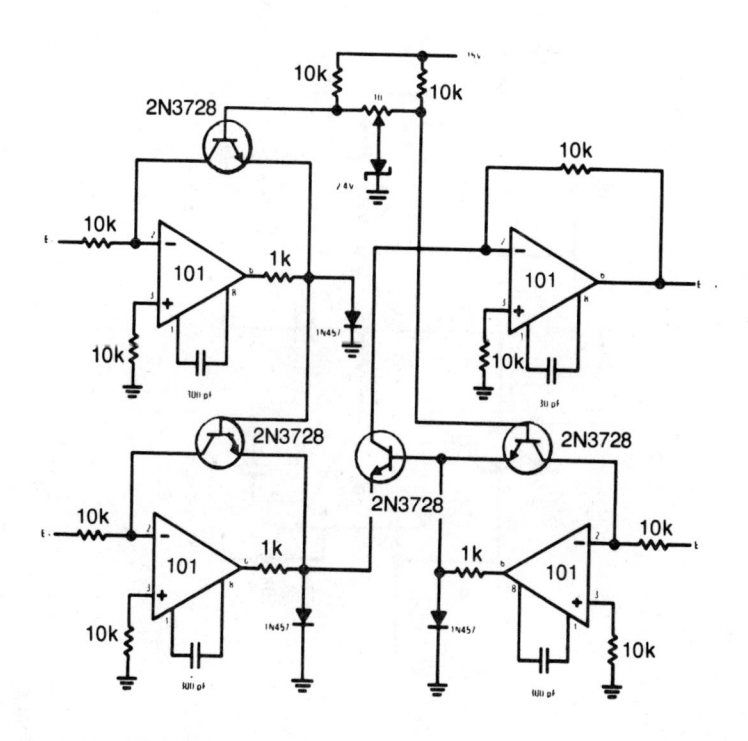

Analog multiplier/divider

Root extractor

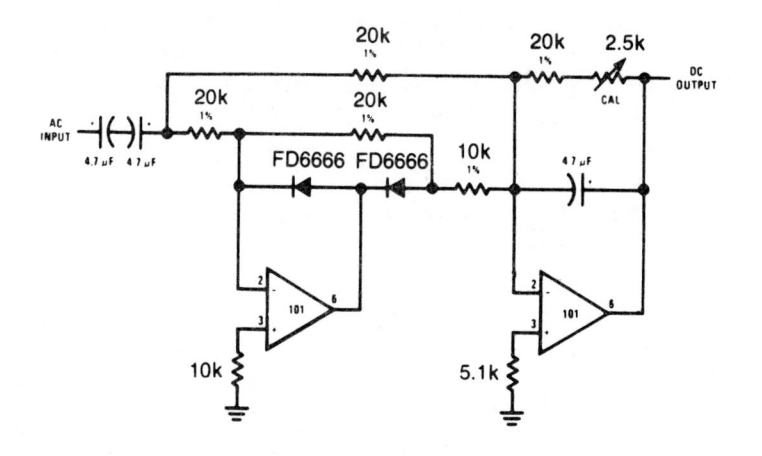

Full-wave rectifier and averaging filter

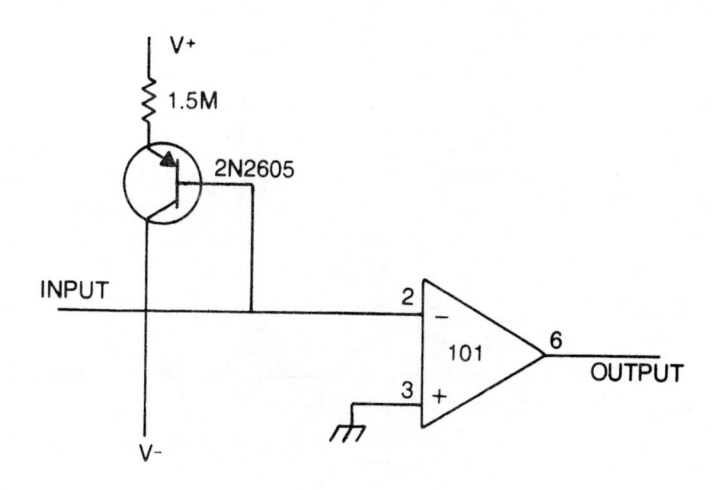

Summing amplifier with bias-current compensation and improved temperature stability

15

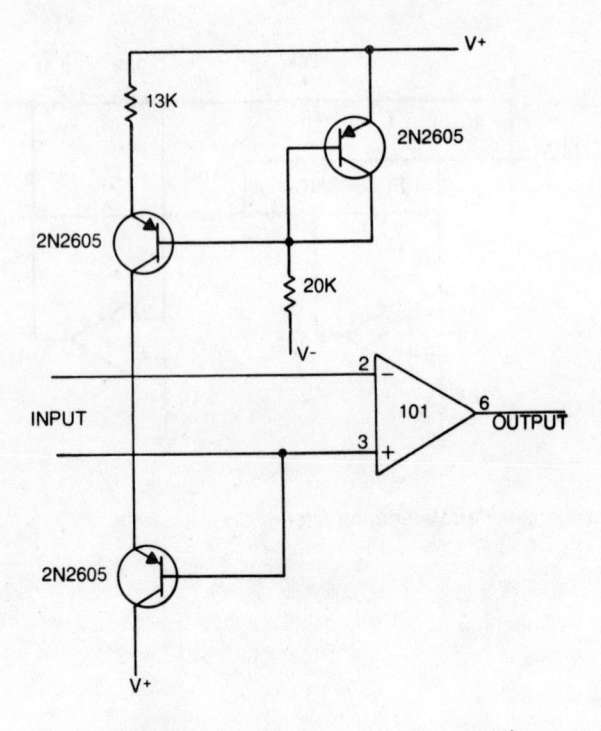

Bias-current compensated noninverting amplifier operating over large common mode range

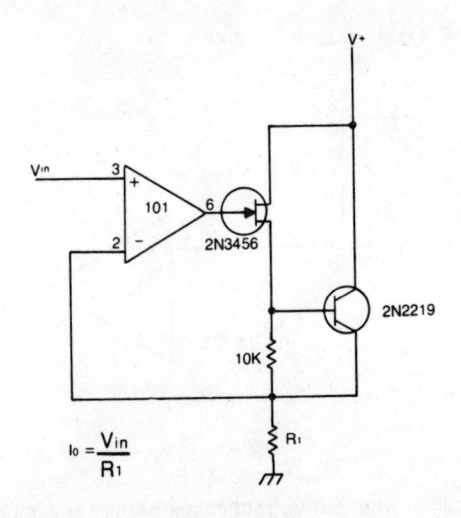

$$I_0 = \frac{V_{in}}{R_1}$$

Precision current sink

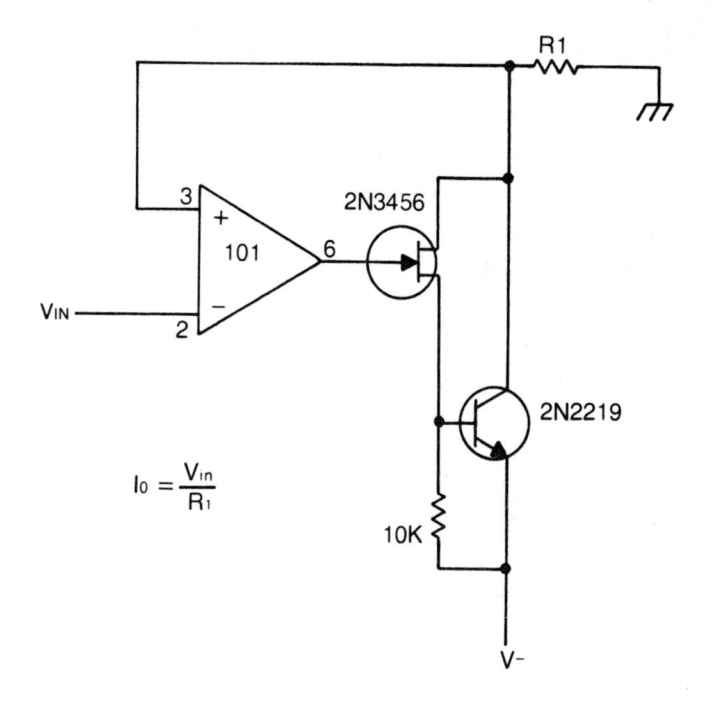

$$I_0 = \frac{V_{in}}{R_1}$$

Precision current source

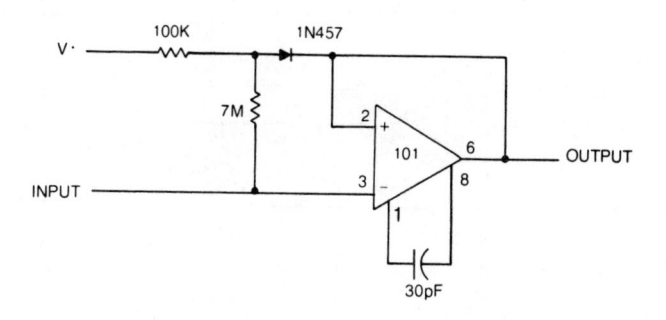

Voltage follower with bias-current compensation

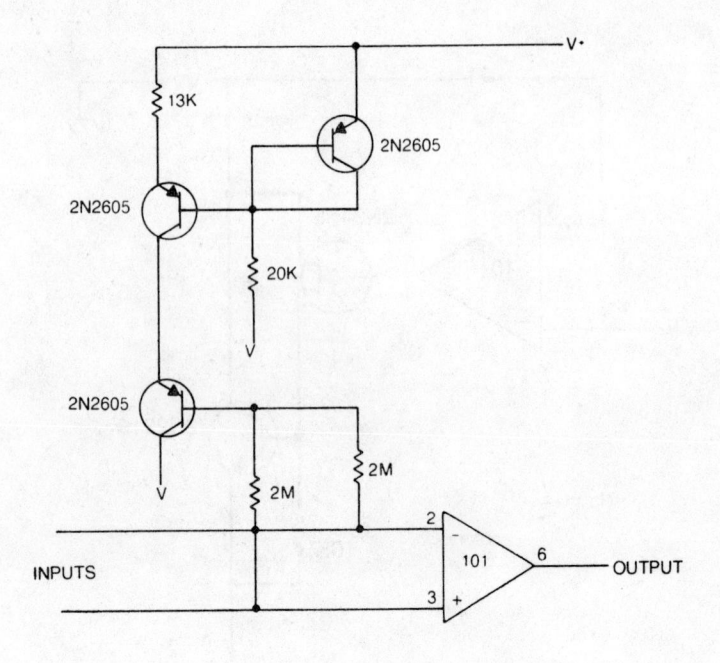

Summing amplifier with bias-current compensation for differential inputs

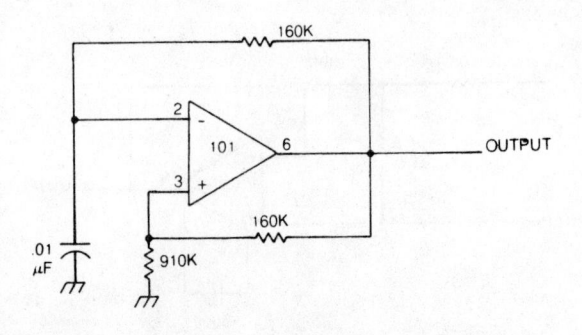

Low frequency free running multivibrator

18

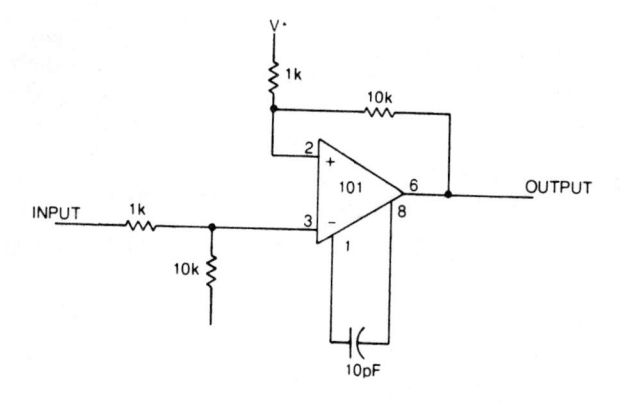

Level shifting differential amplifier

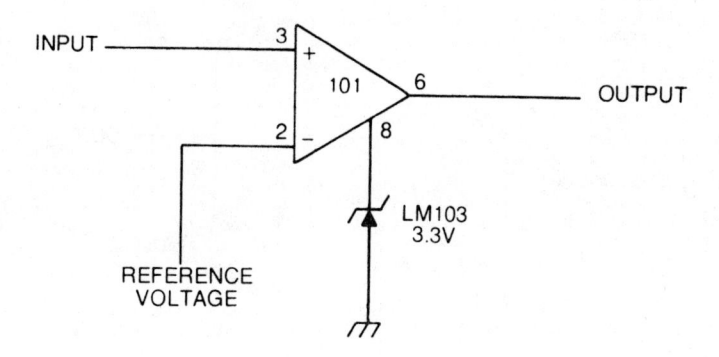

Voltage comparator for driving TTL ICs

19

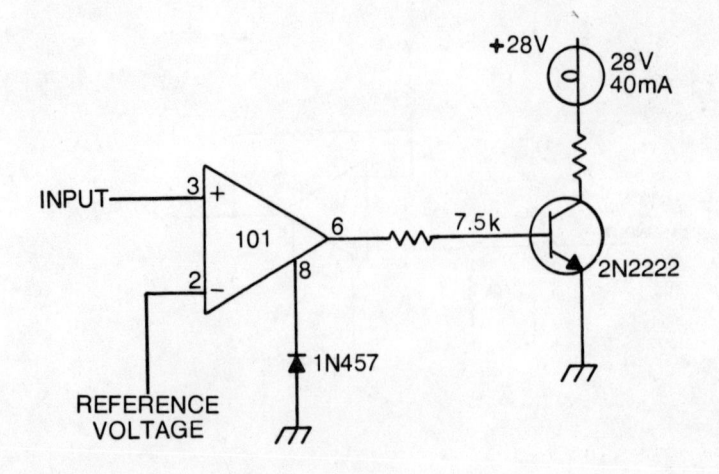

Voltage comparator and lamp driver

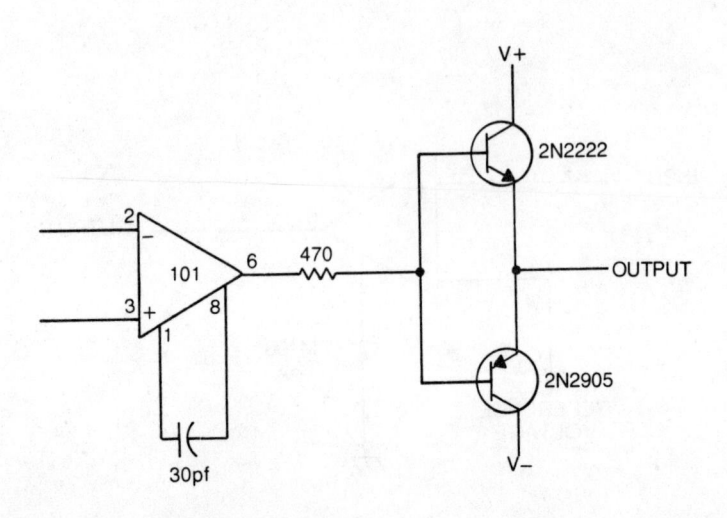

High current output buffer

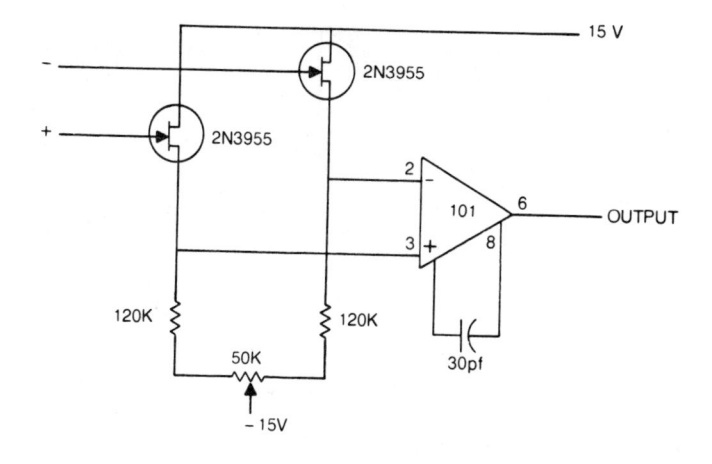

Summing amplifier with FET source followers

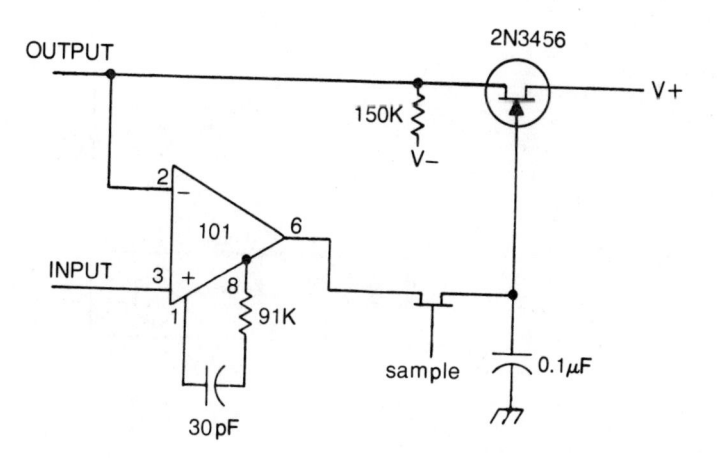

Low drift sample and hold

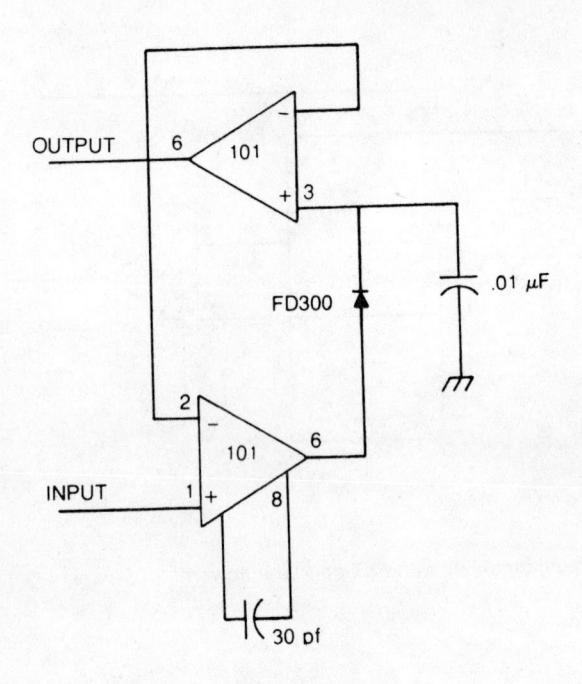

Positive peak detector with buffered output

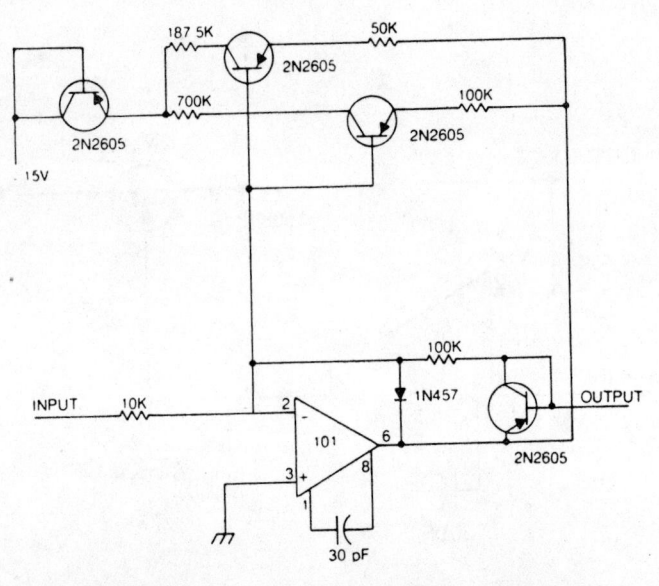

Nonlinear amplifier with temperature compensated breakpoints

22

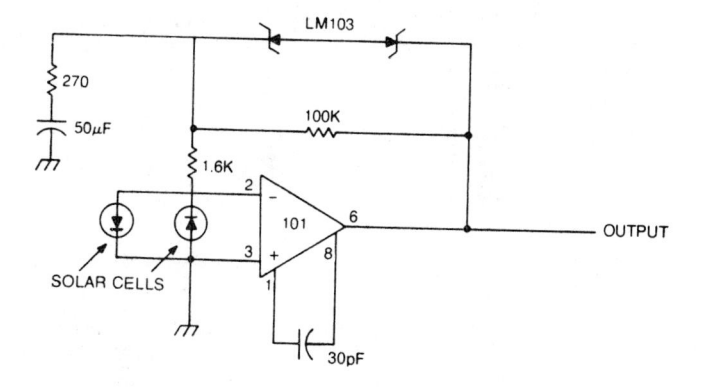

Saturating servo control preamplifier with rate feedback and solar cell sensors

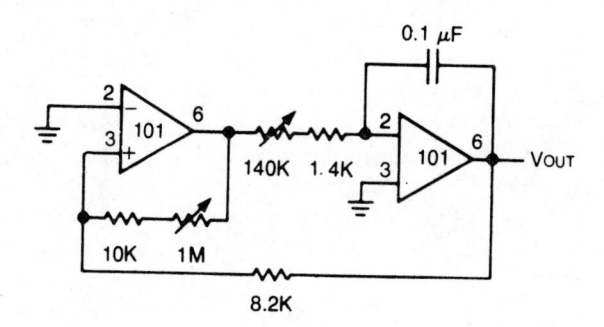

Triangular-wave generator

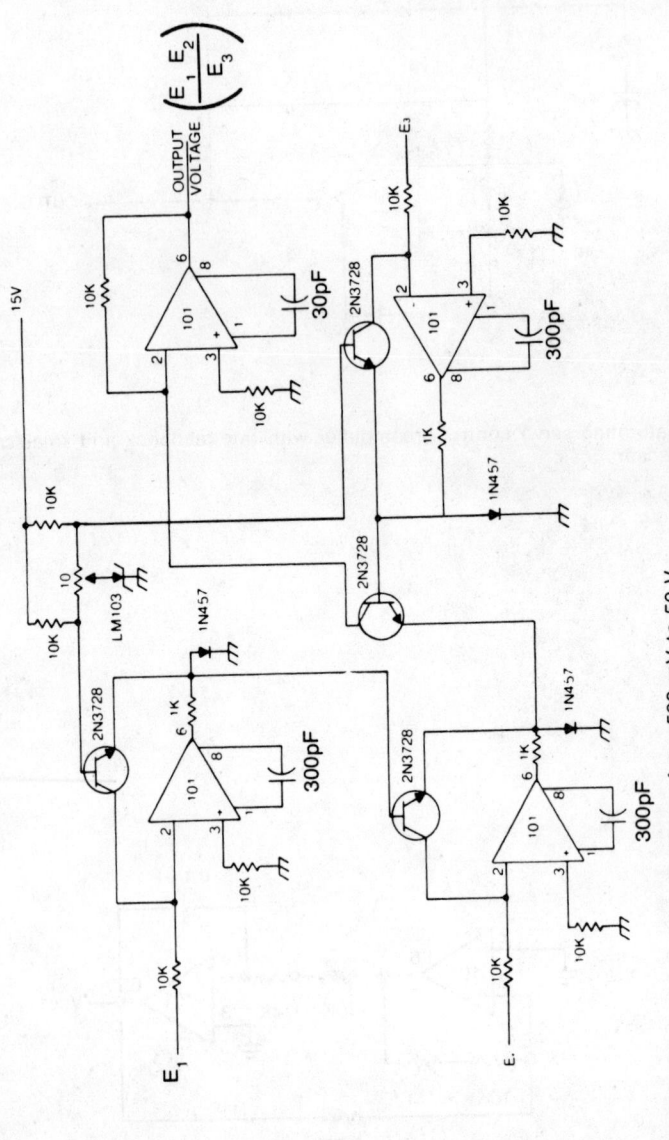

Analog multiplier/divider for input voltages from 500 mV to 50 V

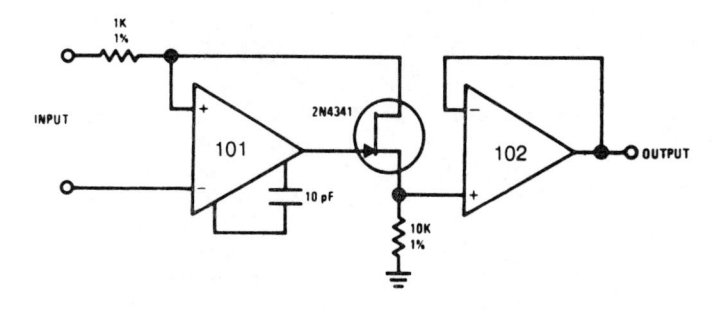

Level-shifting isolation amplifier

101A—LOW INPUT CURRENT GENERAL PURPOSE OPERATIONAL AMPLIFIER

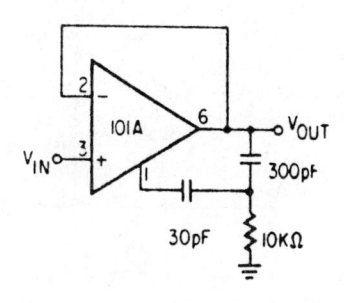

Fast voltage follower

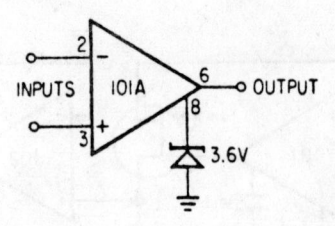

Voltage comparator for driving DTL or TTL integrated circuits

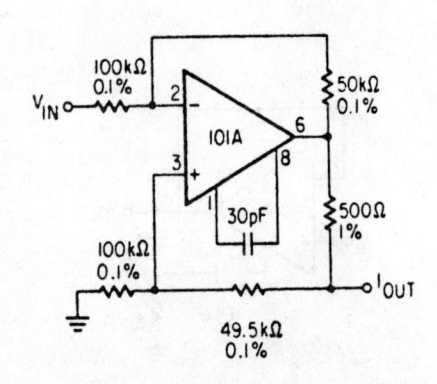

Bilateral current source

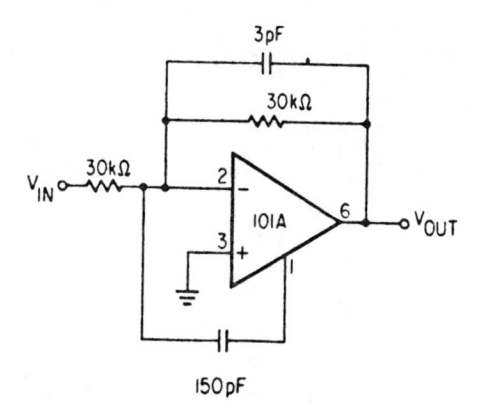

Fast summing amplifier

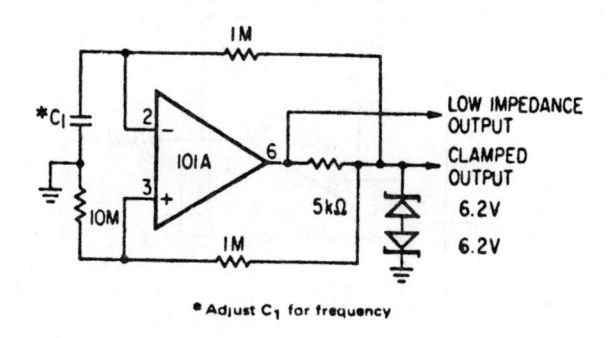

Low frequency square wave generator

V_{OUT} = 4.6V FOR $V_L < V_{IN} < V_U$
V_{OUT} = 0V FOR $V_{IN} < V_L$ OR $V_{IN} < V_U$

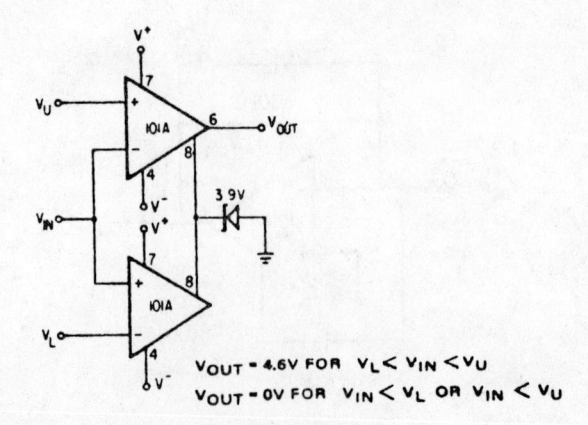

Double ended limit detector

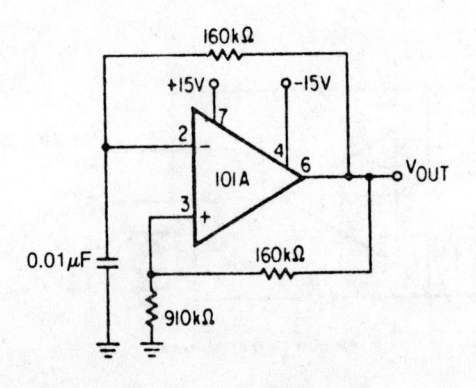

Free-running multivibrator

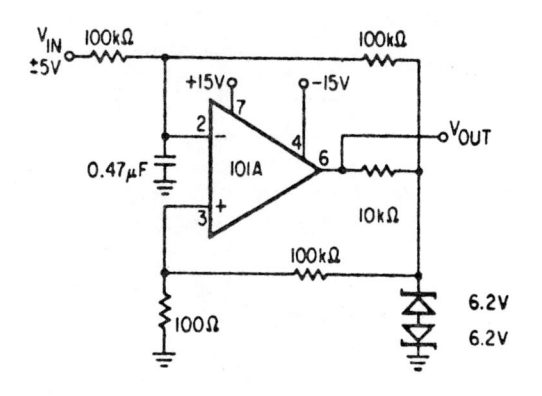

Pulse width modulator

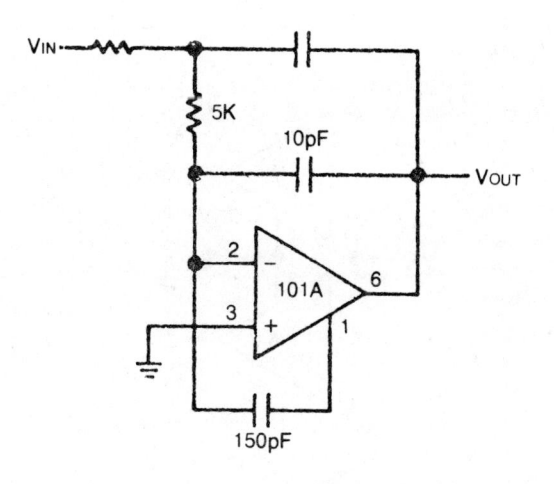

Fast integrator

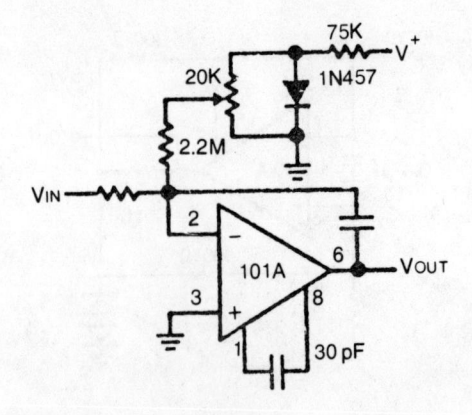

Integrator with bias current compensation

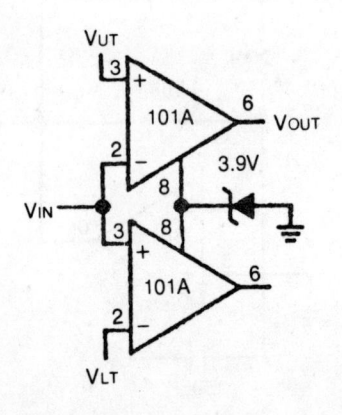

Double-ended limit detector

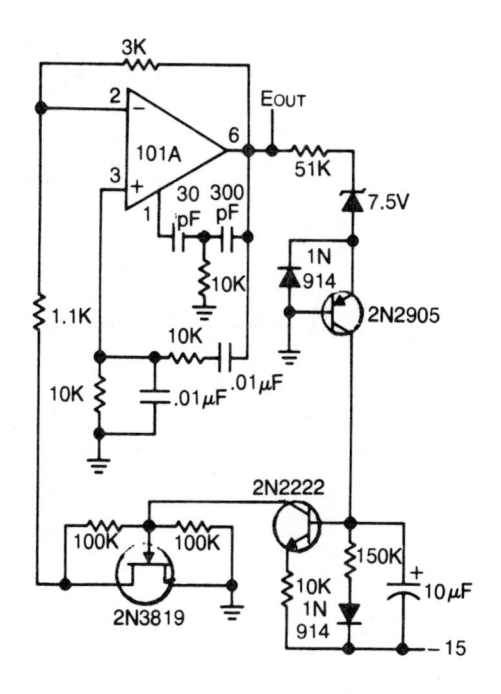

Wien bridge oscillator with FET amplitude stabilization

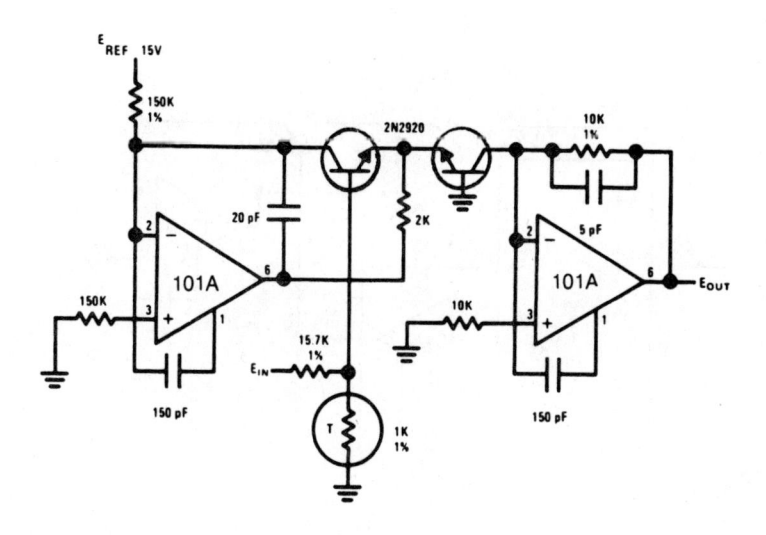

Anti-log generator

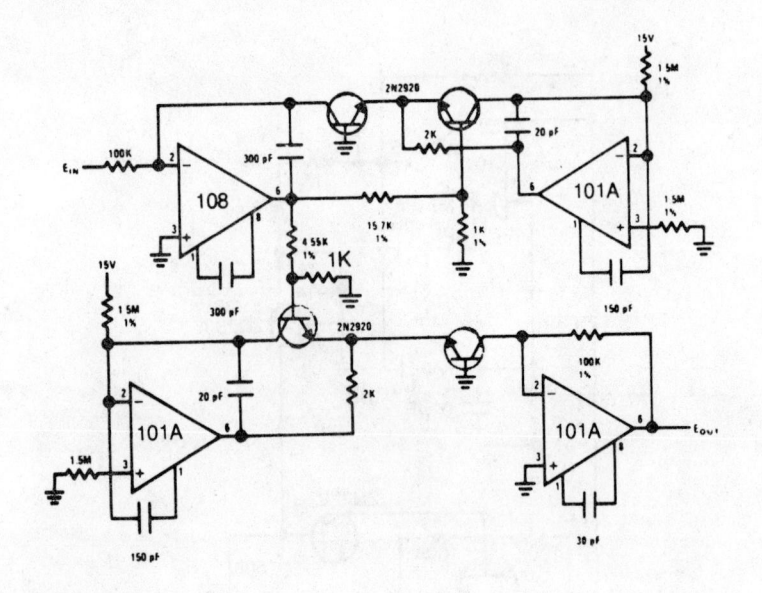

Cube generator

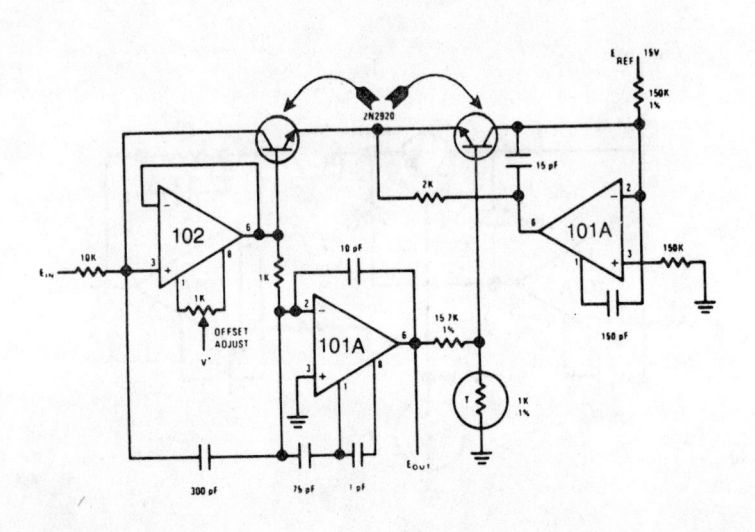

Fast log generator

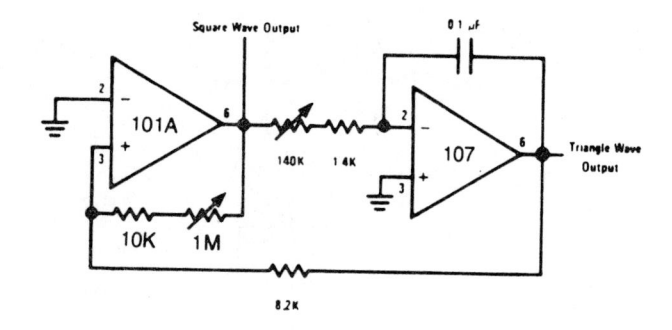

Function generator

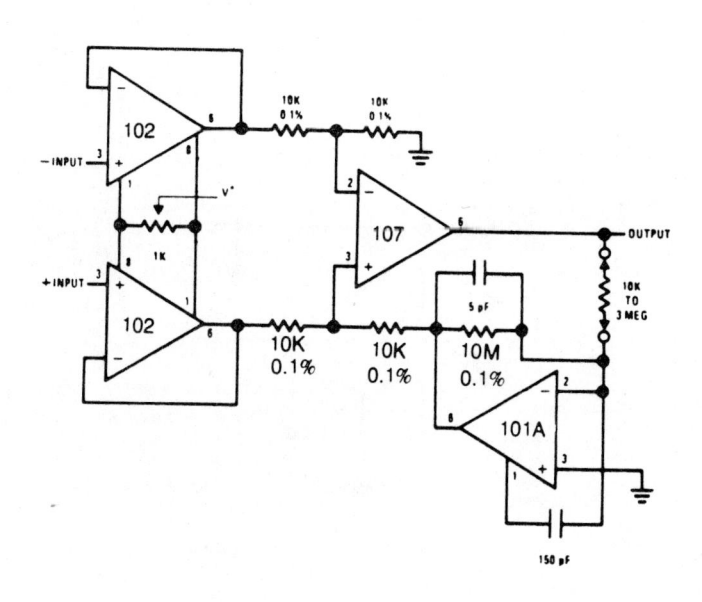

Variable gain, differential-input instrumentation amplifier

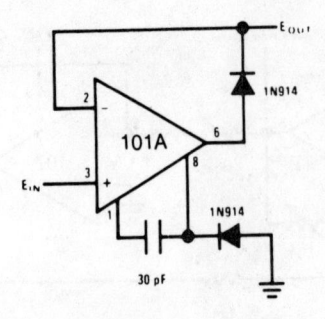

Precision diode

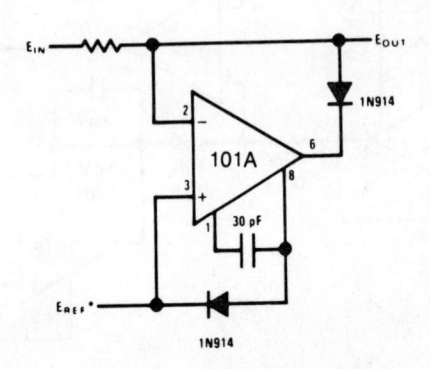

Precision clamp

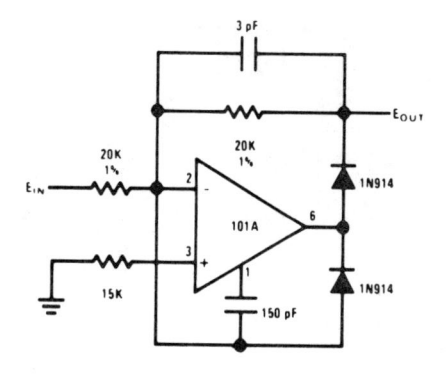

Fast half wave rectifier

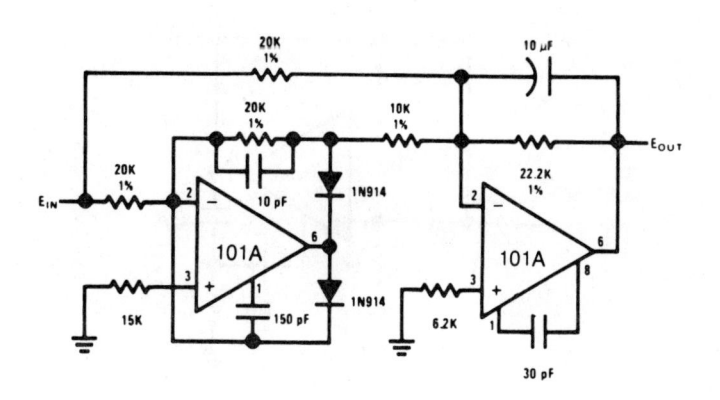

Precision ac to dc converter

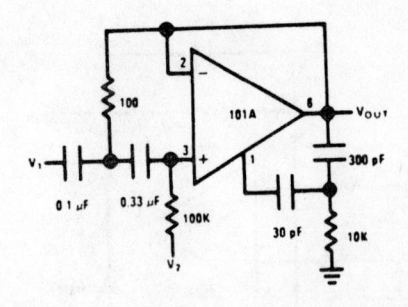

Tuned circuit

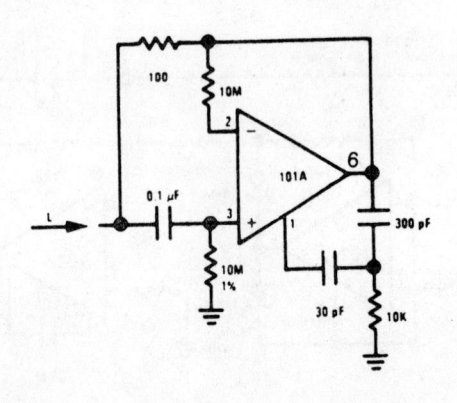

Simulated inductor

36

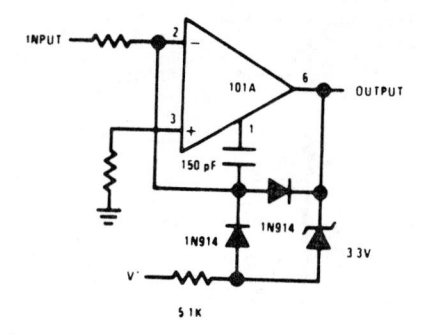

Fast zero crossing detector

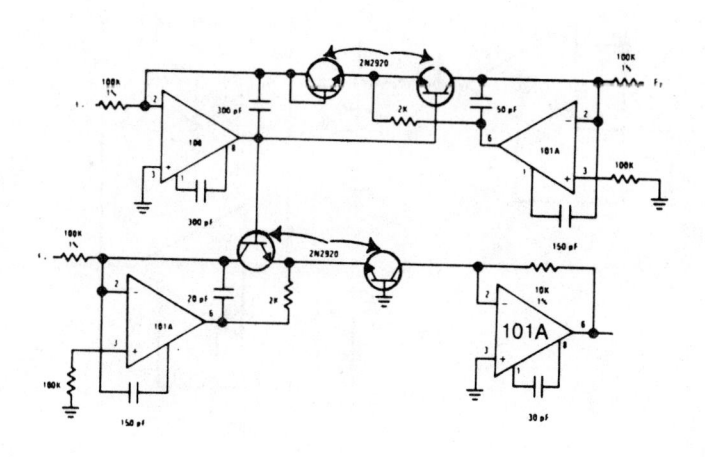

Multiplier/divider

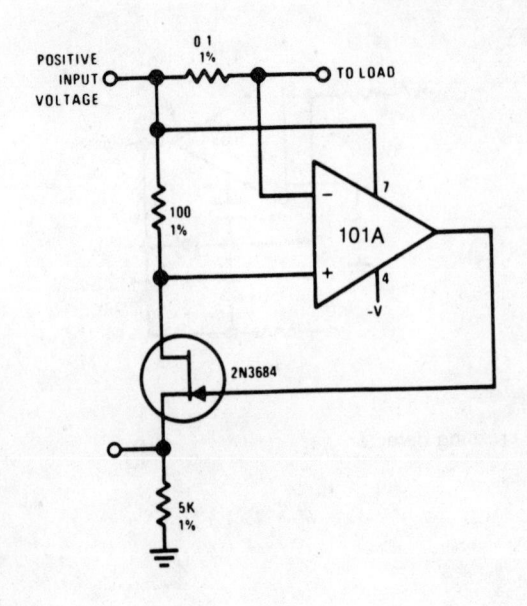

Current Monitor

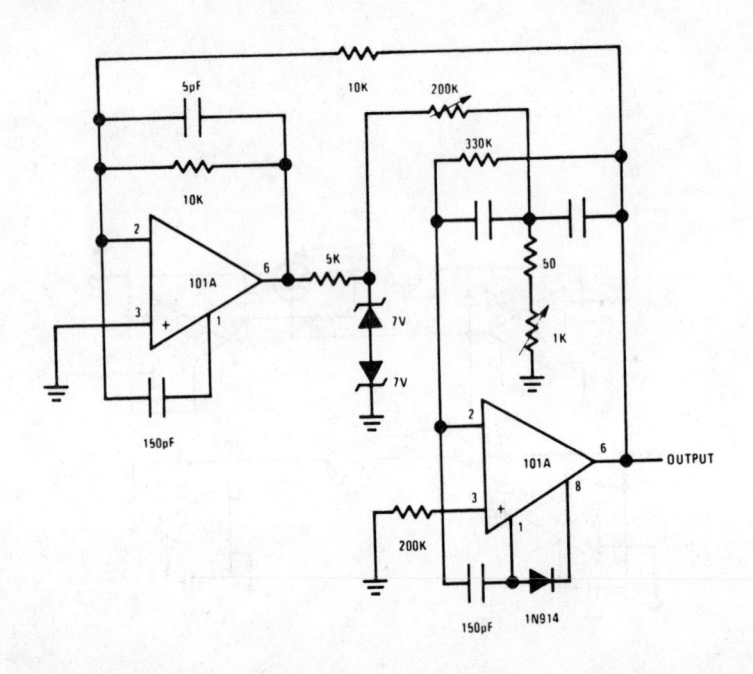

Low distortion sine wave oscillator

38

102—HIGH SPEED OPERATIONAL AMPLIFIER

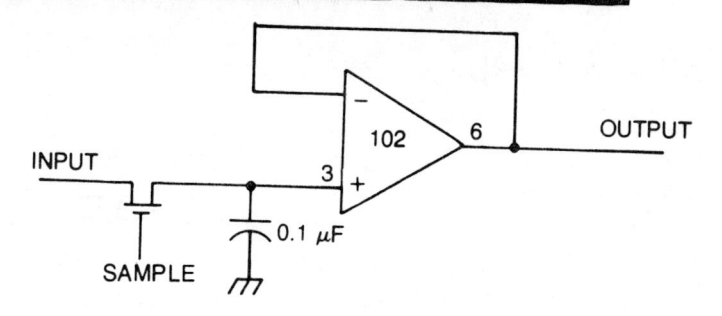

Sample and hold circuit

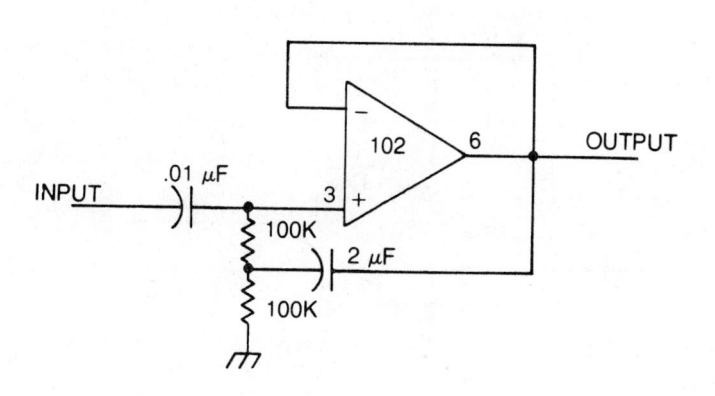

High input impedance ac amplifier

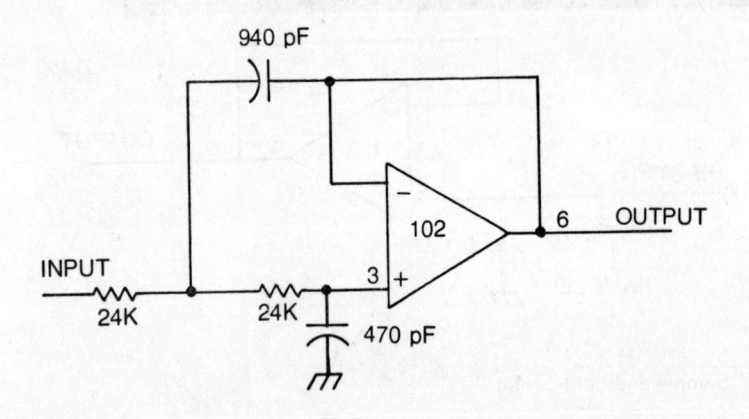

Low pass active filter

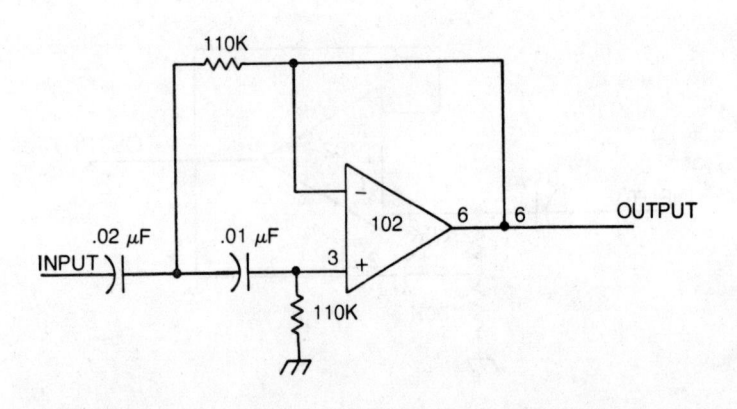

High pass active filter

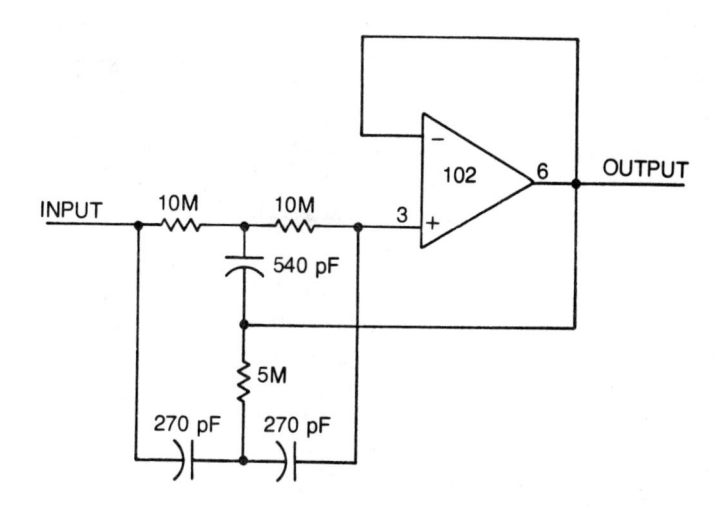

High Q notch filter

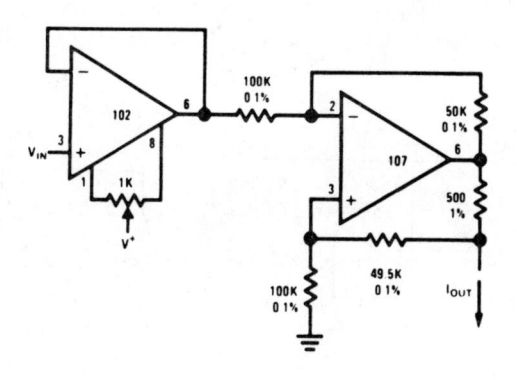

Bilateral current source

41

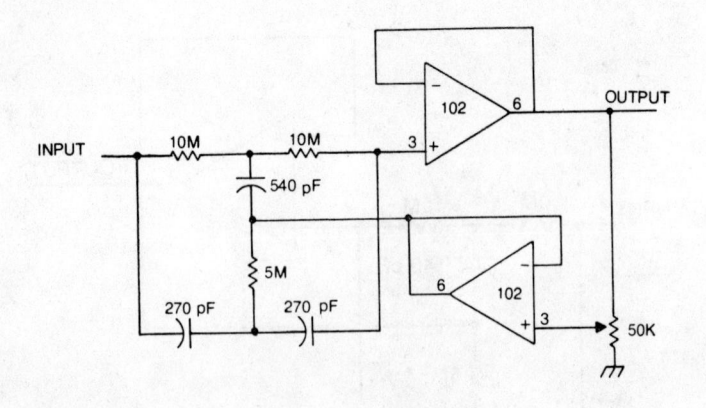

Variable Q notch filter

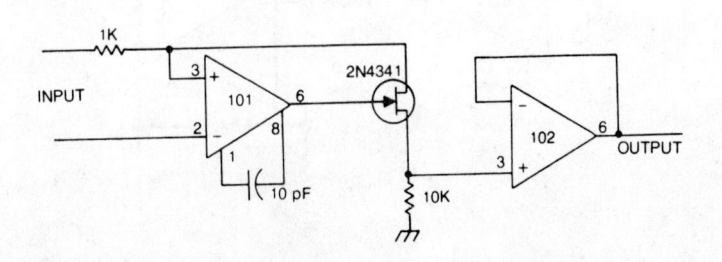

Level shifting isolation amplifier

42

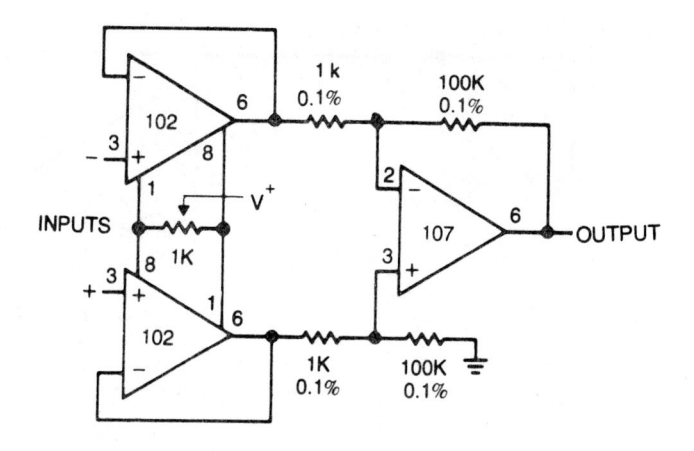

Differential-input instrumentation amplifier

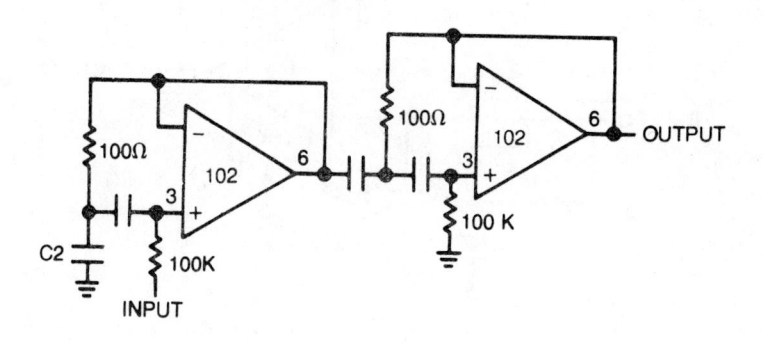

Two-stage tuned circuit

43

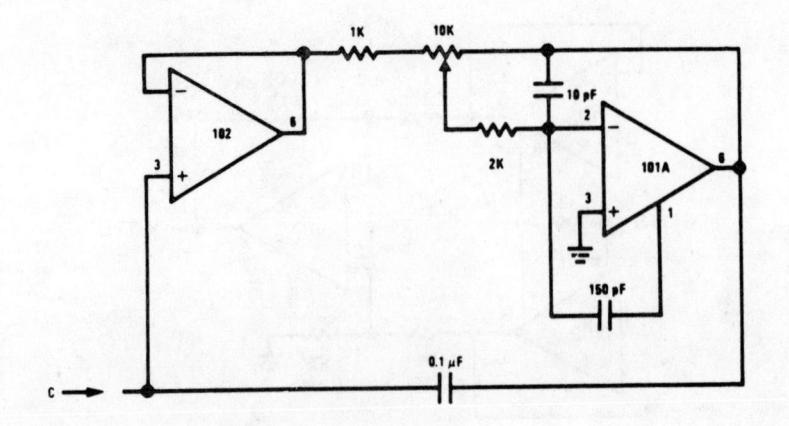

Variable capacitance multiplier

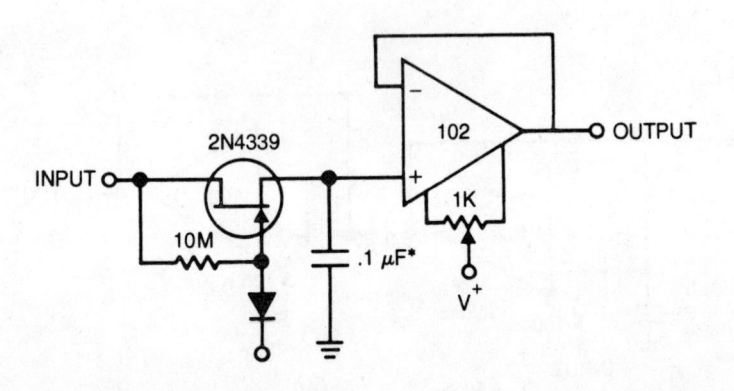

Sample and hold with offset adjustment

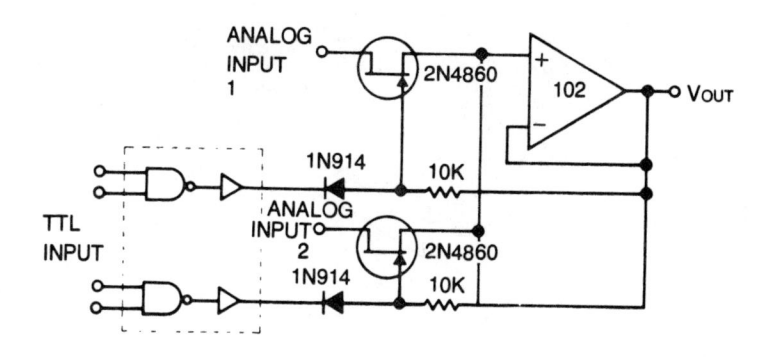

TTL controlled buffered analog switch

106—VOLTAGE COMPARATOR

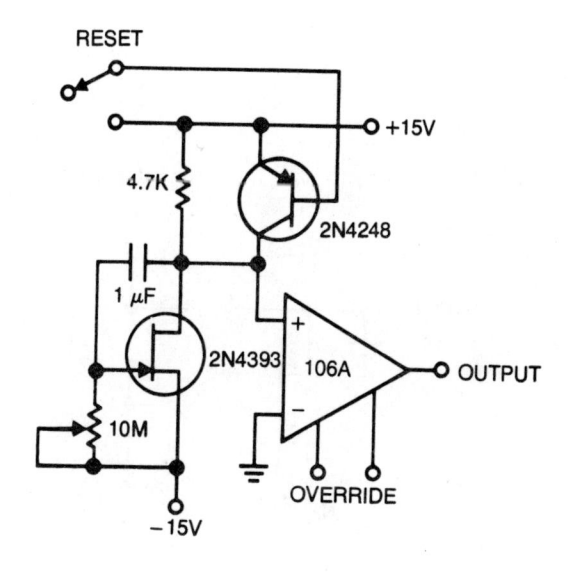

Long time comparator

45

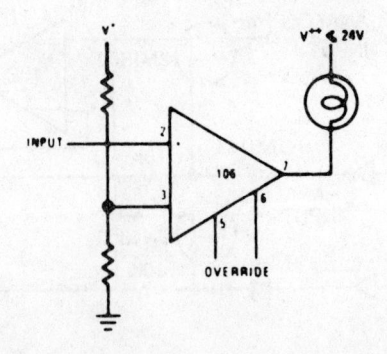

Level detector and lamp driver

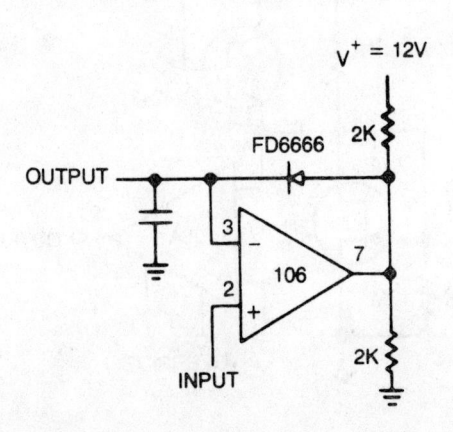

Fast response peak detector

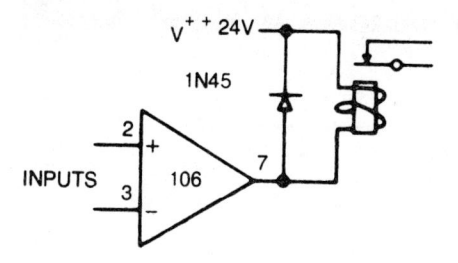

Relay driver

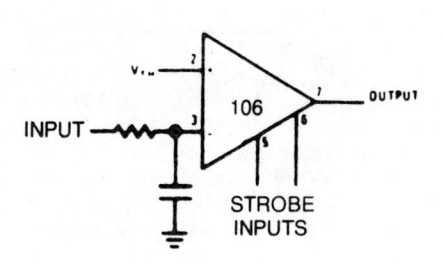

Adjustable threshold line receiver

47

107 — FREQUENCY COMPENSATED OPERATIONAL AMPLIFIER

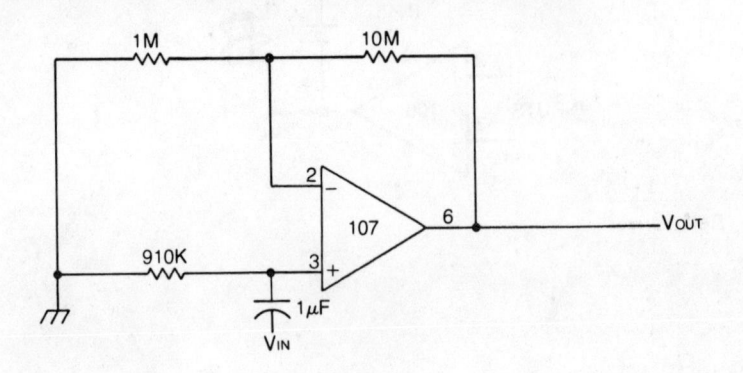

Noninverting ac amplifier

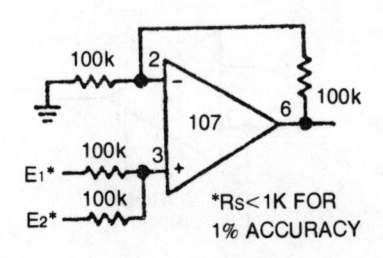

Noninverting summing amplifier

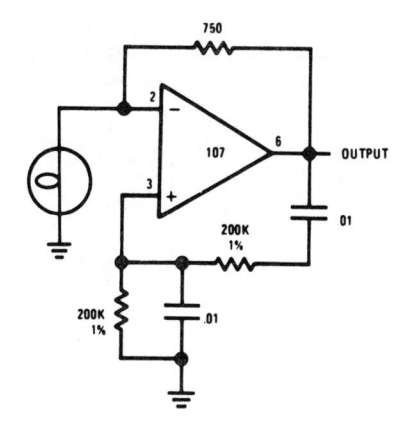

Wien bridge sine wave oscillator

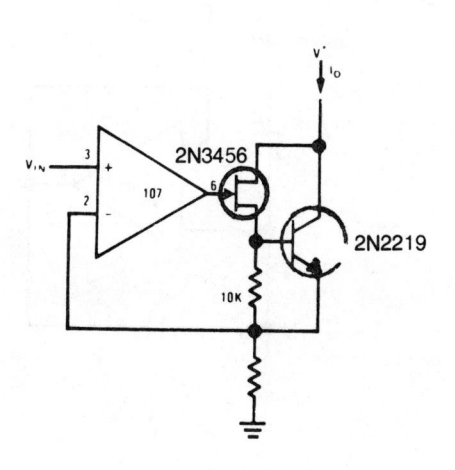

Precision current sink

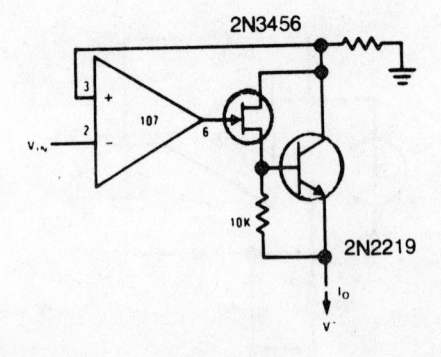

Precision current source

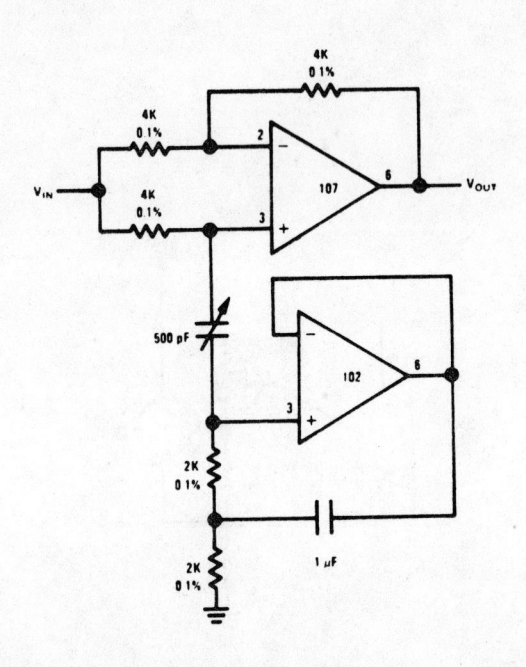

Easily tuned notch filter

50

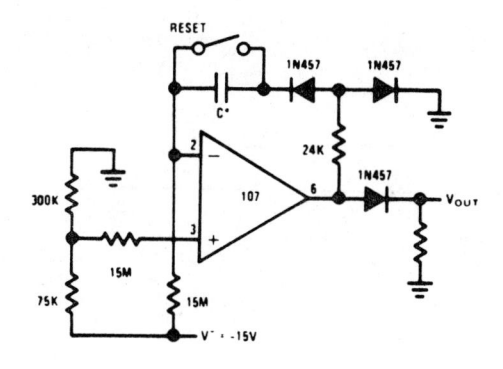

Long interval timer

108 — LOW INPUT CURRENT OPERATIONAL AMPLIFIER

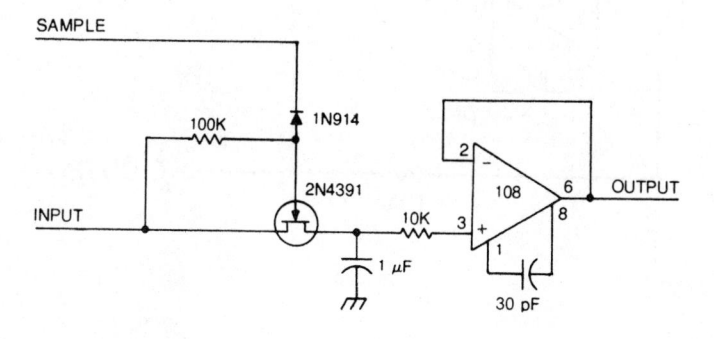

Sample and hold circuit

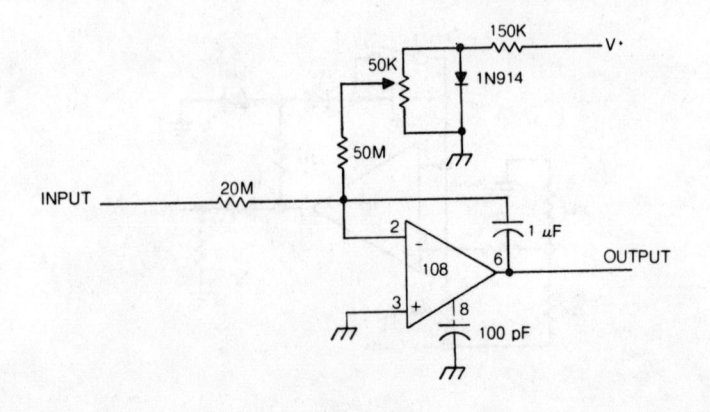

Integrator with bias current compensation

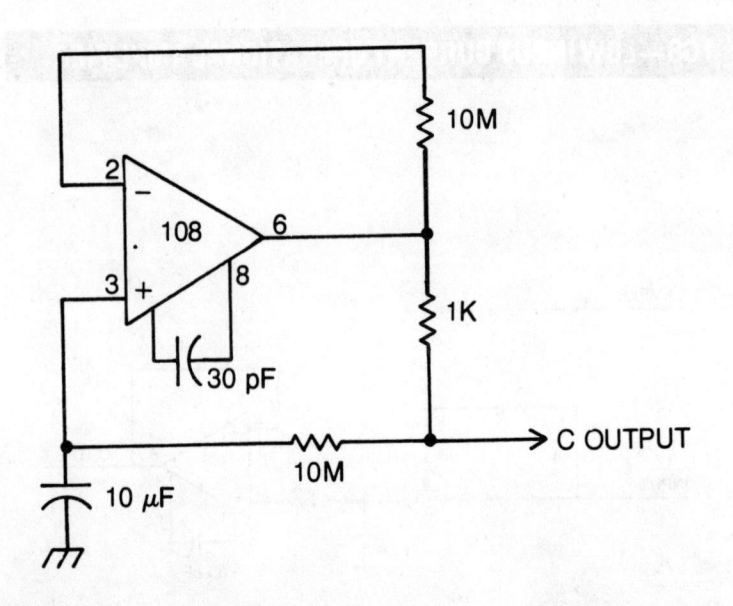

Capacitance multiplier

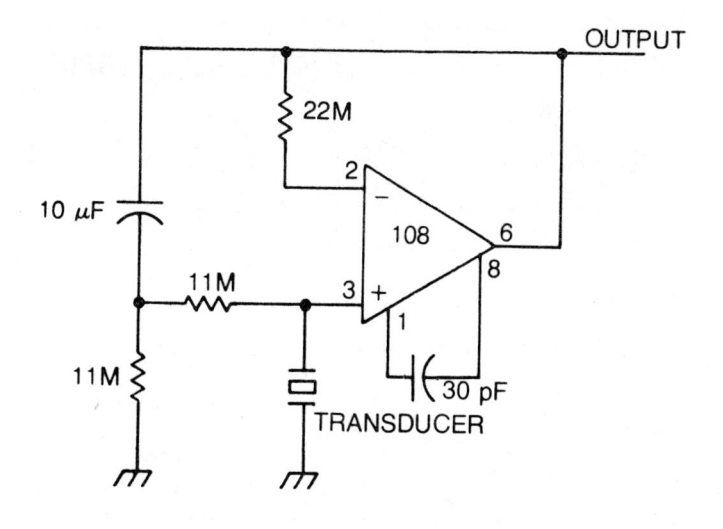

Amplifier for piezoelectric transducers

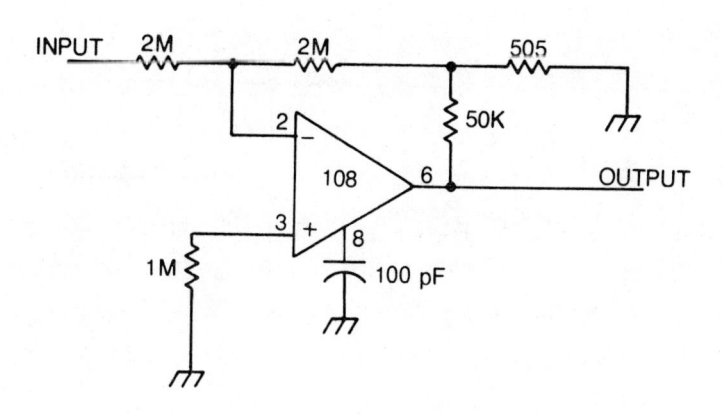

High input impedance inverting amplifier

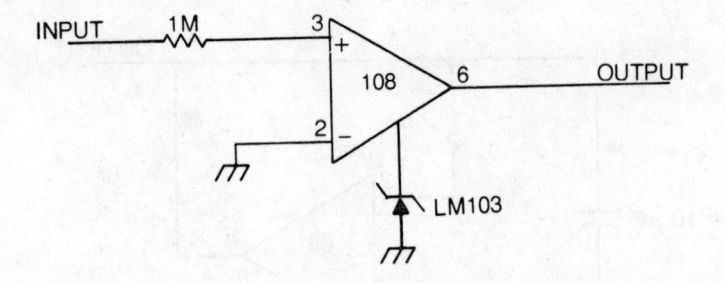

Zero crossing detector voltage comparator

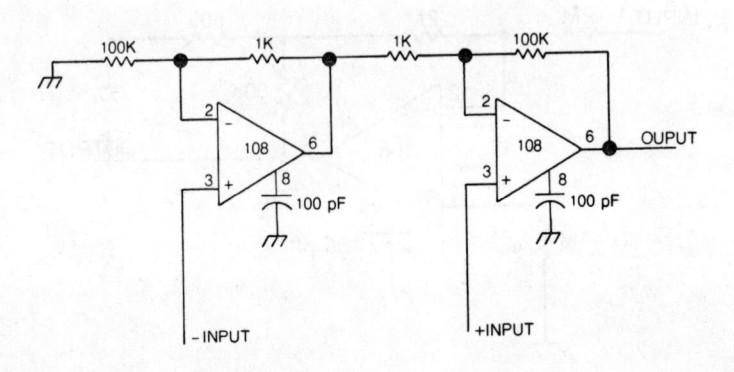

Differential input instrumentation amplifier

54

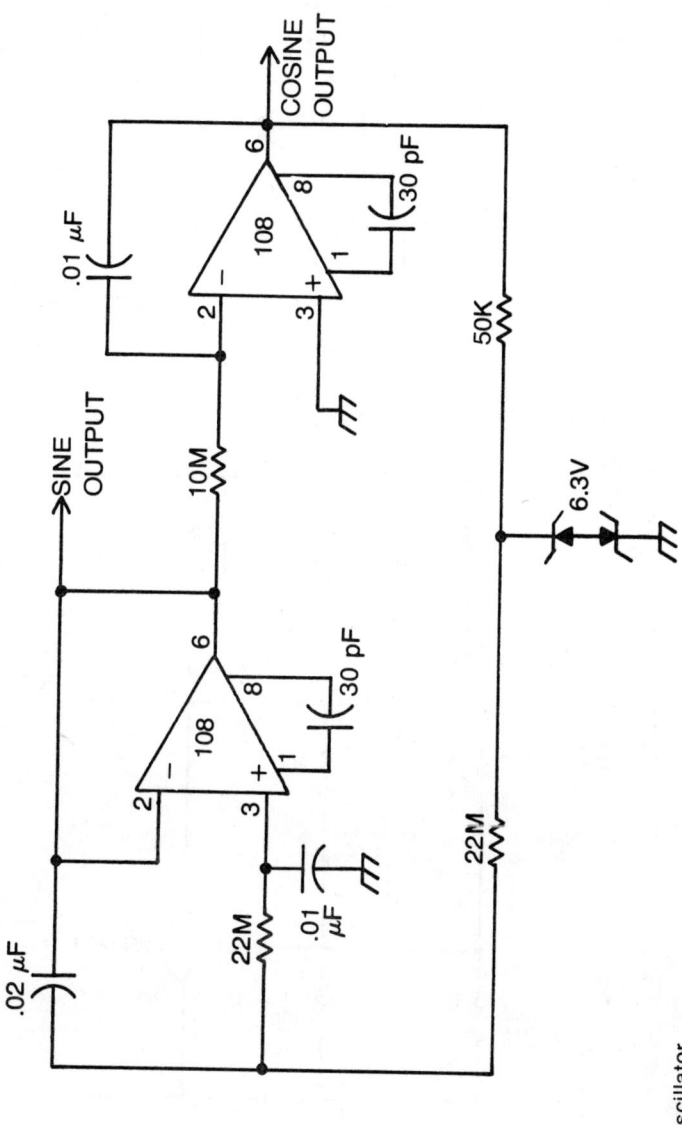

Sine wave oscillator

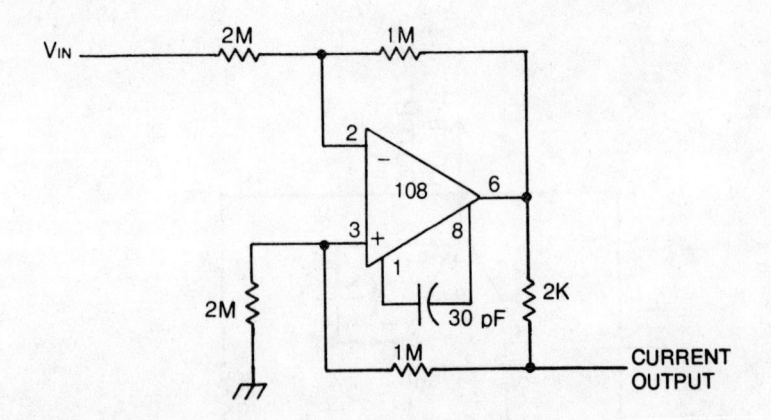

Bilateral current source

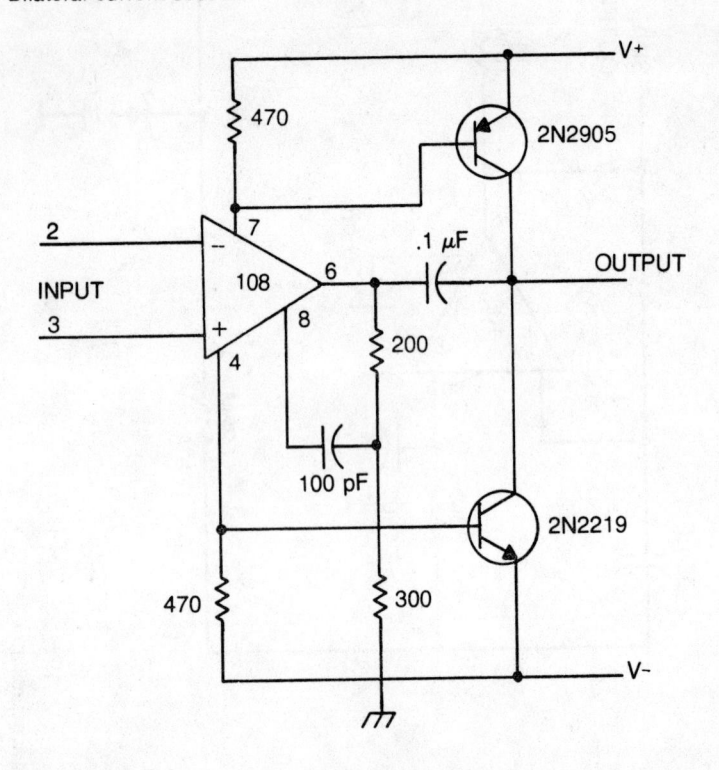

Power booster

Power op amp

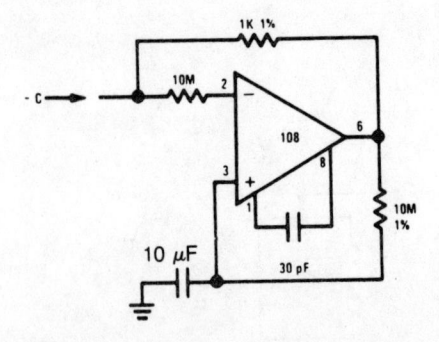

Negative capacitance multiplier

110—UNITY GAIN OPERATIONAL AMPLIFIER/BUFFER

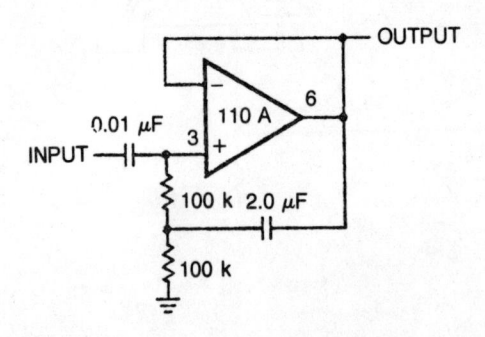

High input impedance ac amplifier

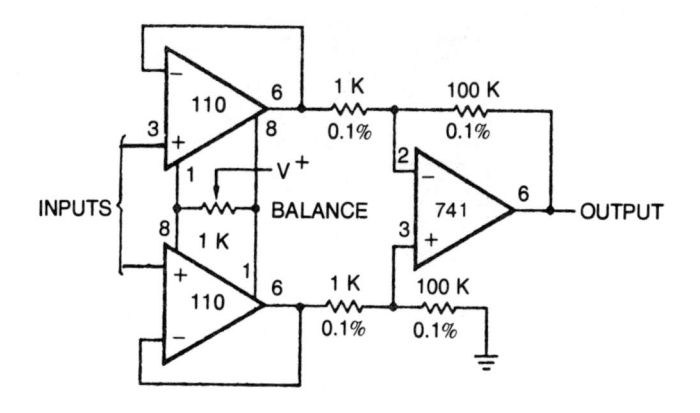

Differential input instrumentation amplifier

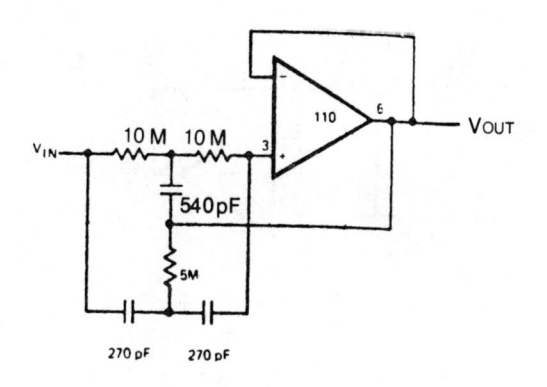

High Q notch filter

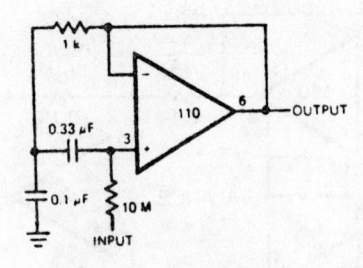

Bandpass filter

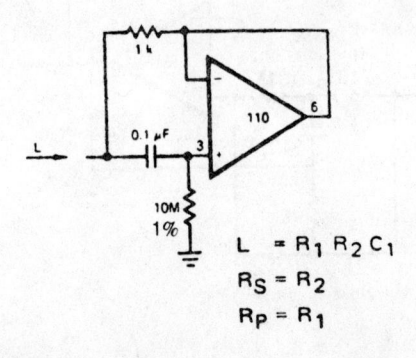

Simulated inductor

$L = R_1 R_2 C_1$
$R_S = R_2$
$R_P = R_1$

60

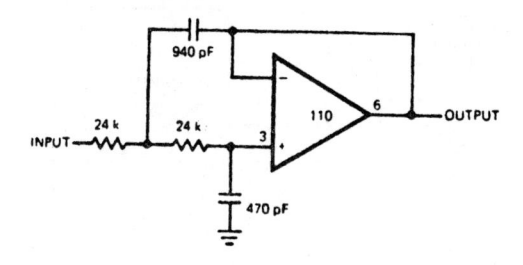

Low pass active filter
(10 kHz CUTOFF)

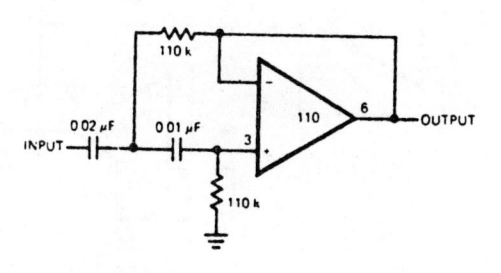

High pass active filter
(100 Hz CUTOFF)

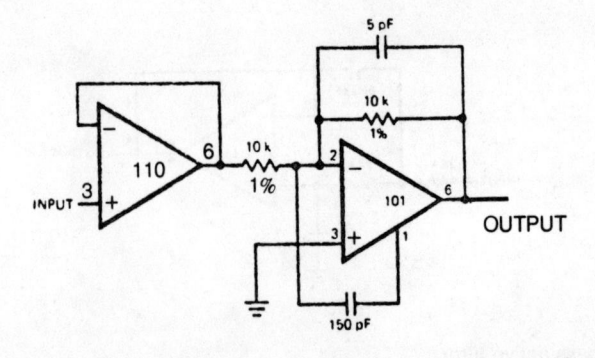

Fast inverting amplifier with high input impedance

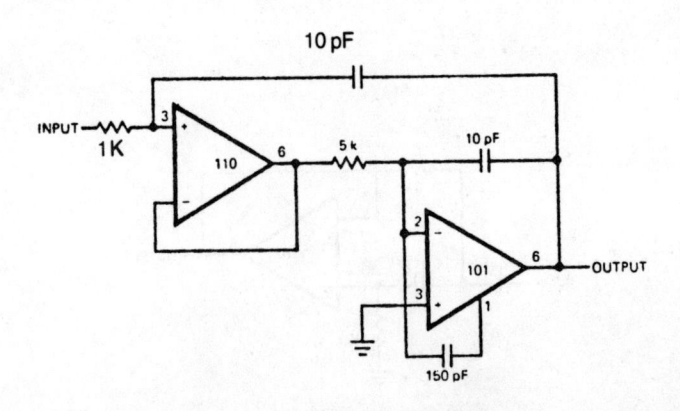

Fast integrator with low input current

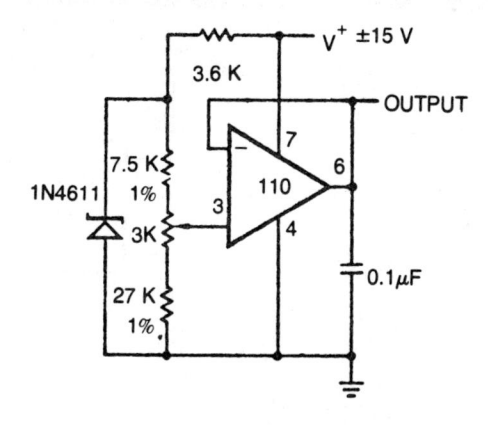

Buffered reference source

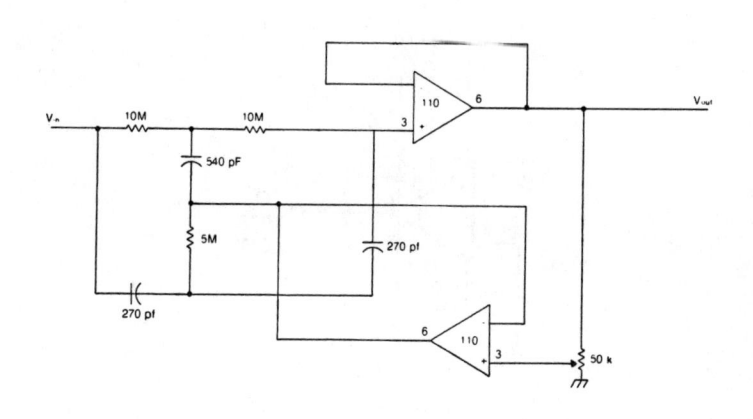

Adjustable Q notch filter

111—INTEGRATED VOLTAGE COMPARATOR

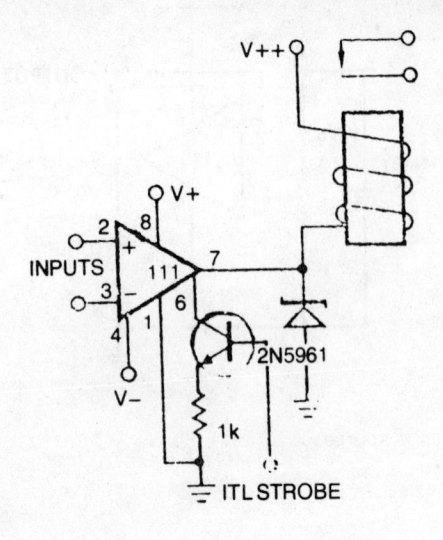

Relay driver with strobe

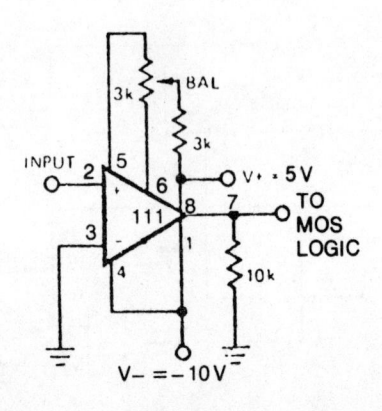

Zero crossing detector driving MOS logic

64

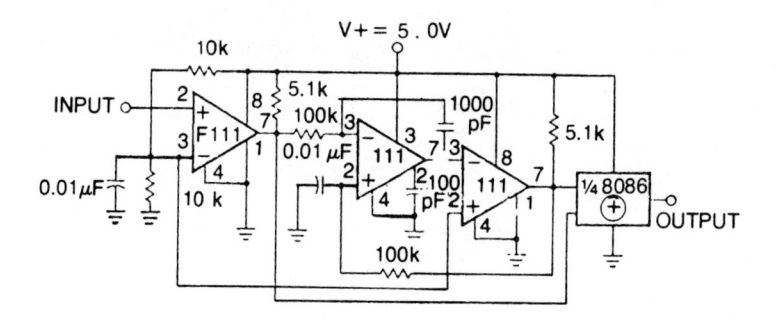

Frequency doubler

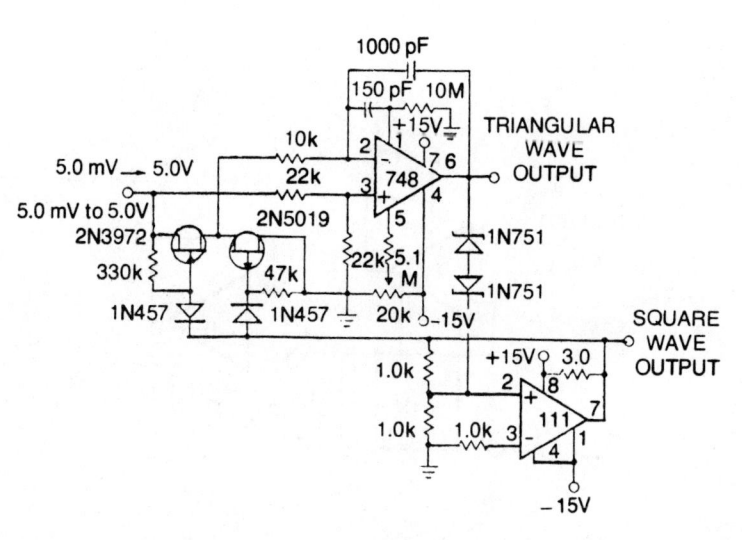

10 Hz to 10 kHz voltage controlled oscillator

65

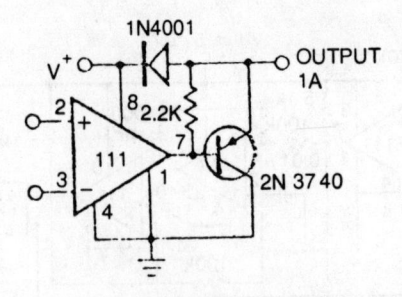

Comparator and solenoid driver

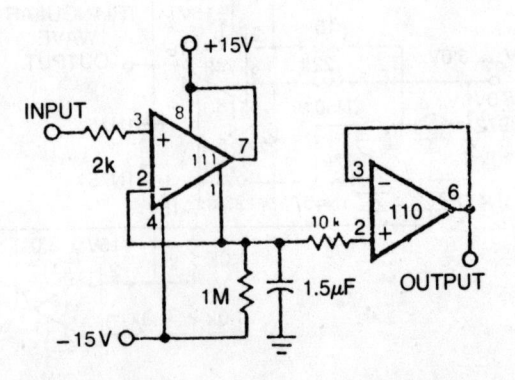

Positive peak detector

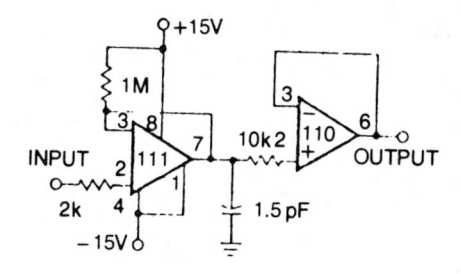

Negative peak detector

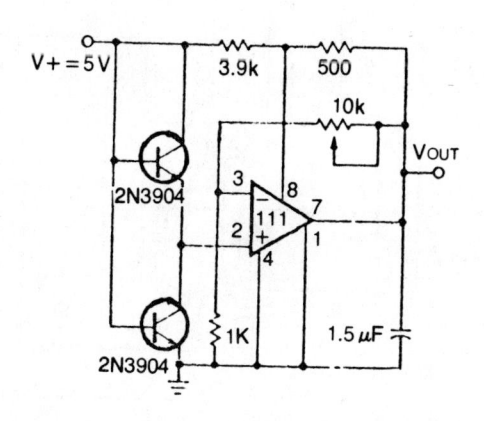

Adjustable low voltage reference supply

Switching power amplifier

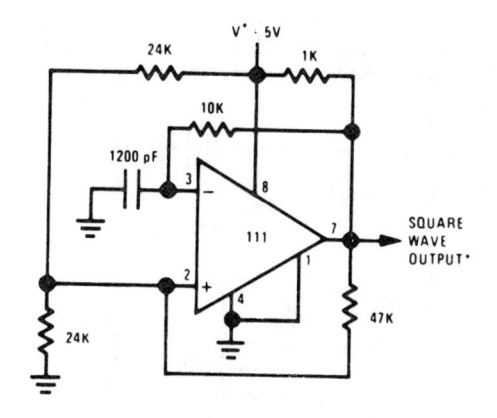

Free-running multivibrator

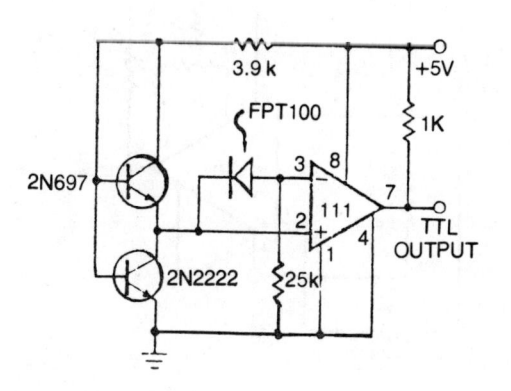

Precision photodiode comparator

69

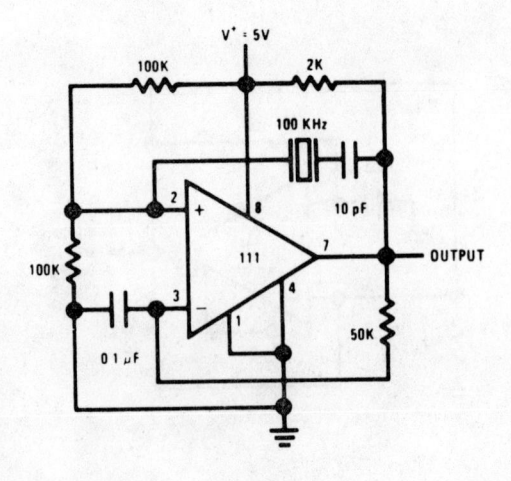

Crystal-controlled oscillator

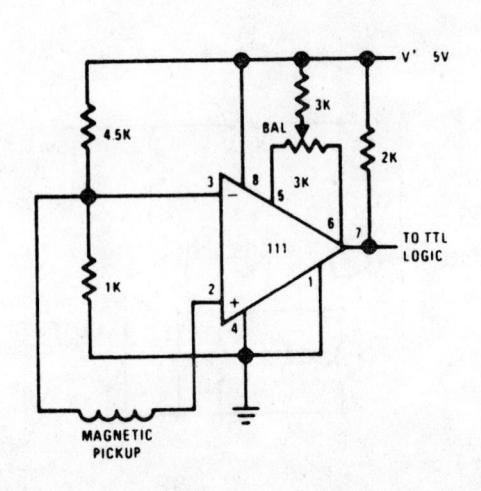

Zero crossing detector for magnetic transducer

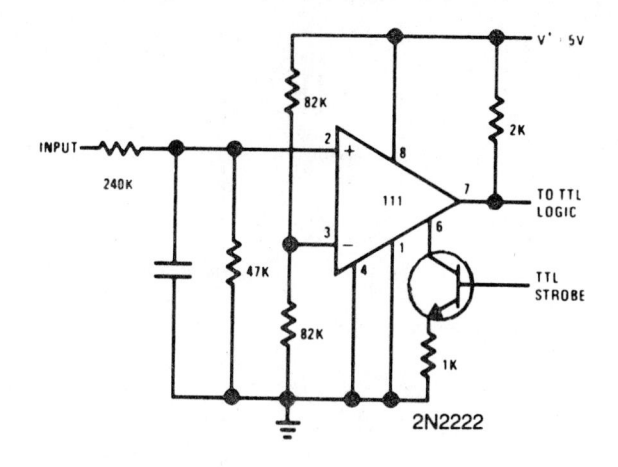

Circuit for transmitting data between high-level logic and TTL

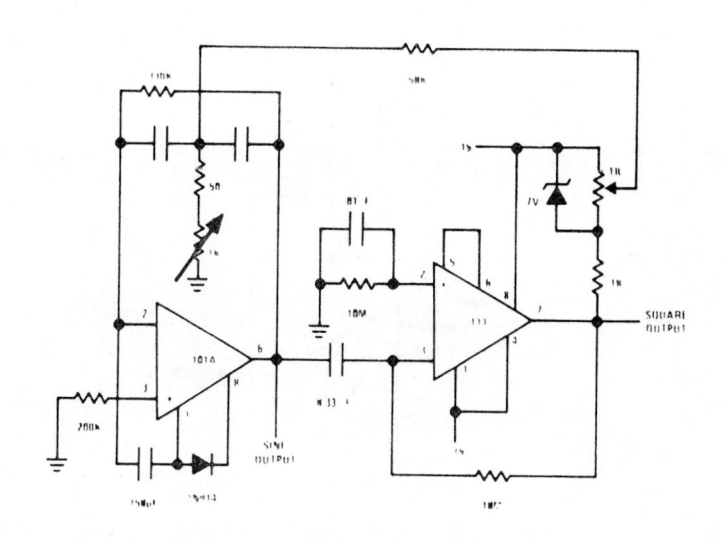

Easily tuned sine wave oscillator

118 — HIGH SPEED OPERATIONAL AMPLIFIER

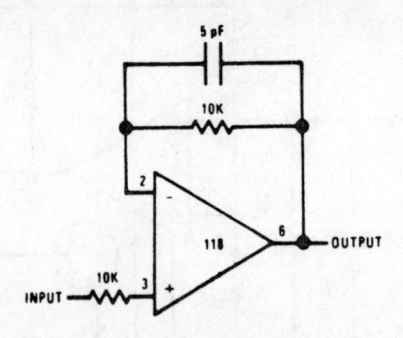

Fast voltage follower

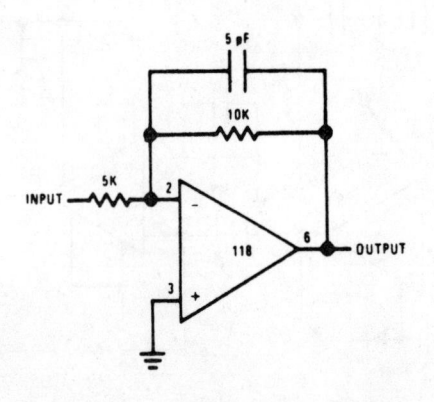

Fast summing amplifier

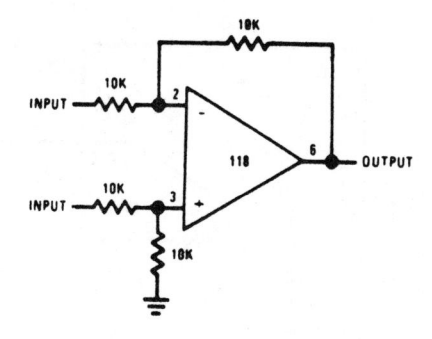

Differential amplifier

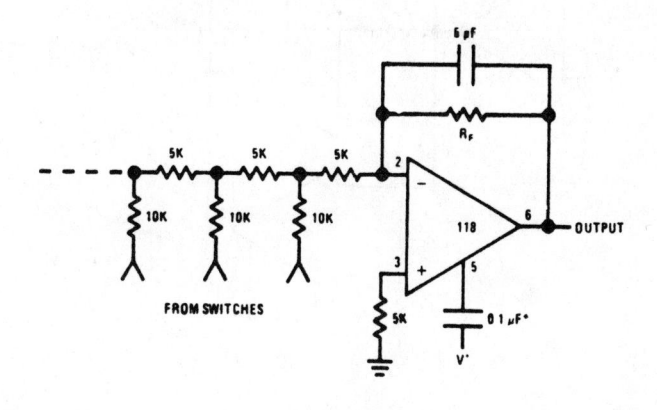

D/A converter using ladder network

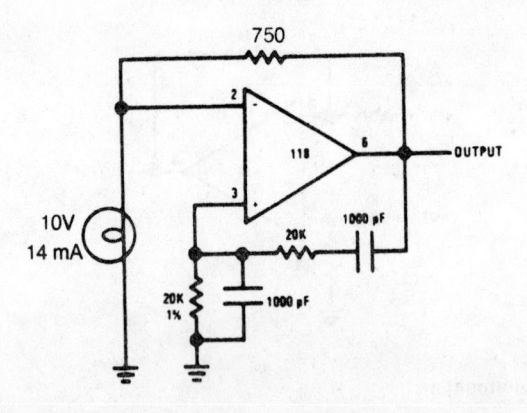

Wien bridge sine wave oscillator

119—HIGH SPEED DUAL COMPARATOR

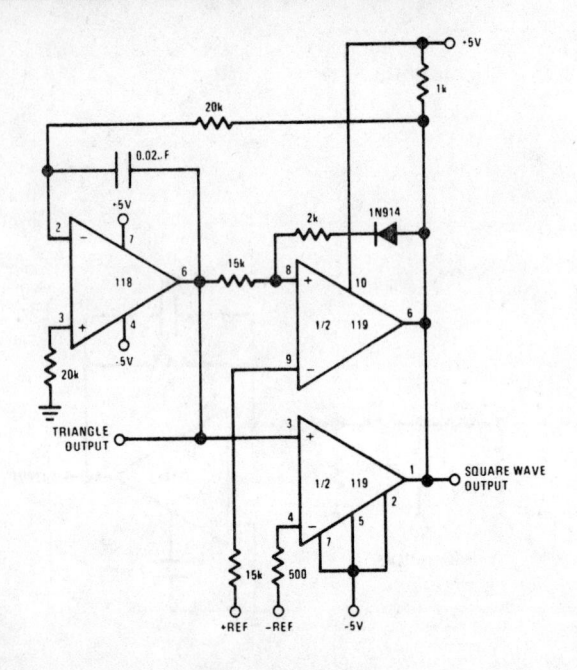

Precision triangle generator

74

121—PRECISION PREAMPLIFIER

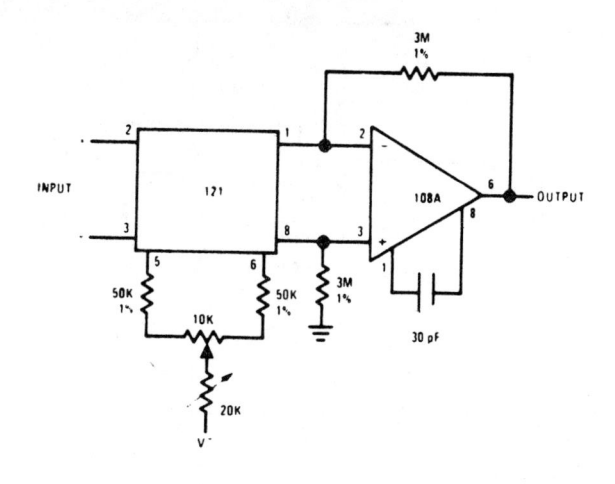

Gain of 1000 instrumentation amplifier

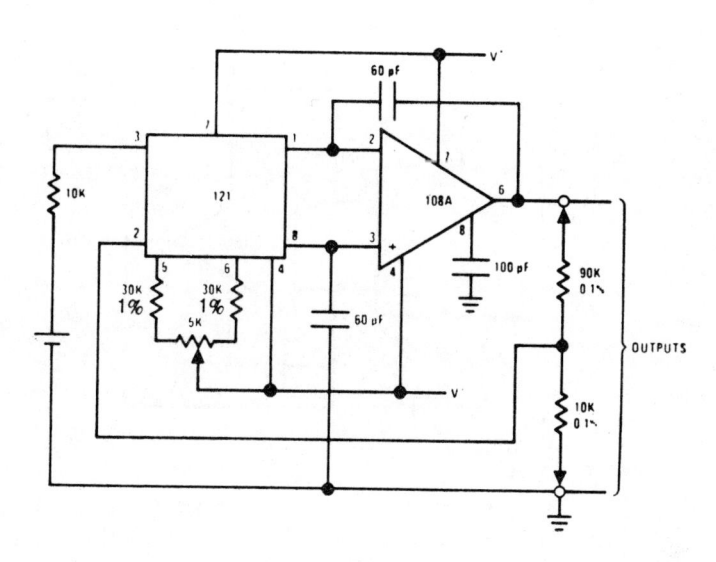

10V reference

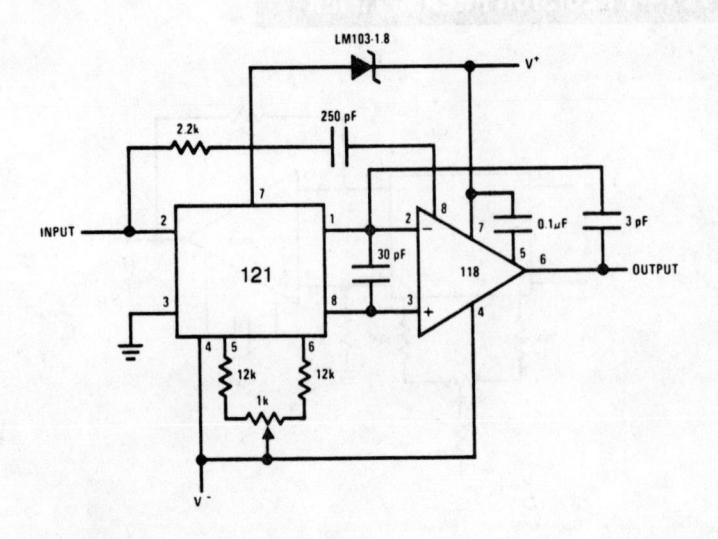

High speed inverting amplifier with low drift

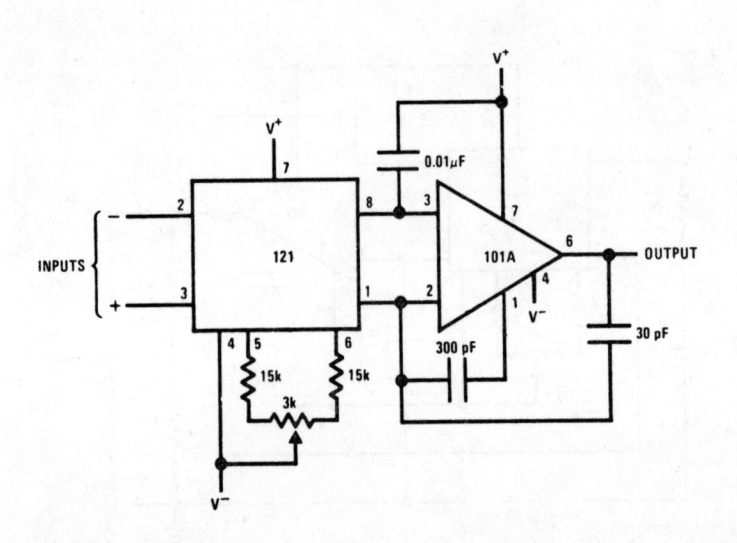

Medium speed general purpose amplifier

76

122 — PRECISION TIMER

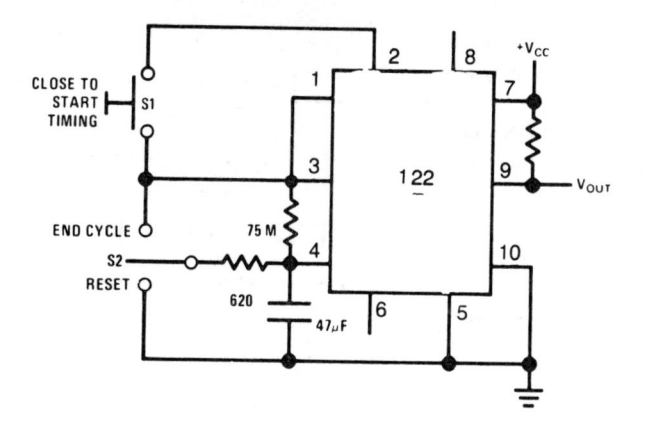

One hour timer with reset and manual cycle end

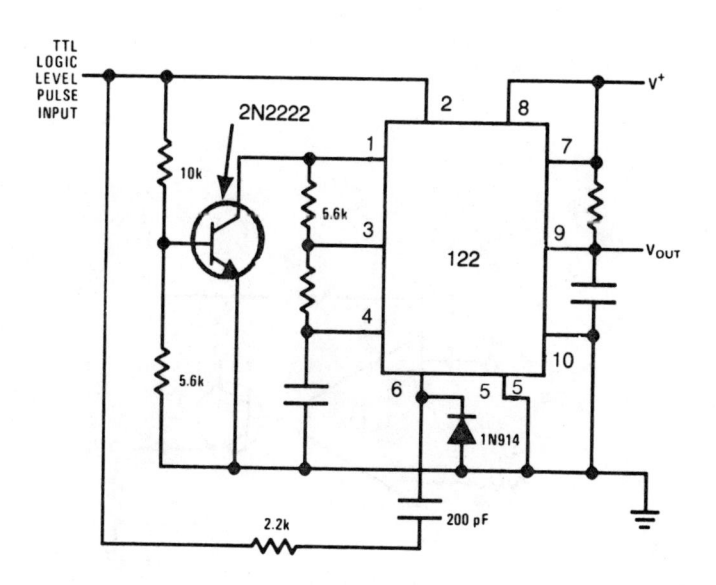

Pulse width detector

124—QUAD OPERATIONAL AMPLIFIER

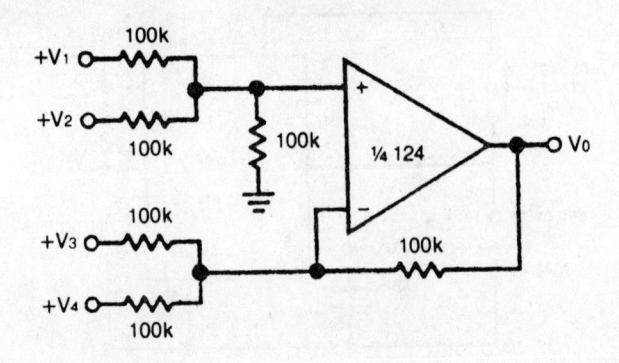

DC summing amplifier

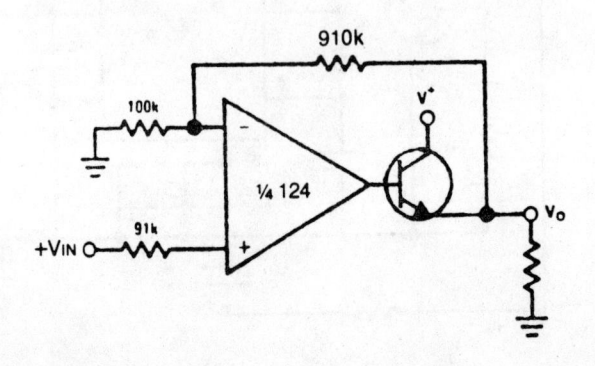

Power amplifier

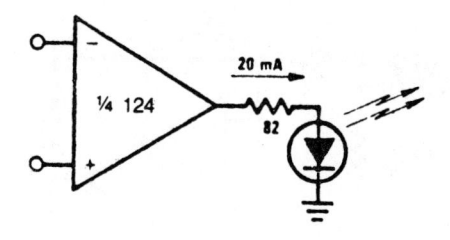

LED driver

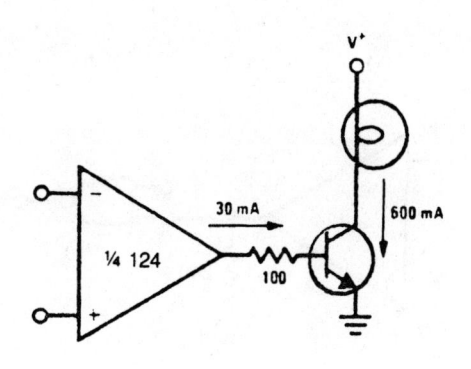

Lamp driver

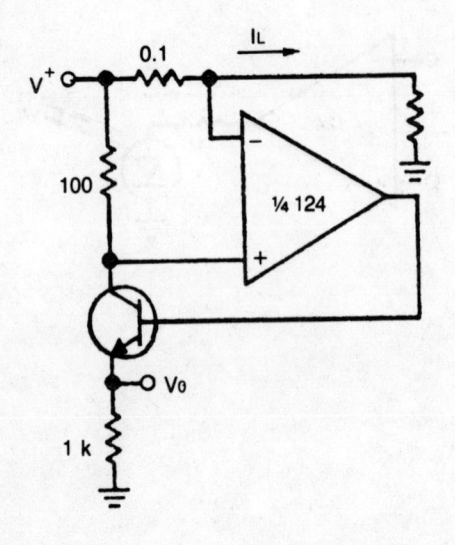

Current monitor

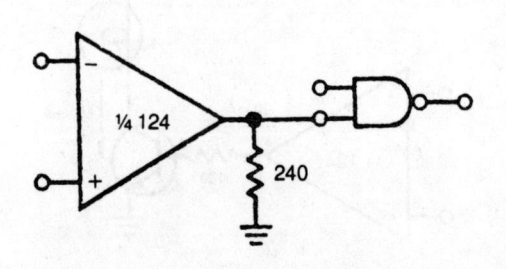

Driving TTL

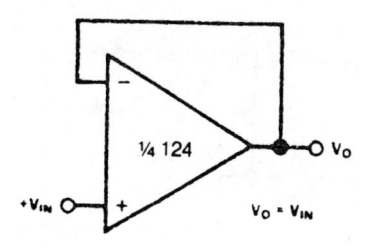

Voltage follower

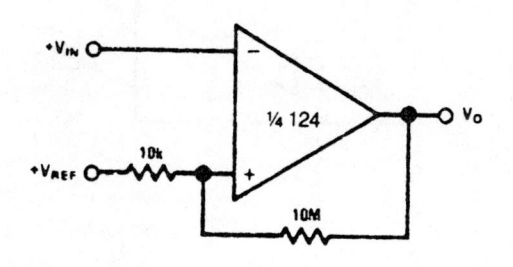

Comparator with hysteresis

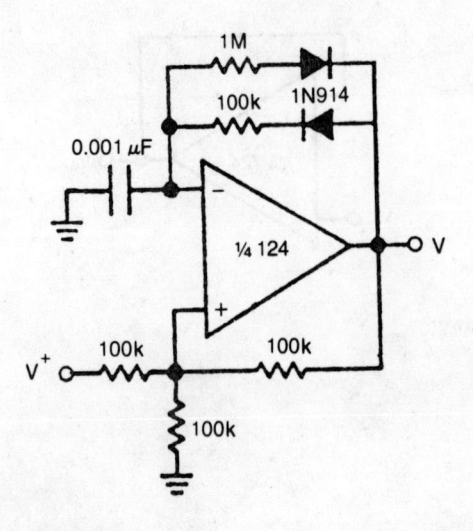

Pulse generator

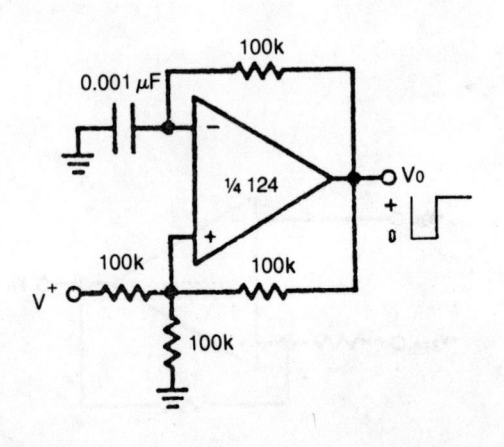

Squarewave oscillator

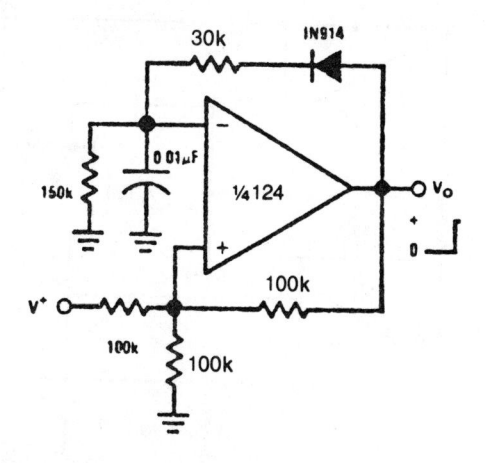

Pulse generator

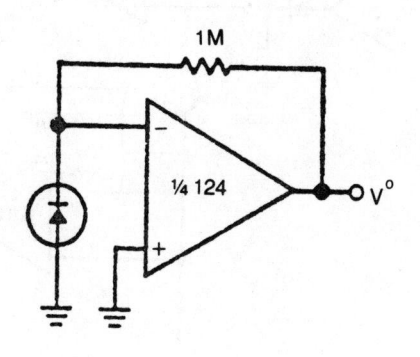

Photo voltaic-cell amplifier

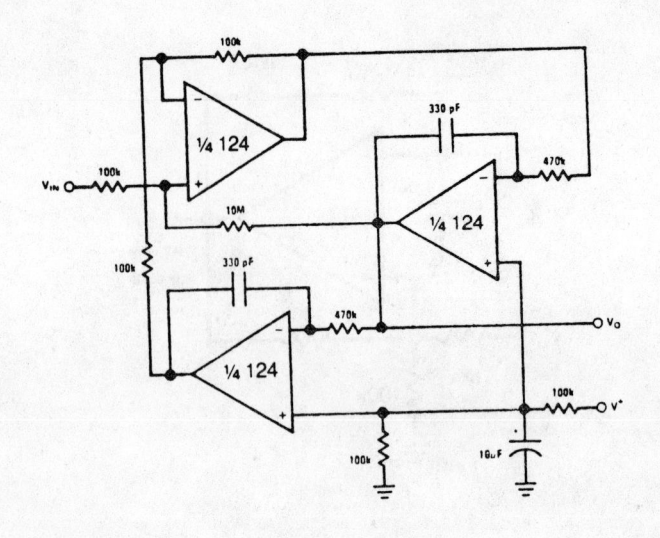

"BI-QUAD" RC active bandpass filter

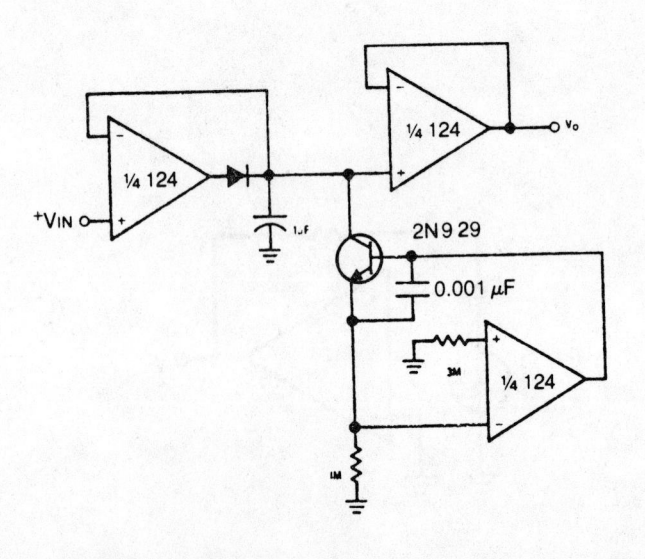

Low drift peak detector

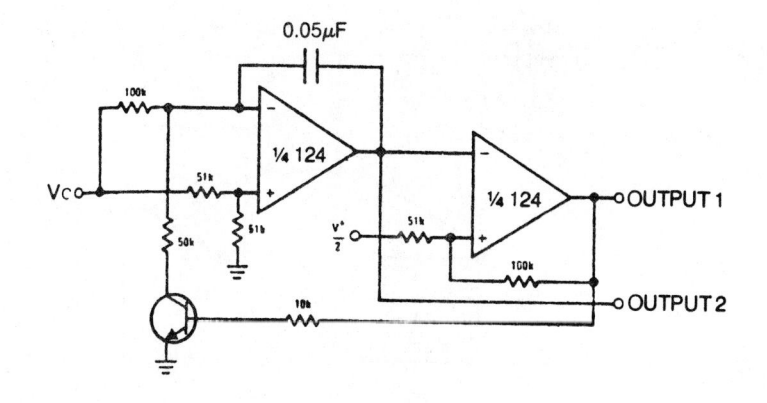

Voltage controlled oscillator circuit

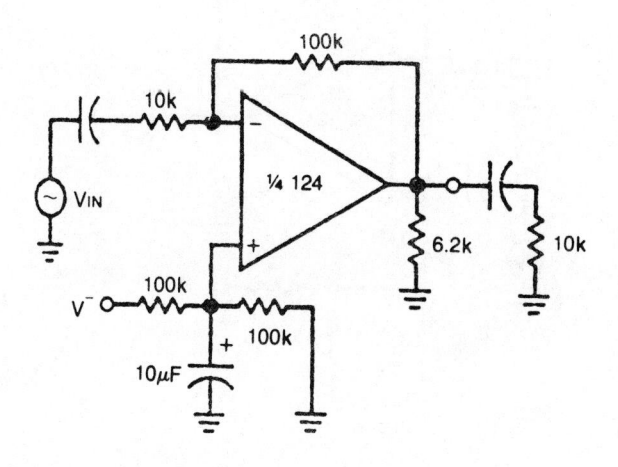

AC coupled inverting amplifier

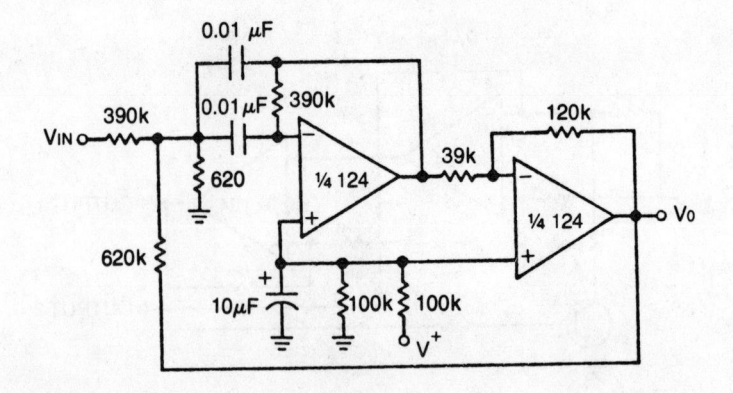

Bandpass active filter

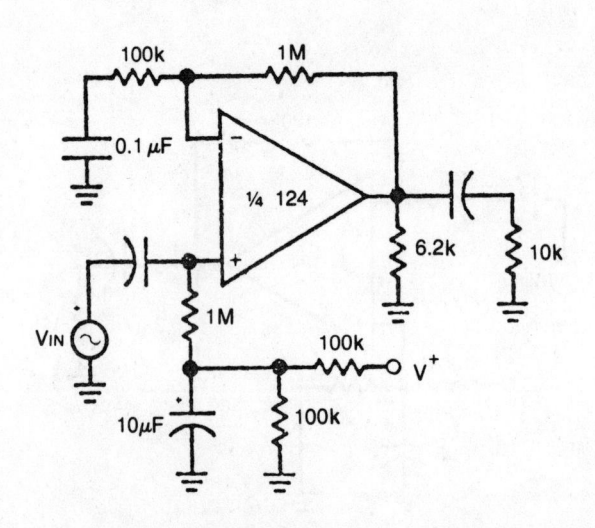

AC coupled noninverting amplifier

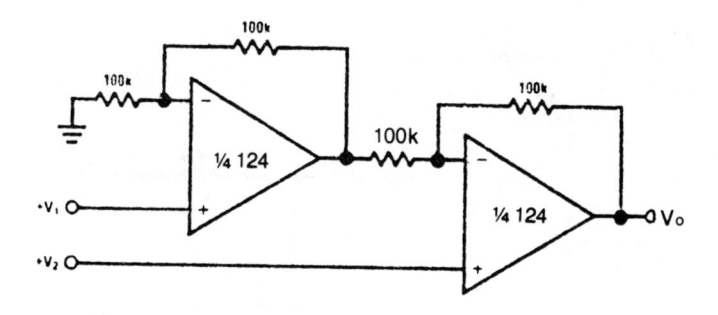

High input Z, dc differential amplifier

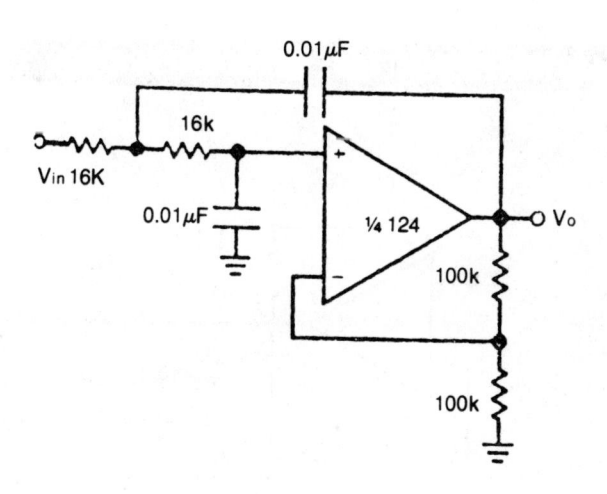

DC coupled low-pass RC active filter

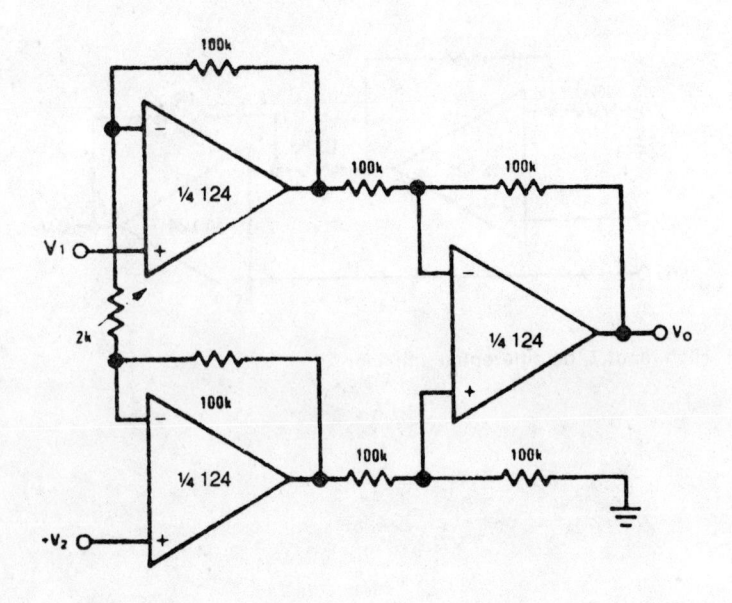

High input Z adjustable-gain dc instrumentation amplifier

139 — LOW OFFSET VOLTAGE QUAD COMPARATOR

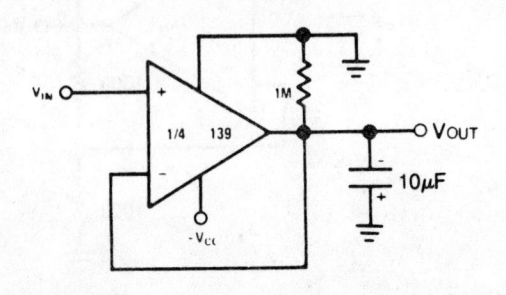

Negative peak detector

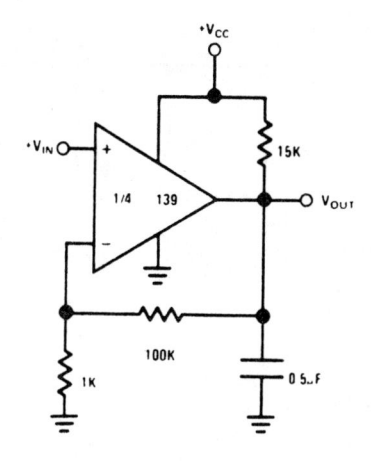

Noninverting amplifier

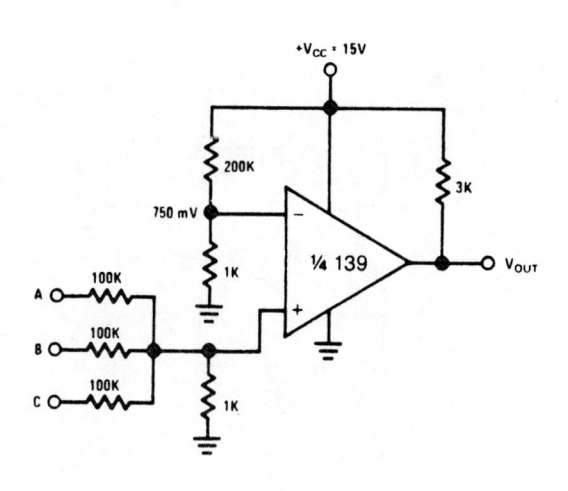

Three input OR gate

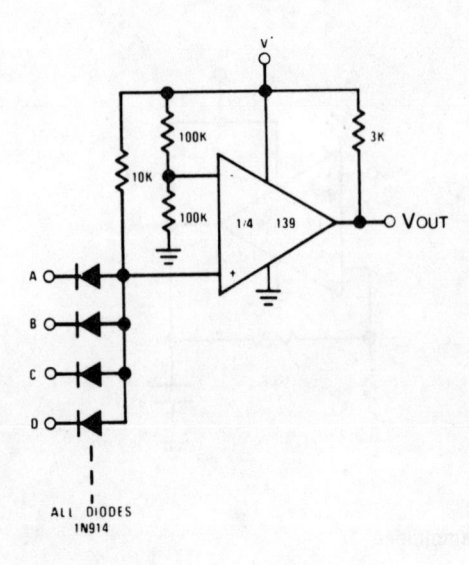

AND gate with large fan-in

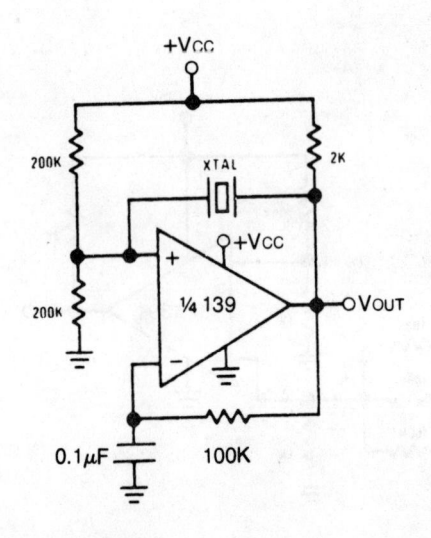

Crystal controlled oscillator

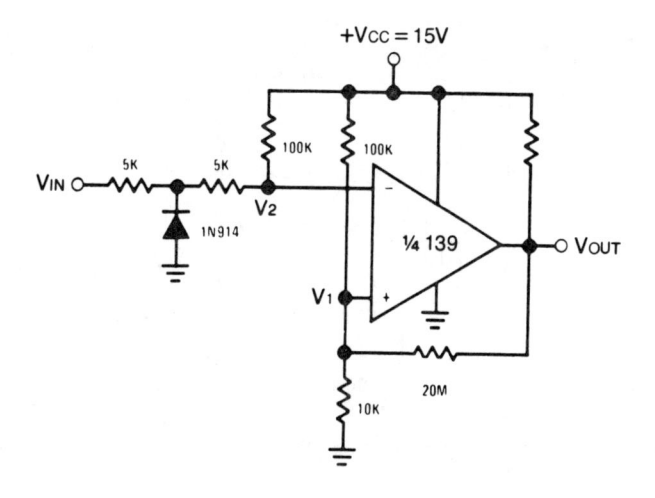

Zero crossing detector

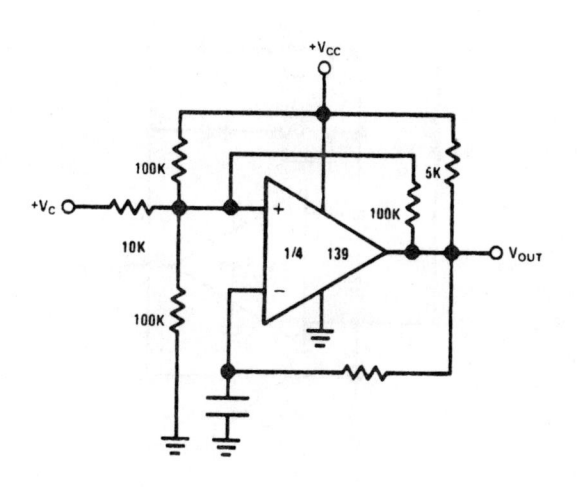

Pulse width modulator

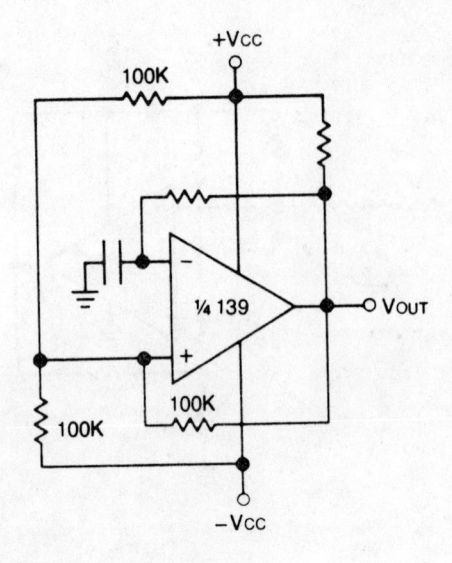

Squarewave generator using dual supplies

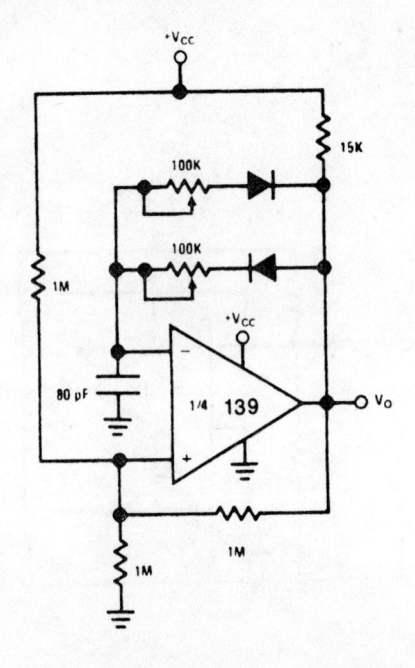

Pulse generator with variable duty cycle

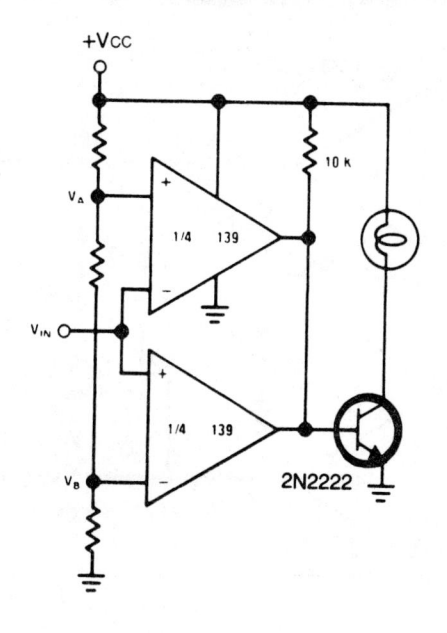

Limit comparator with lamp driver

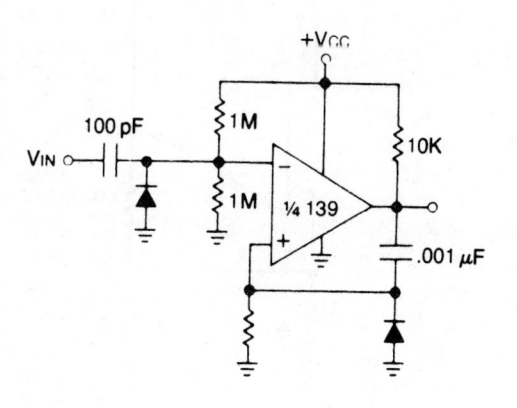

One shot multivibrator

93

143—HIGH VOLTAGE OPERATIONAL AMPLIFIER

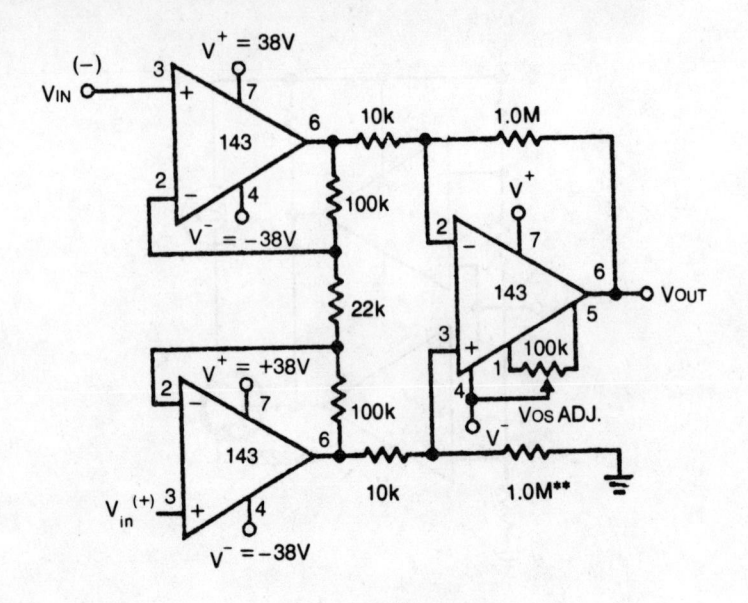

±34V common-mode instrumentation amplifier

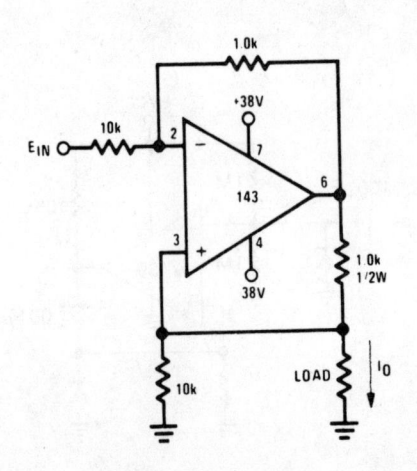

High-compliance current source

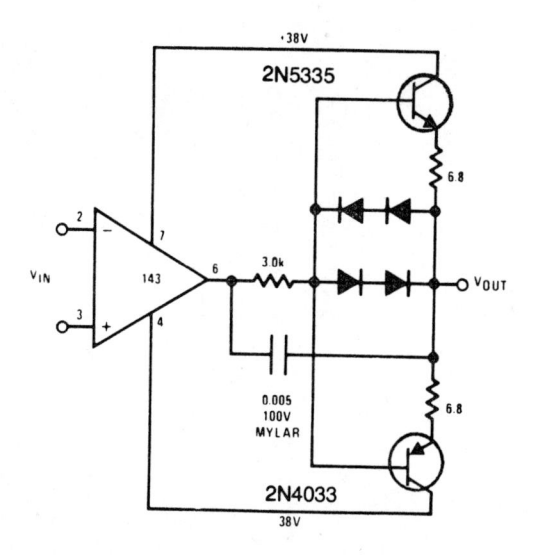

100 mA current boost circuit

All diodes are 1N914

144—HIGH VOLTAGE, HIGH SLEW RATE OPERATIONAL AMPLIFIER

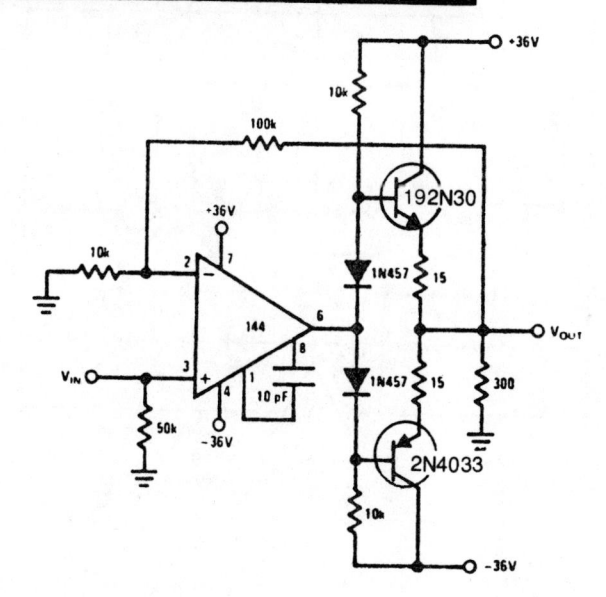

Large power bandwidth, current boosted audio line driver

90W audio power amplifier

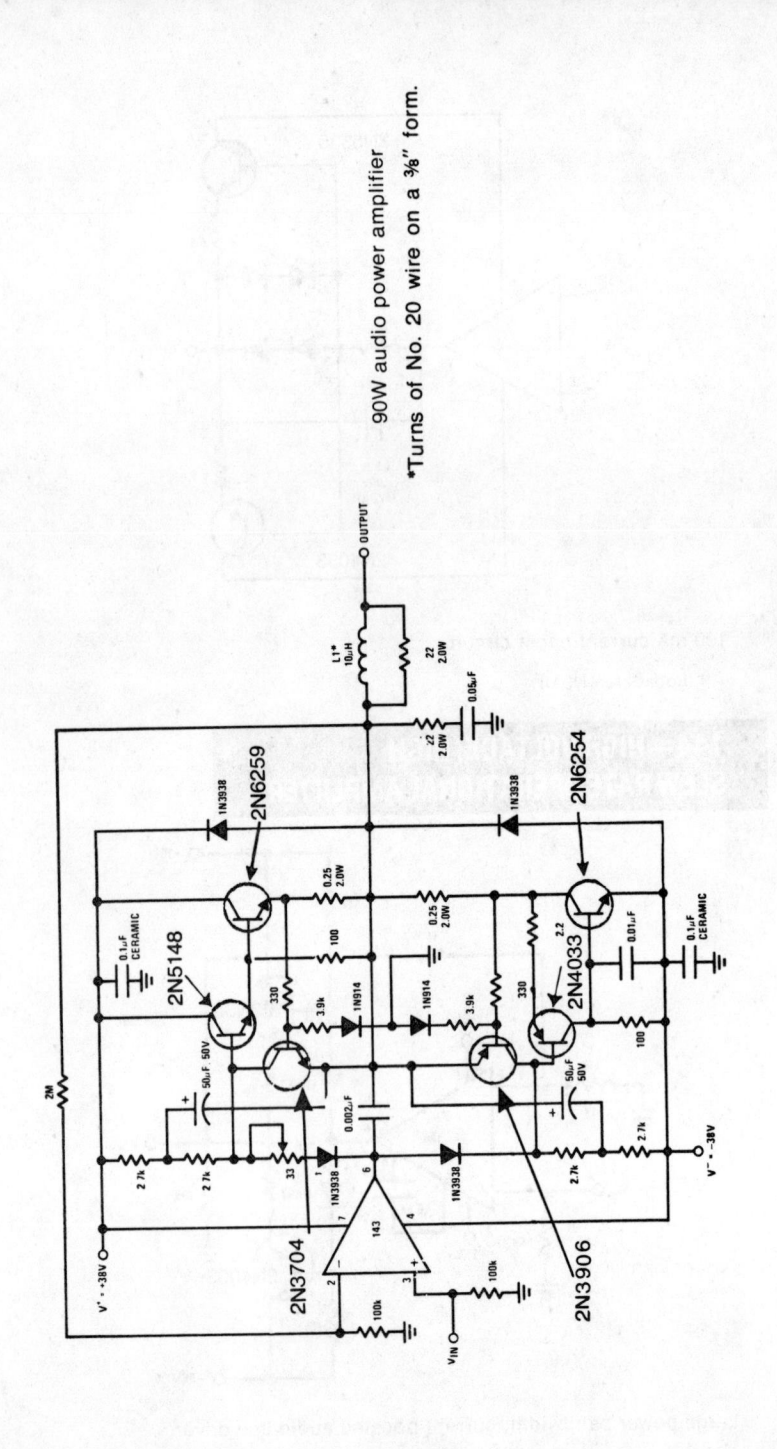

*Turns of No. 20 wire on a ⅜" form.

146—PROGRAMMABLE QUAD OPERATIONAL AMPLIFIER

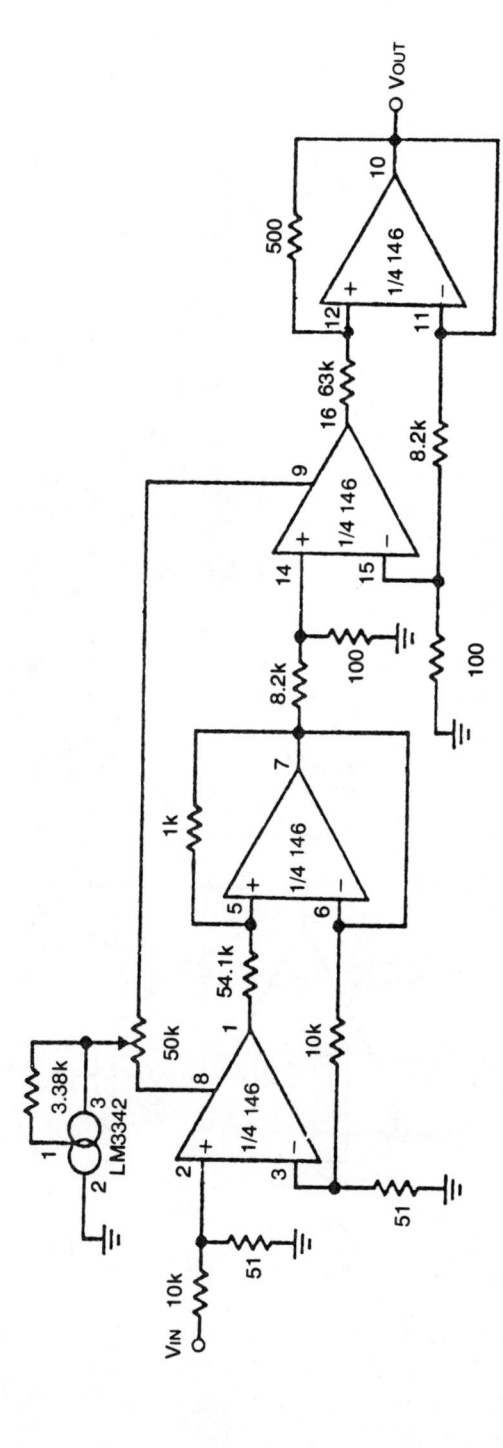

A 4th order Butterworth low pass capacitorless filter

Voice activated switch and amplifier

148 — QUAD 741 OPERATIONAL AMPLIFIER

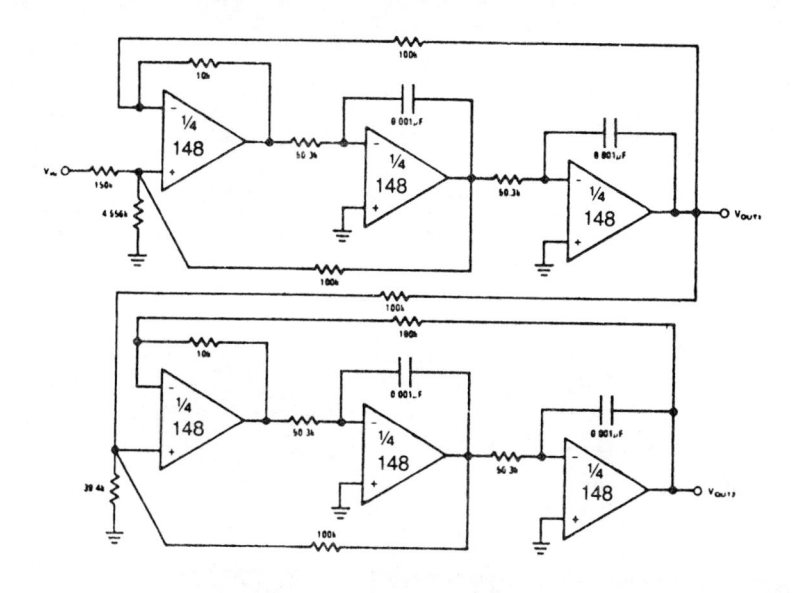

A 1 kHz 4 pole Butterworth

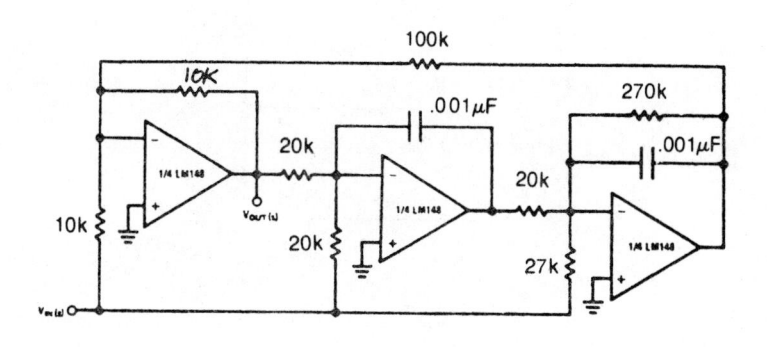

A 3 amplifier bi-quad notch filter

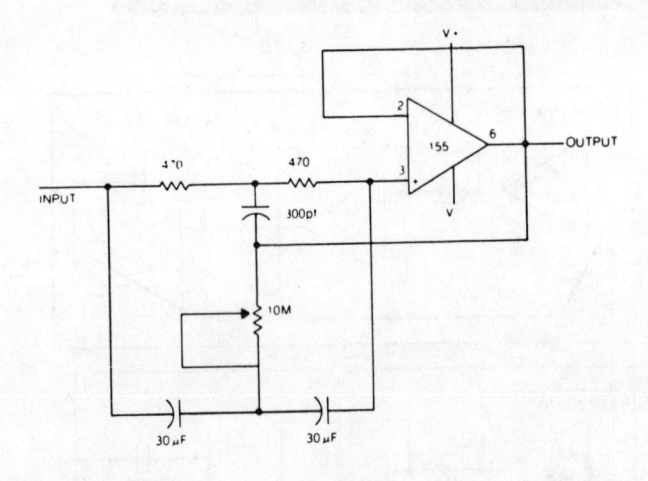

High Q notch filter

155—JFET INPUT OPERATIONAL AMPLIFIER

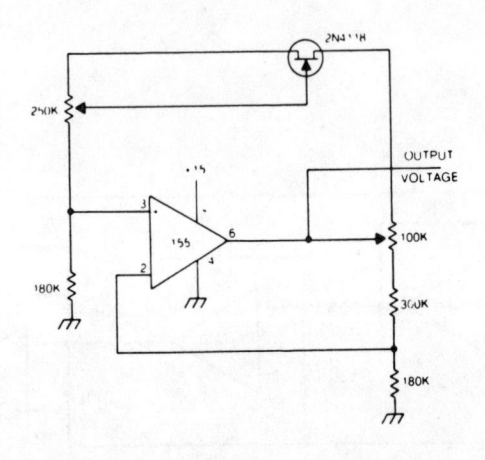

Adjustable voltage reference

High Q bandpass filter

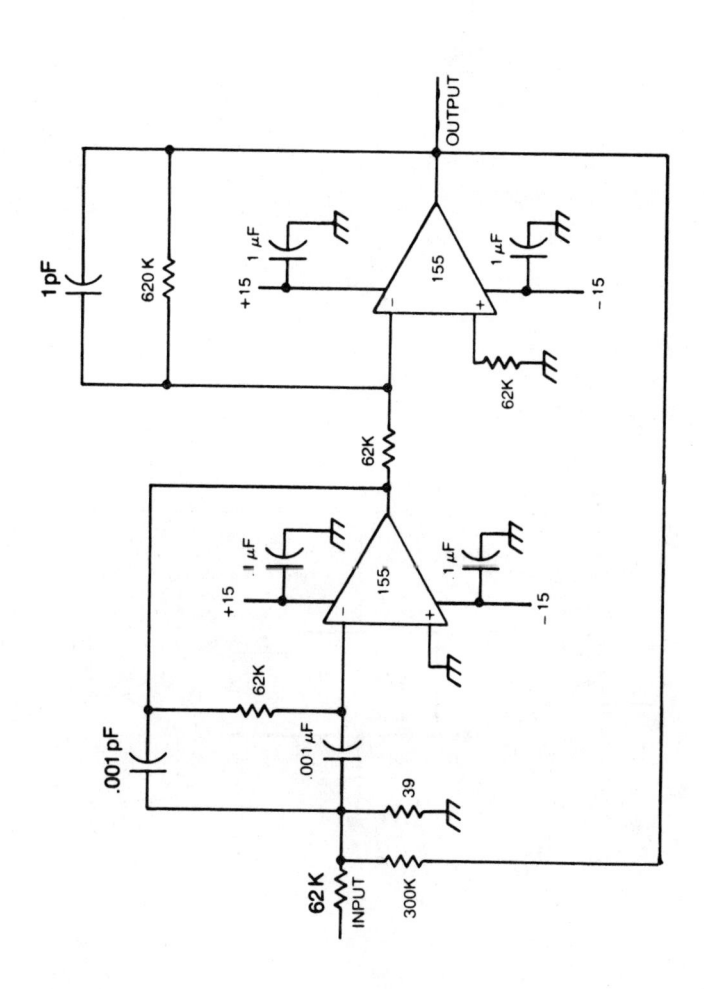

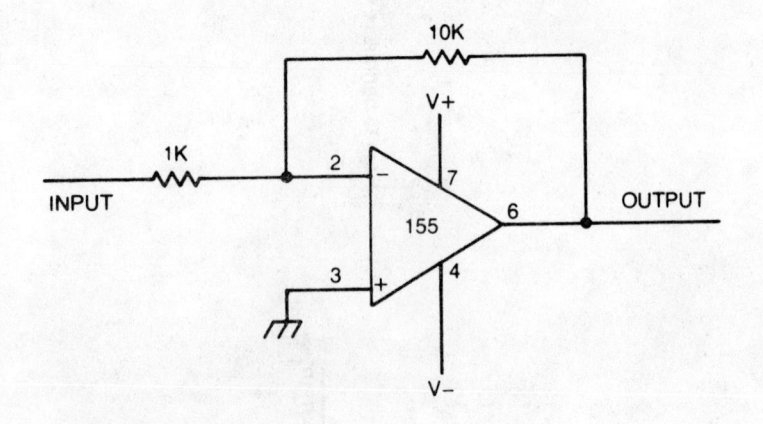

Large bandwidth amplifier

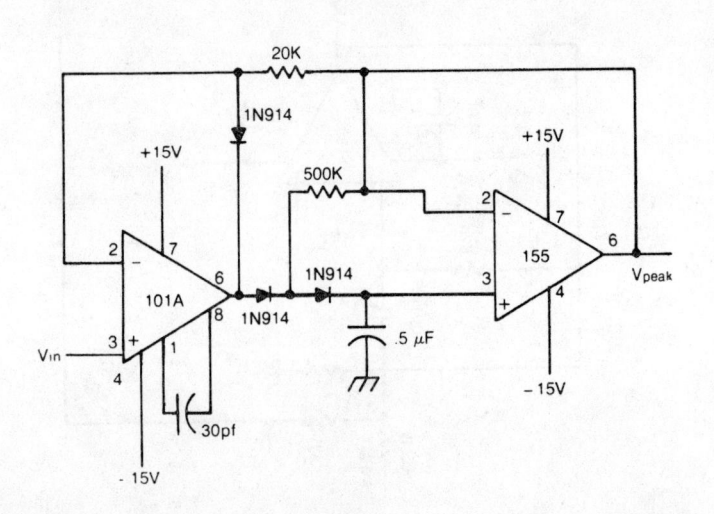

Peak voltage detector

Current boosted amplifier

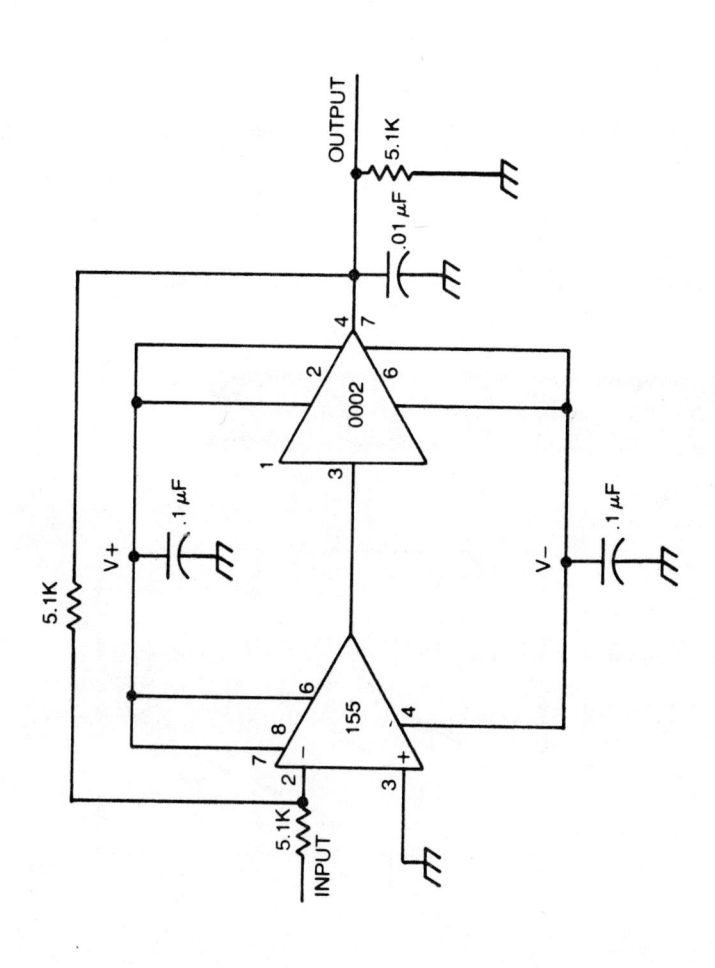

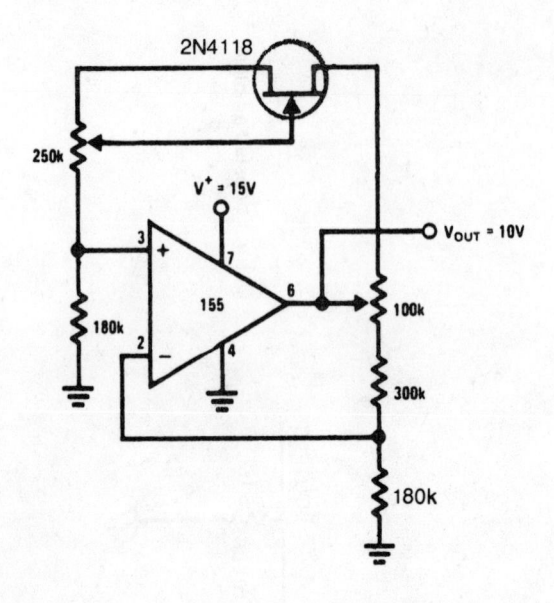

Low drift adjustable voltage reference

157 — WIDE BAND DECOMPENSATED JFET INPUT OPERATIONAL AMPLIFIER

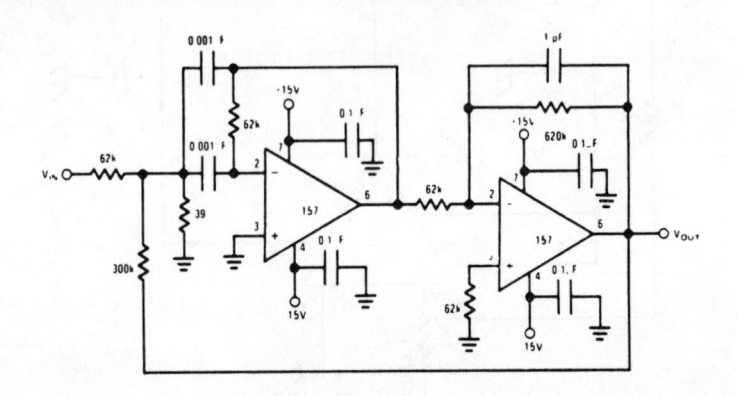

High Q bandpass filter

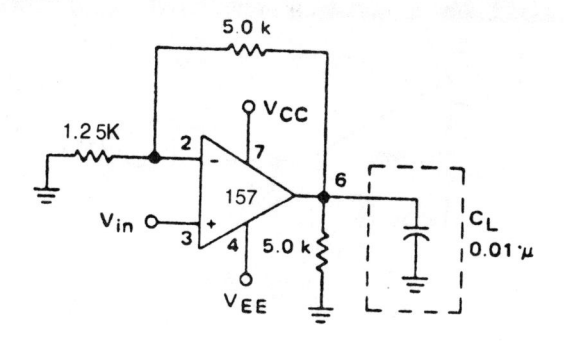

Driving capacitive loads

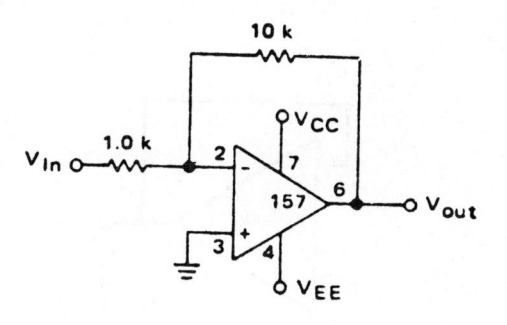

Large power bandwidth amplifier

105

158 — LOW POWER DUAL OPERATIONAL AMPLIFIER

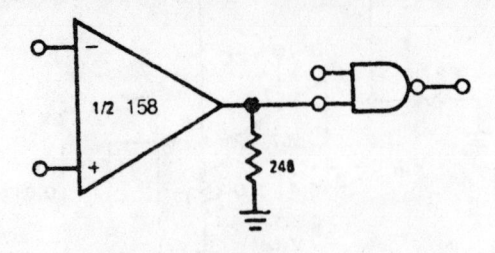

Driving TTL

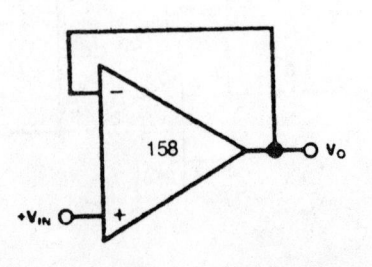

Voltage follower

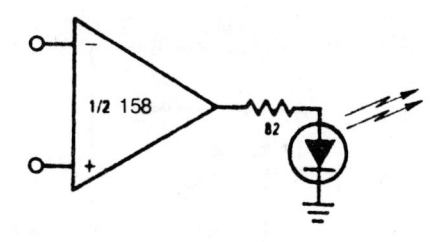

LED driver

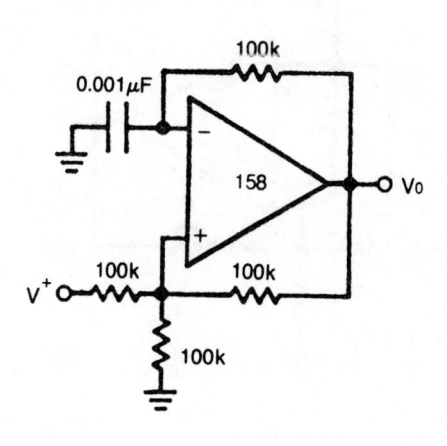

Squarewave oscillator

107

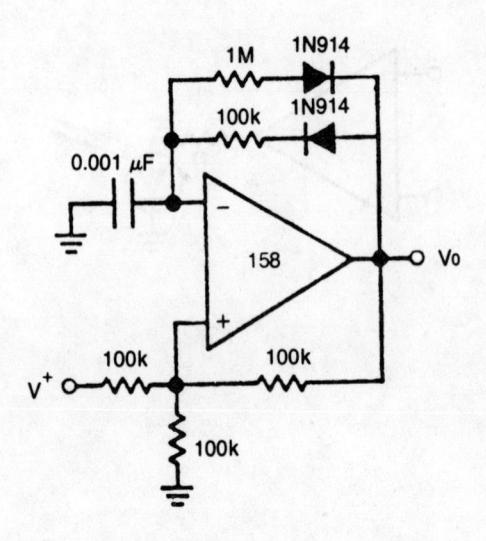

Pulse generator

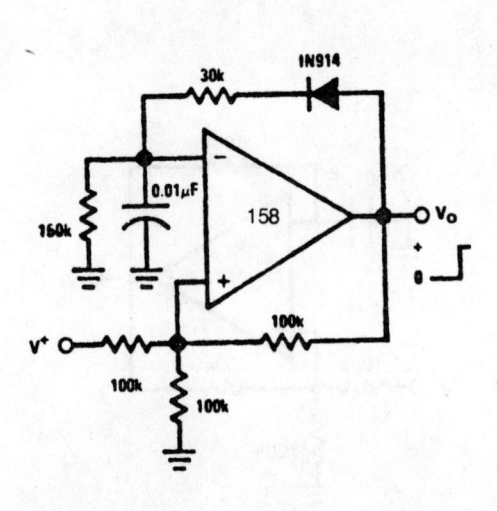

Pulse generator

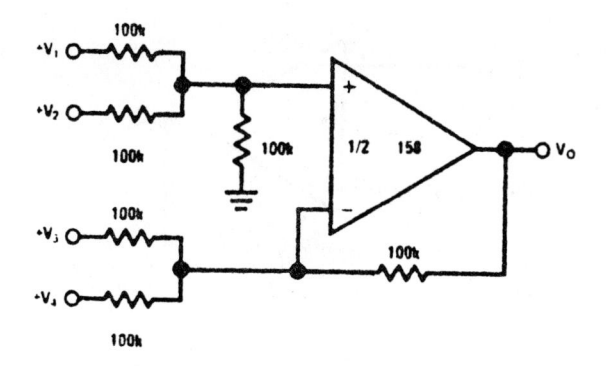

DC summing amplifier

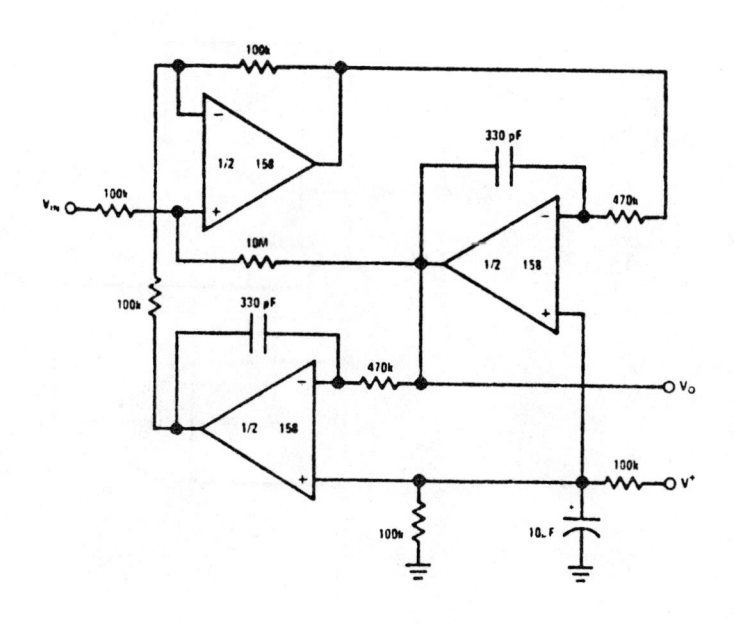

"BI-QUAD" RC active bandpass filter

109

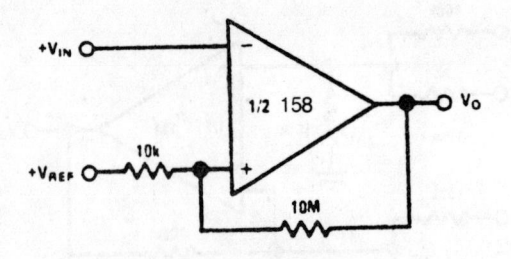

Comparator with hysteresis

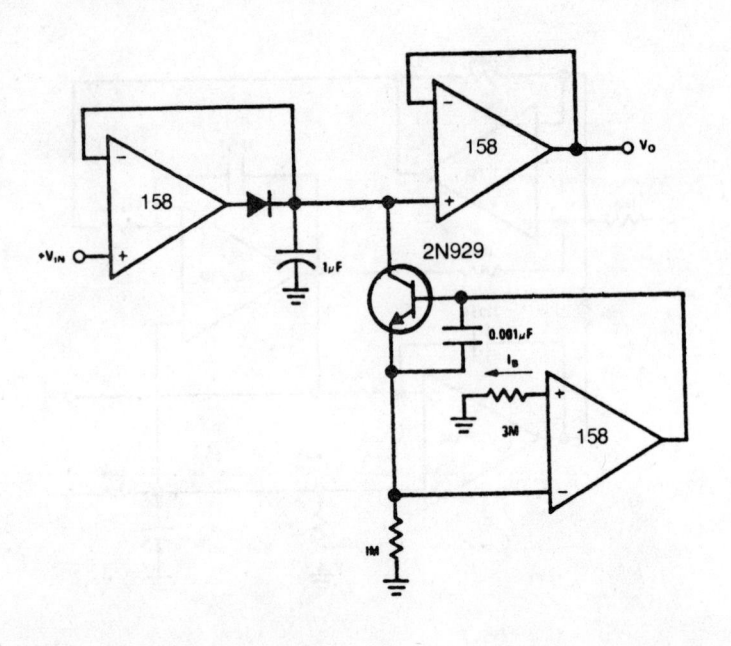

Low drift peak detector

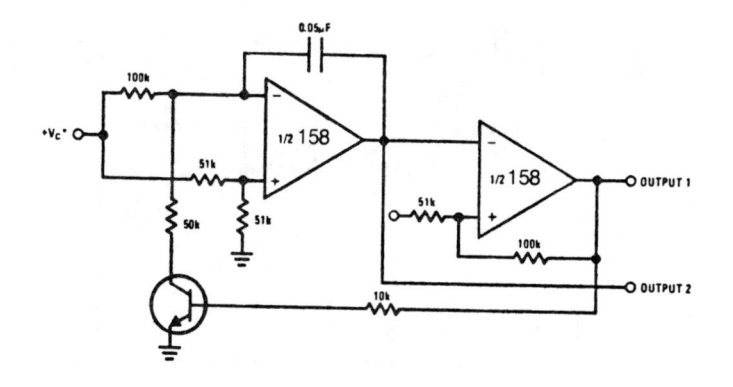

Voltage controlled oscillator (VCO)

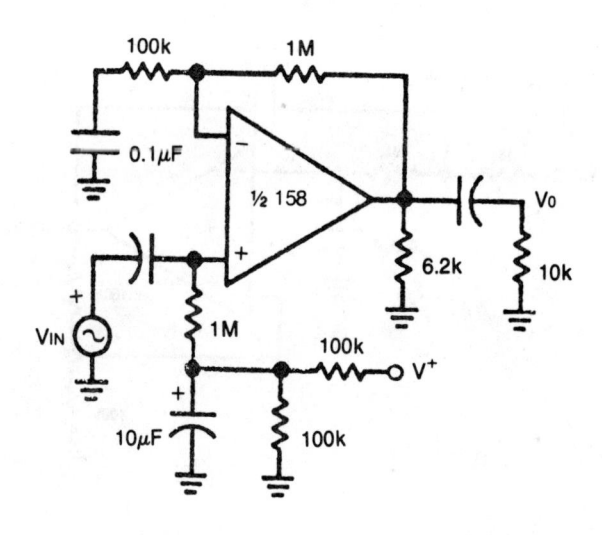

AC coupled noninverting amplifier

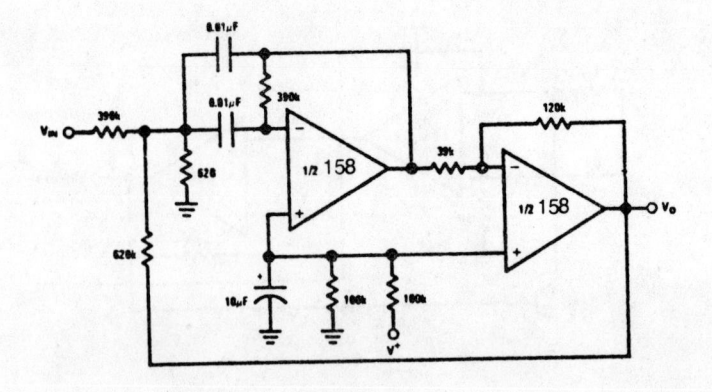

Bandpass active filter

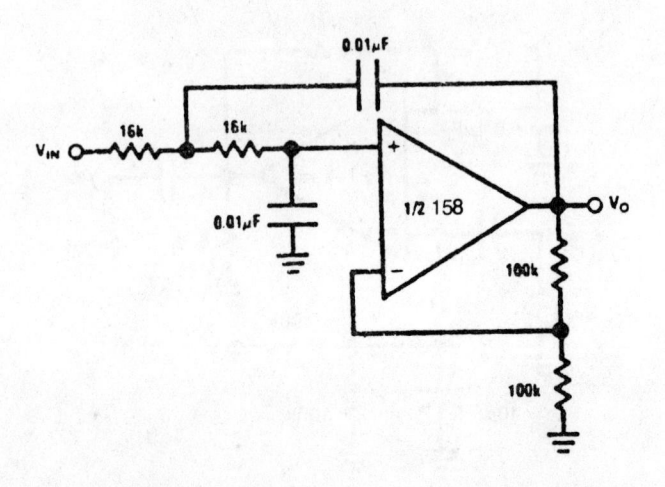

DC coupled low-pass RC active filter

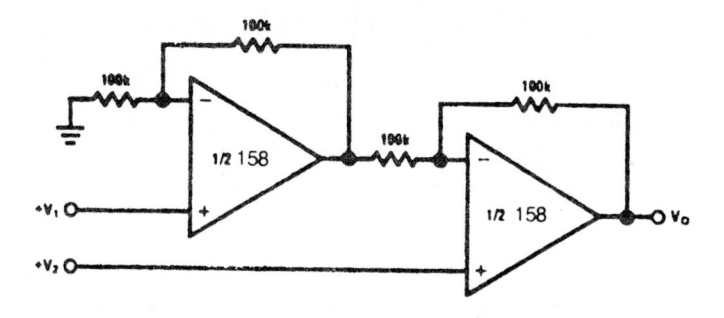

High input Z, dc differential amplifier

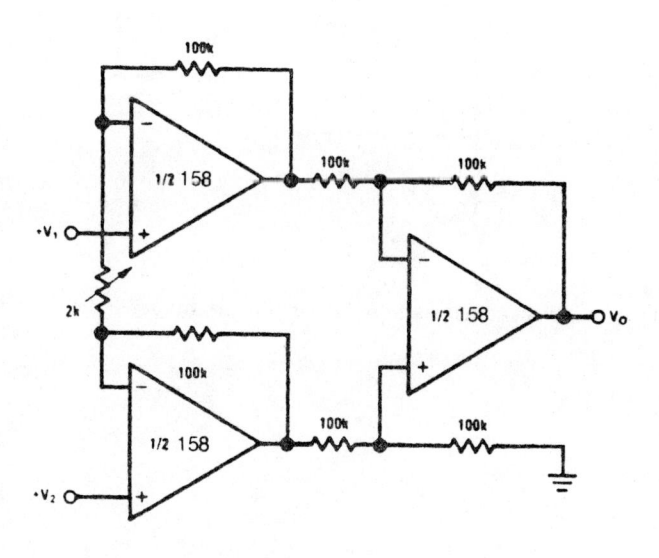

High input Z adjustable-gain dc instrumentation amplifier

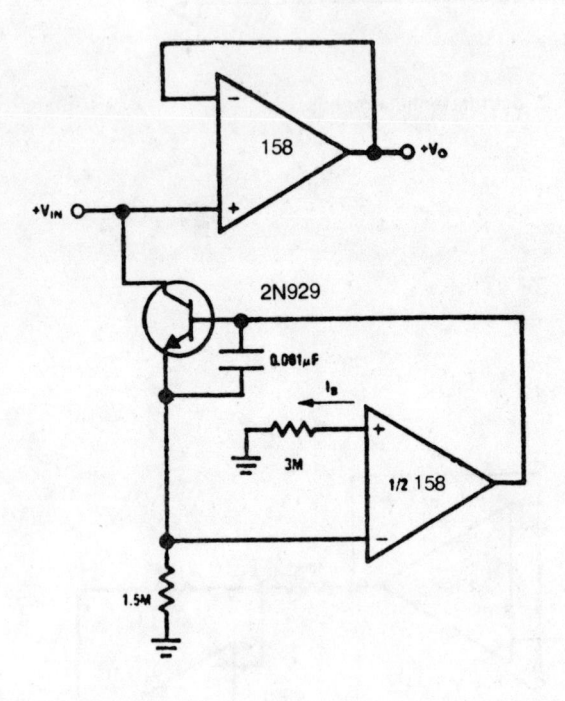

Using symmetrical amplifiers to reduce input current (general concept)

170—INTEGRATED AGC/SQUELCH AMPLIFIER

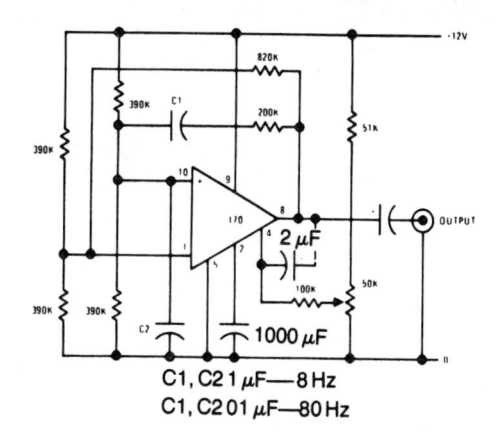

Wien-bridge oscillator

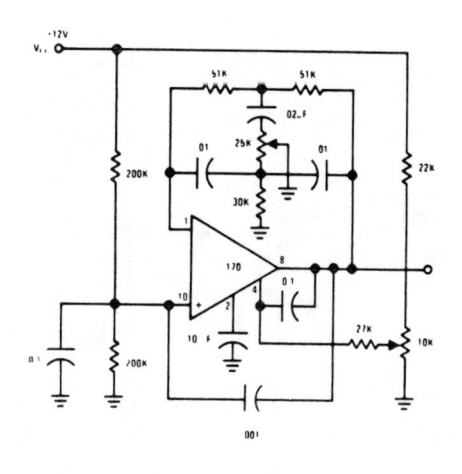

Decade tunable oscillator

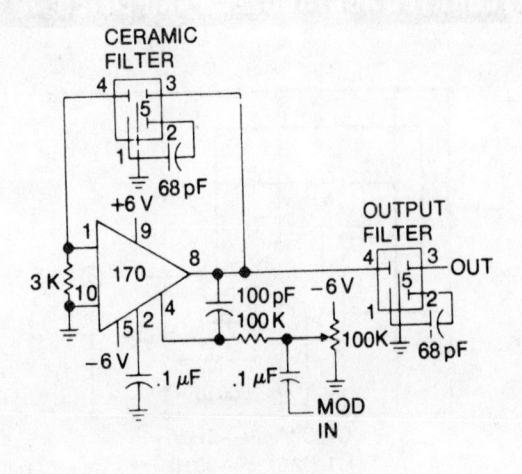

455 kHz modulated constant output oscillator

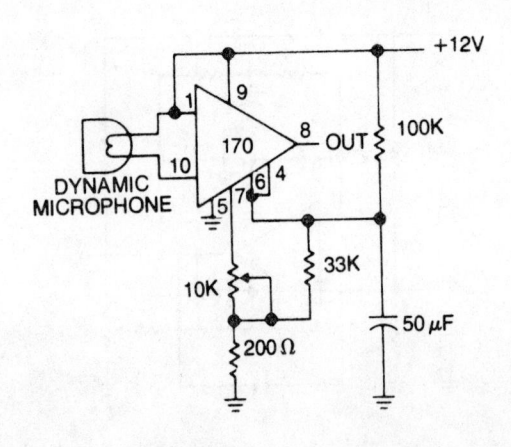

Squelched preamplifier with hysteresis

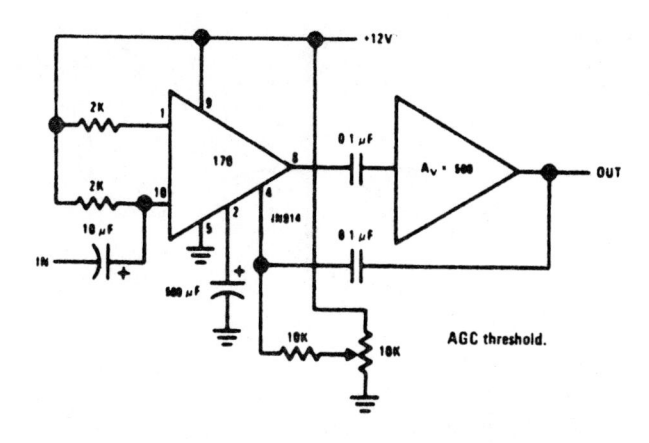

AGC using built-in detection, driven by additional system gain

193—LOW POWER LOW OFFSET VOLTAGE DUAL COMPARATORS

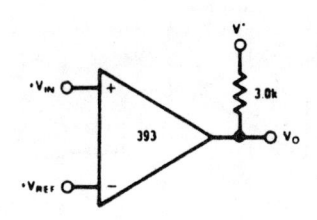

Basic comparator

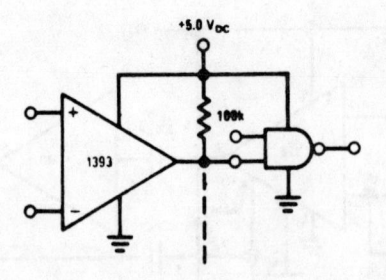

Driving CMOS

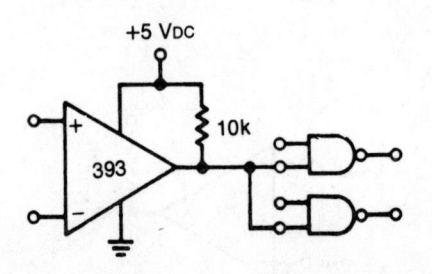

Driving TTL

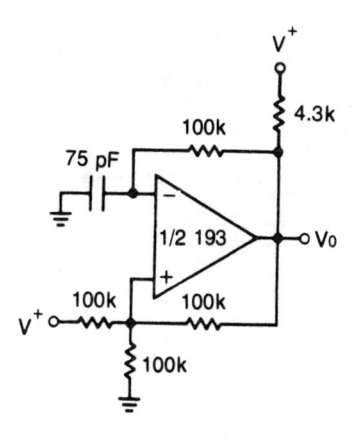

Squarewave oscillator

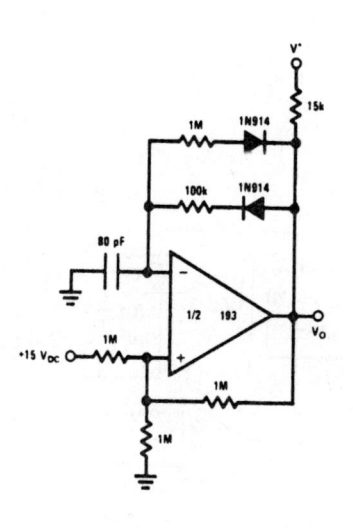

Pulse generator

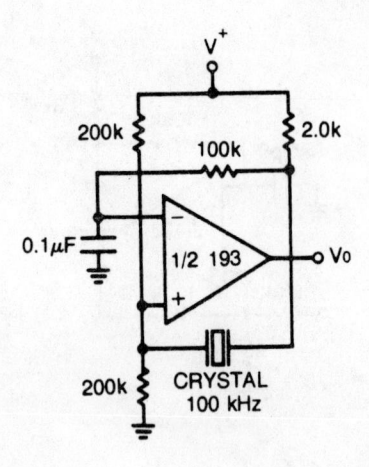

Crystal controlled oscillator

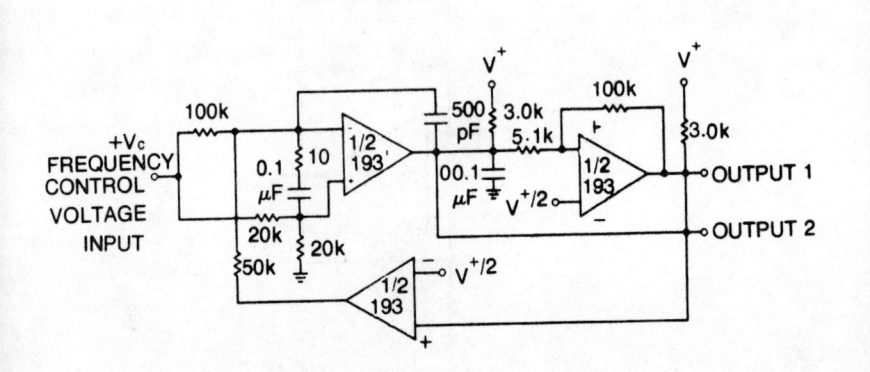

Two-decade high-frequency VCO

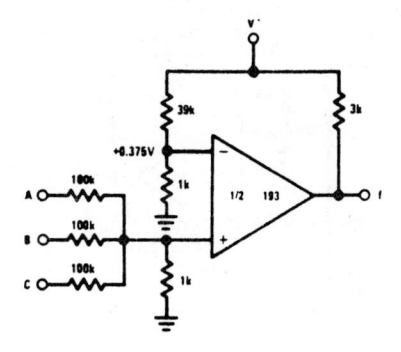

AND gate

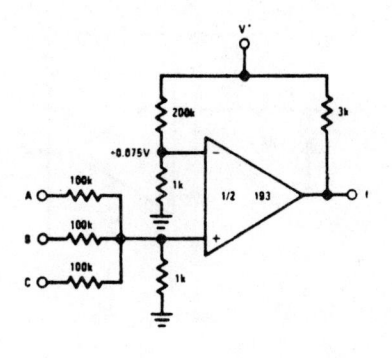

OR gate

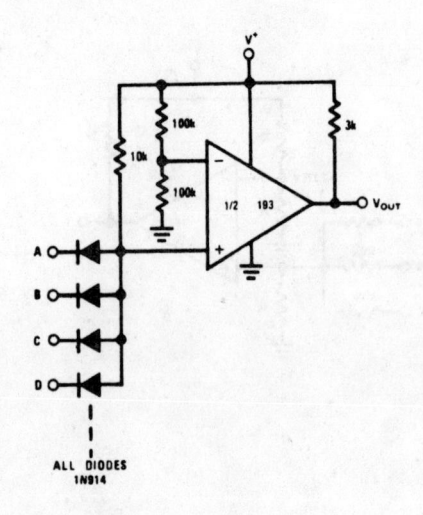

Large fan-in AND gate

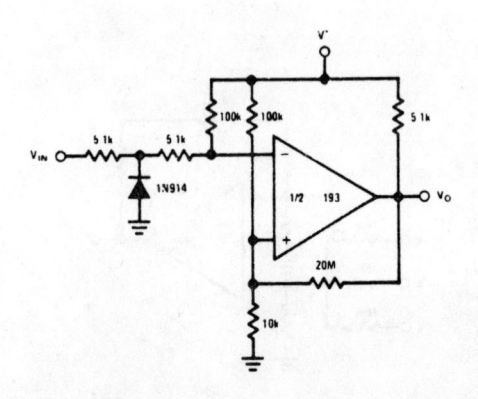

Zero crossing detector (single power supply)

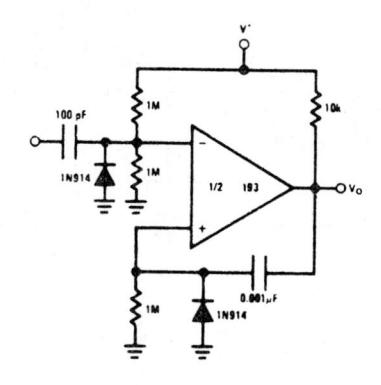

One-shot multivibrator

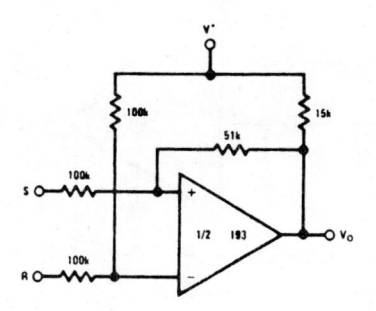

Bi-stable multivibrator

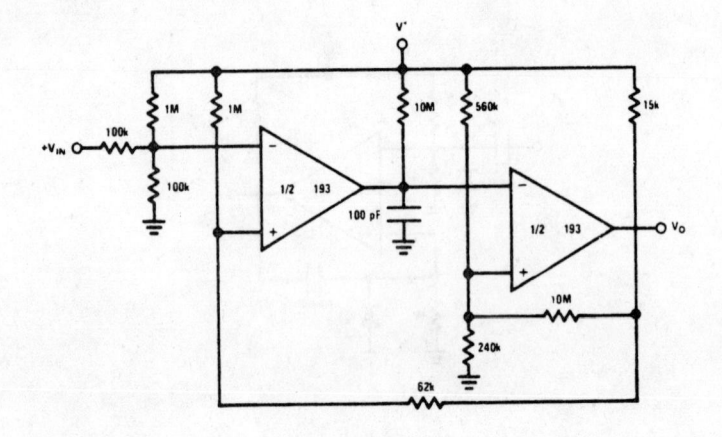

One-shot multivibrator with input lock out

198 — MONOLITHIC SAMPLE AND HOLD CIRCUIT

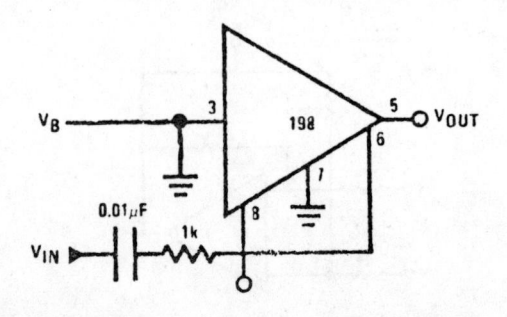

Sample and difference circuit

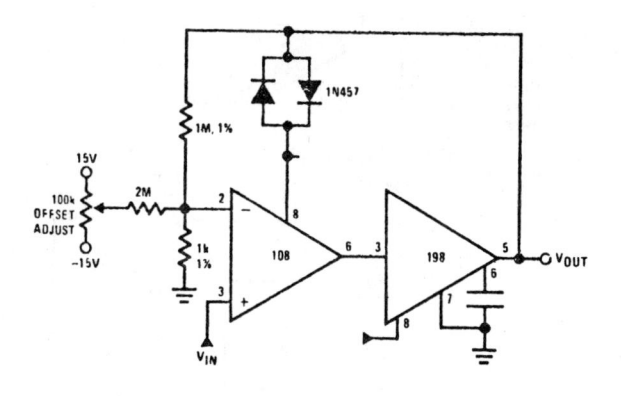

X1000 sample & hold

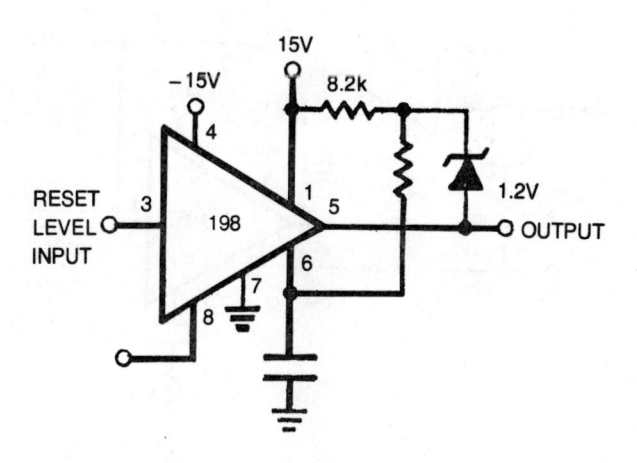

Ramp generator with variable reset level

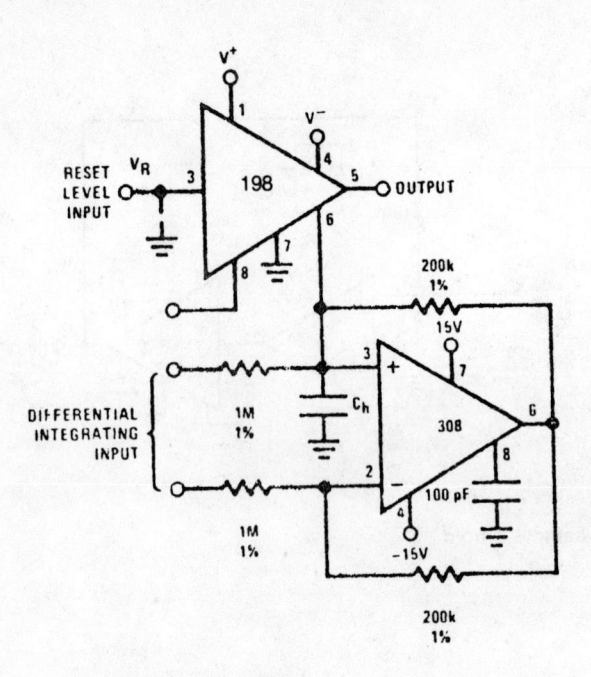

Integrator with programmable reset level

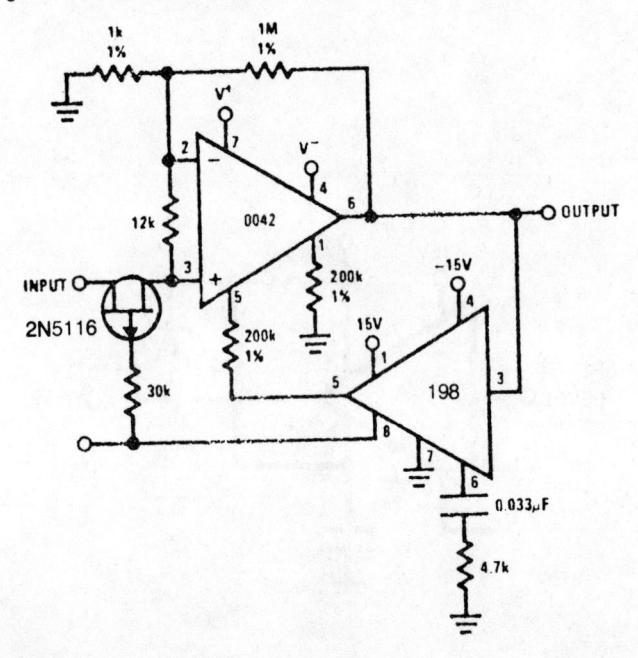

Reset stabilized amplifier (gain of 1000)

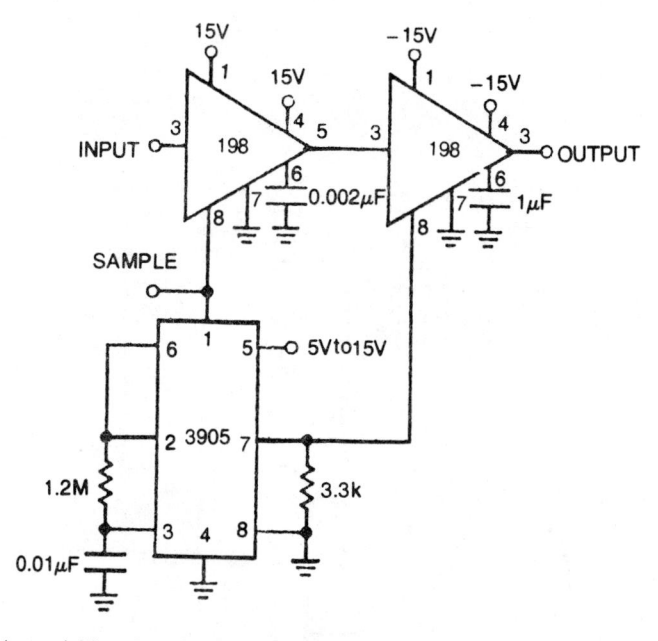

Fast acquisition, low droop sample & hold

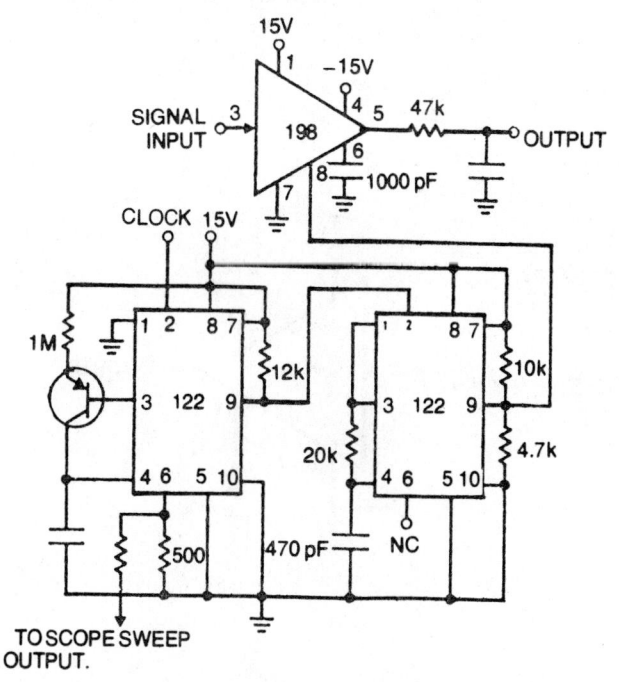

Synchronous correlator for recovering signals below noise level

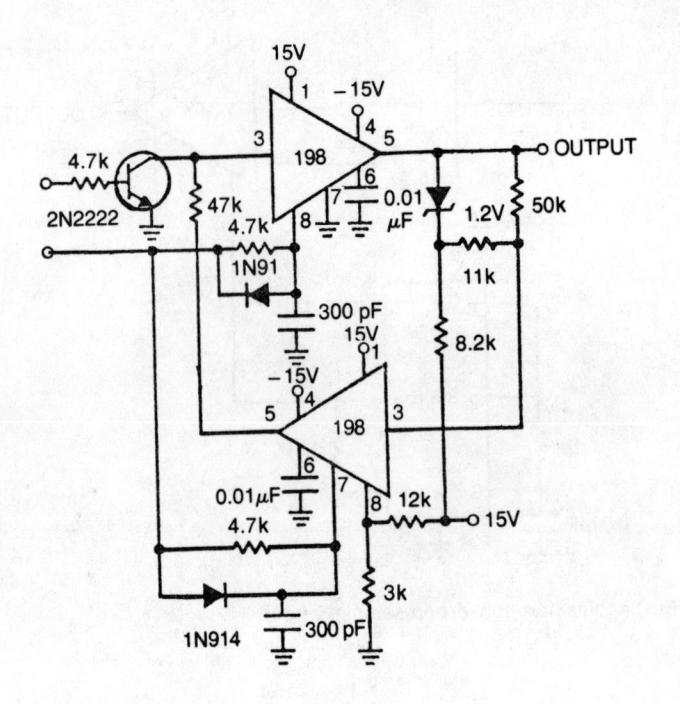

Staircase generator

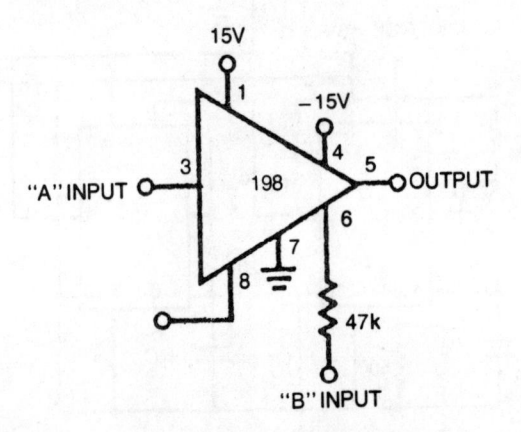

2-channel switch

311 — VOLTAGE COMPARATOR

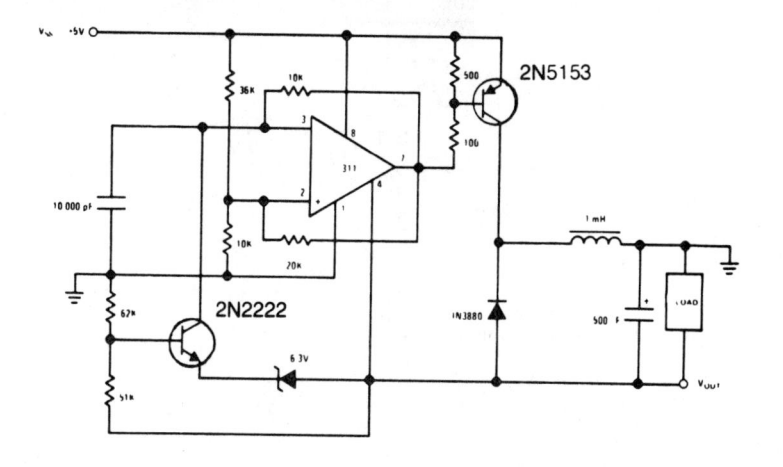

Switching regulator for voltage conversion

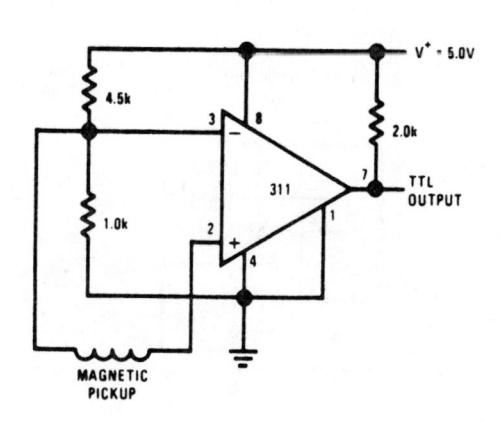

Detector for magnetic transducer

129

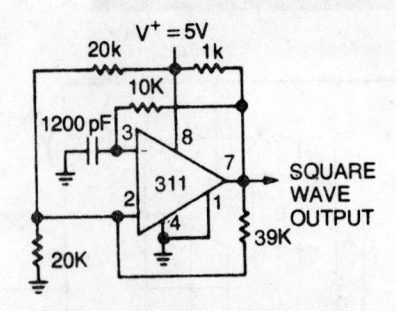

100 kHz free running multivibrator

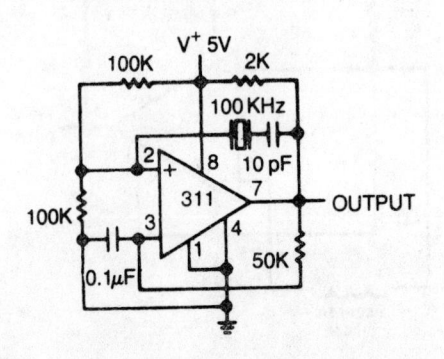

Crystal oscillator

318 — WIDE TEMPERATURE RANGE OPERATIONAL AMPLIFIER

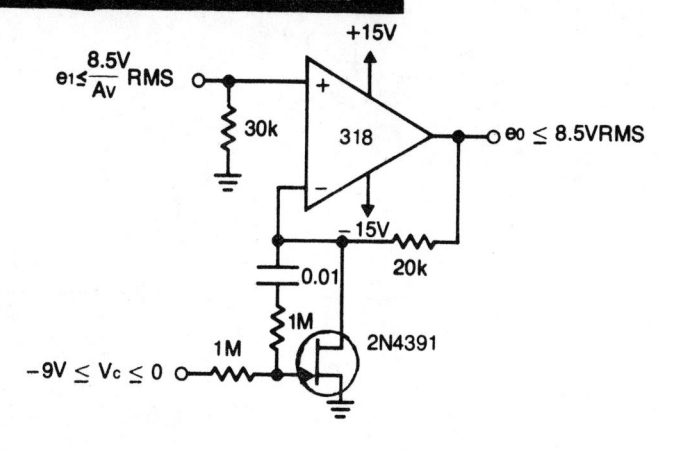

Amplifier with gain range = 1 – 1000

324 — QUAD OPERATIONAL AMPLIFIER

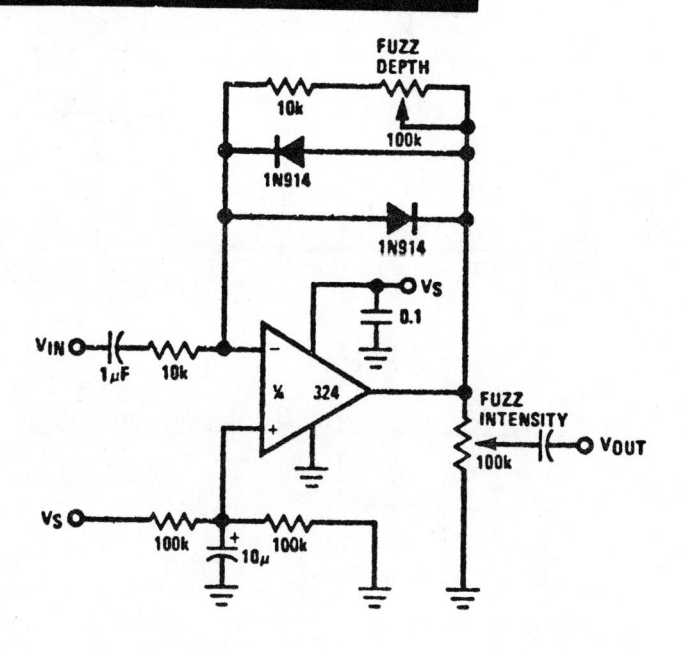

"Fuzz" circuit

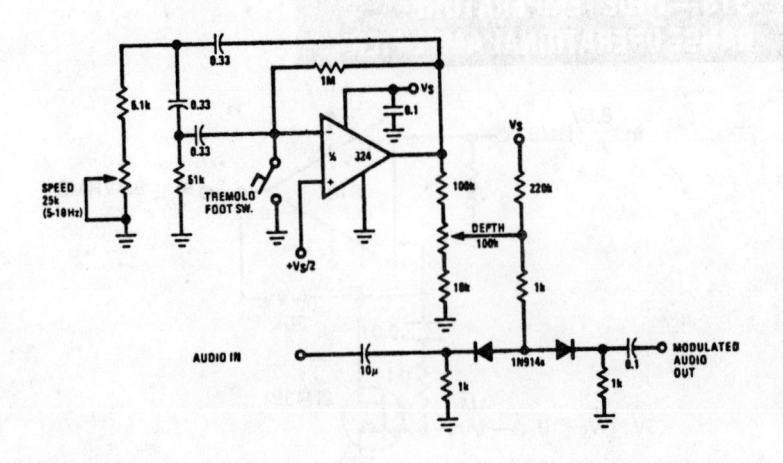

Tremolo circuit

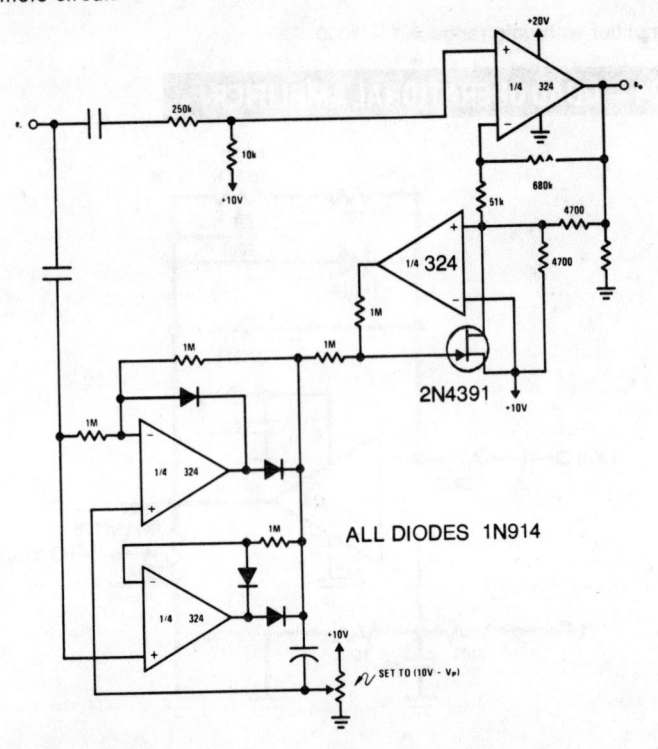

Volume expander circuit

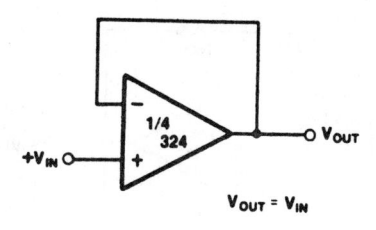

Voltage follower

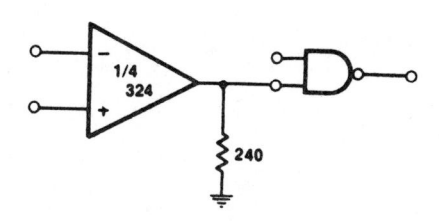

Driving TTL

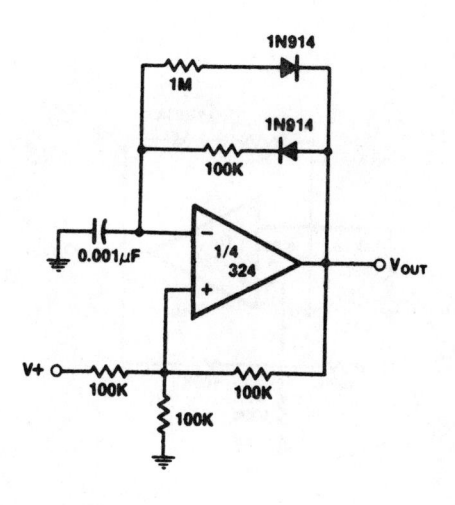

Pulse generator

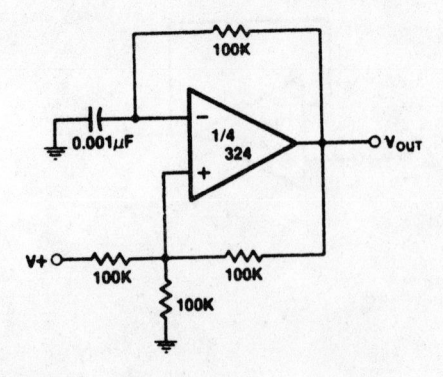

Squarewave oscillator

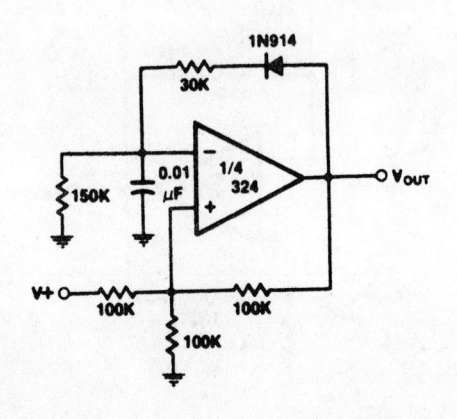

Pulse generator

339 — QUAD VOLTAGE COMPARATOR

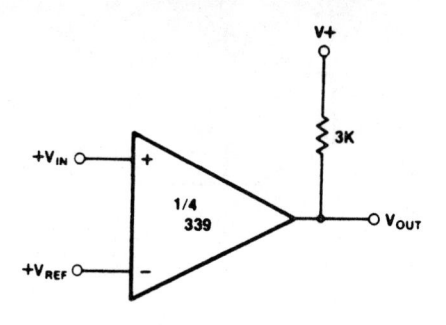

Basic comparator

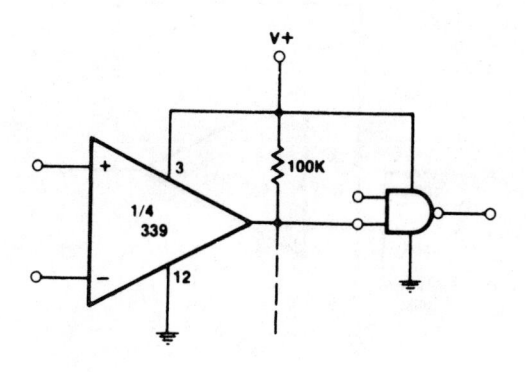

Driving CMOS

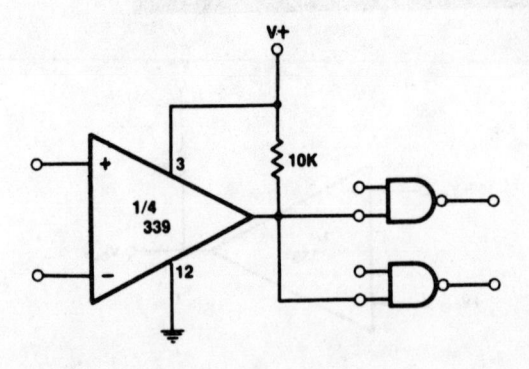

Driving TTL

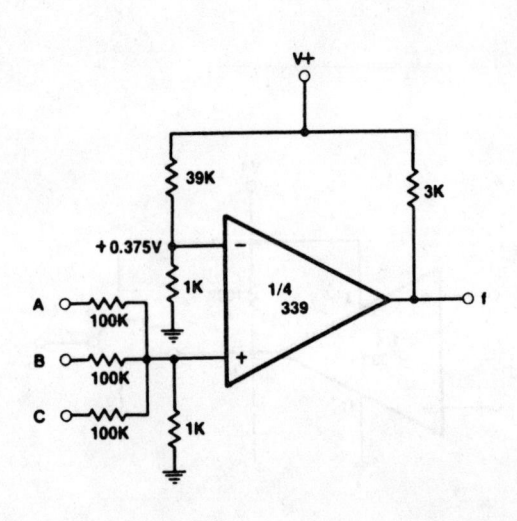

AND gate

136

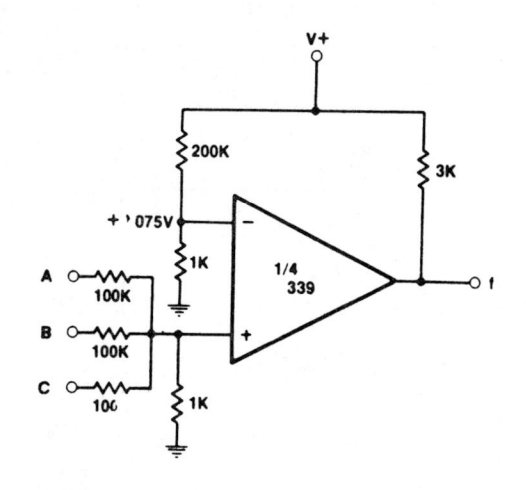

OR gate

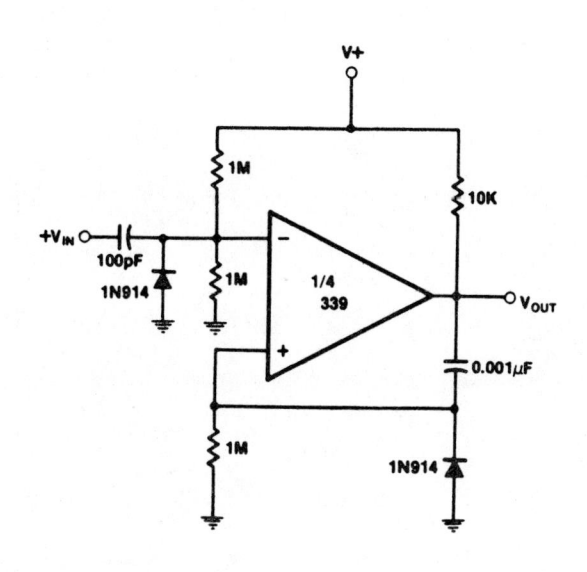

One-shot multivibrator

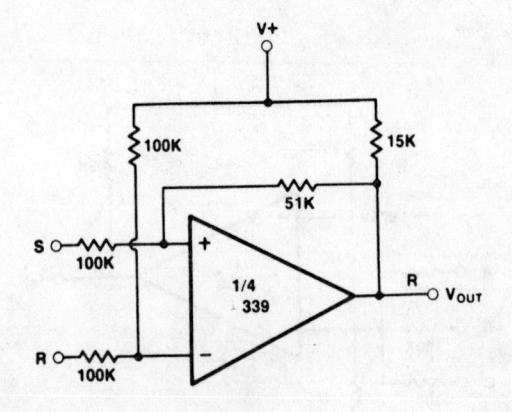

Bi-stable multivibrator

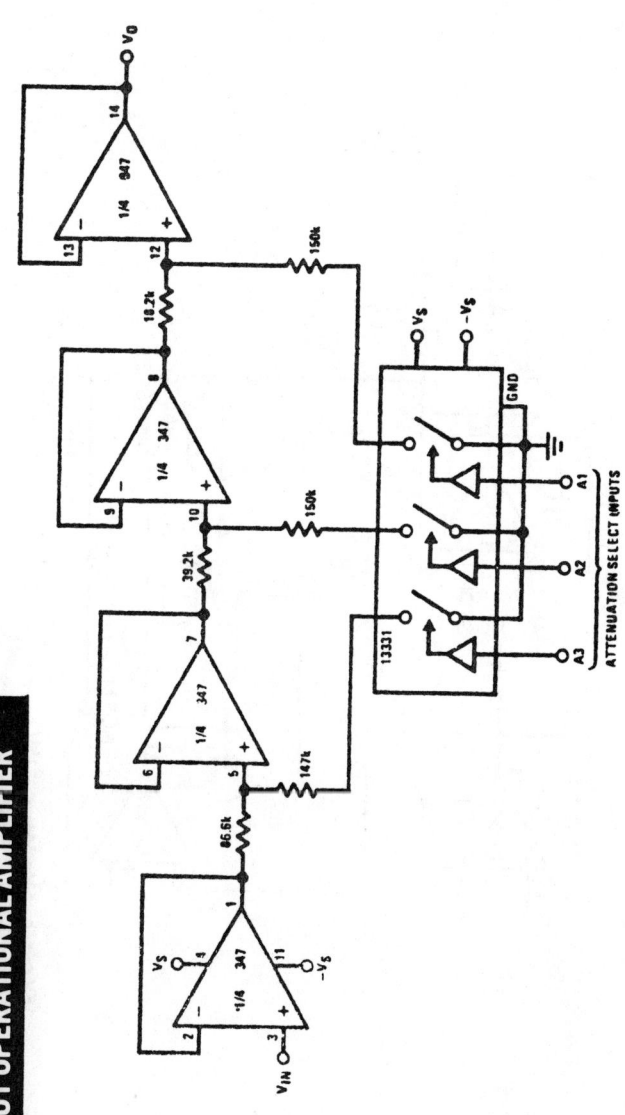

347 — WIDE BANDWIDTH QUAD JFET INPUT OPERATIONAL AMPLIFIER

Digitally selectable precision attenuator

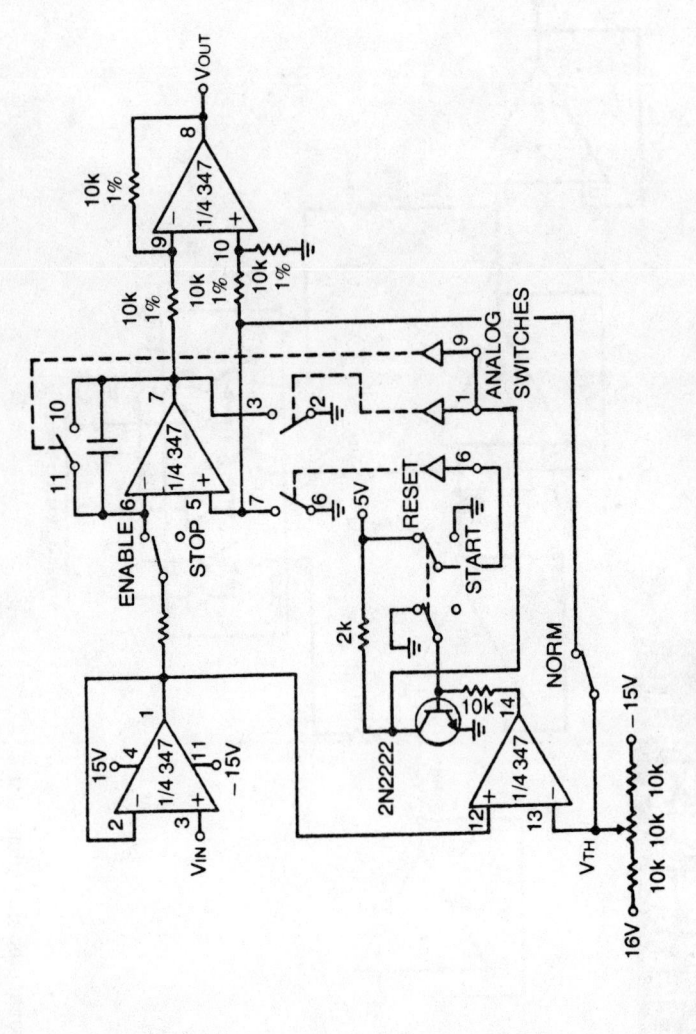

Long time integrator with reset, hold and starting threshold adjustment

Universal state variable filter

348 — QUAD 741 OPERATIONAL AMPLIFIER

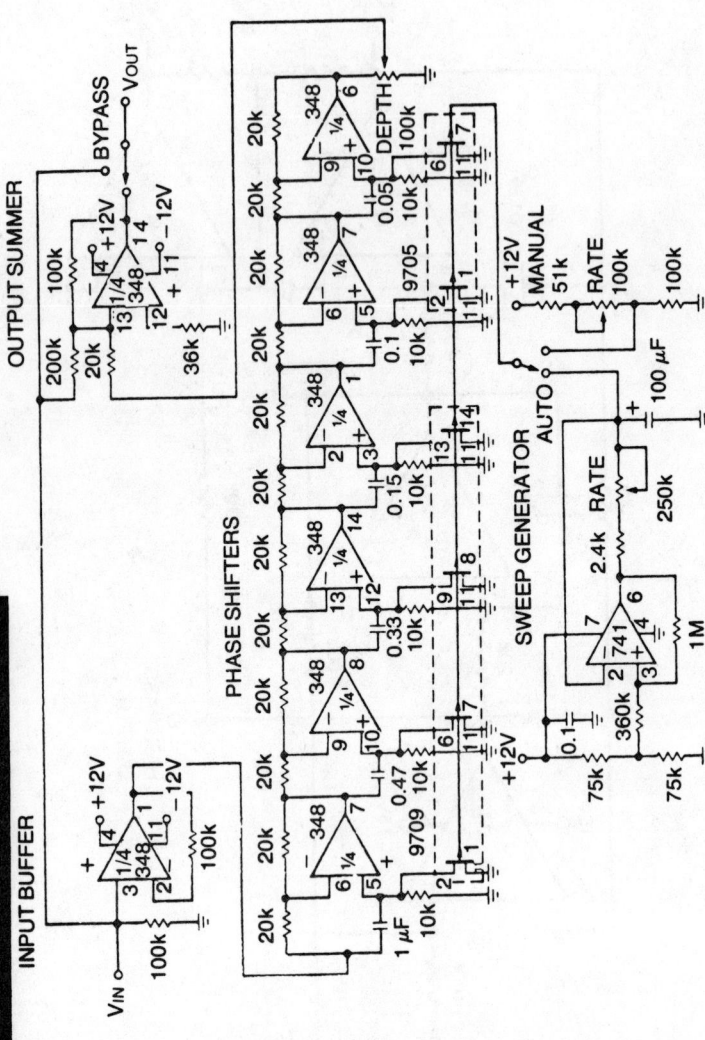

Phase Shifter

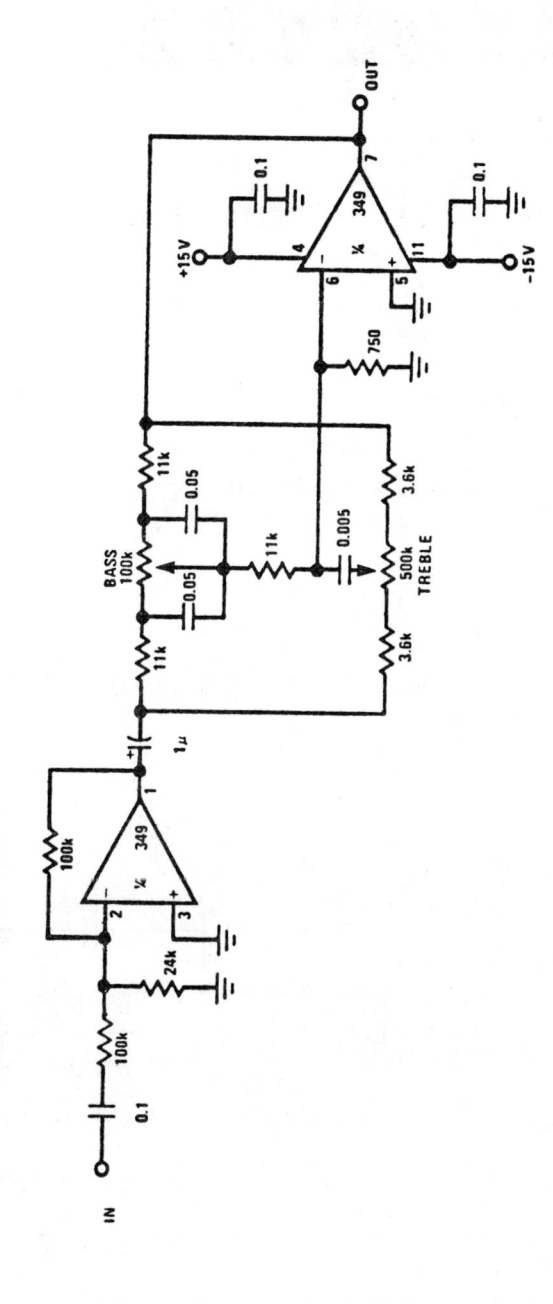

349 — WIDE BAND DECOMPENSATED QUAD 741 OPERATIONAL AMPLIFIER

Active bass & treble tone control with buffer

353 — WIDE BANDWIDTH DUAL JFET INPUT OPERATIONAL AMPLIFIER

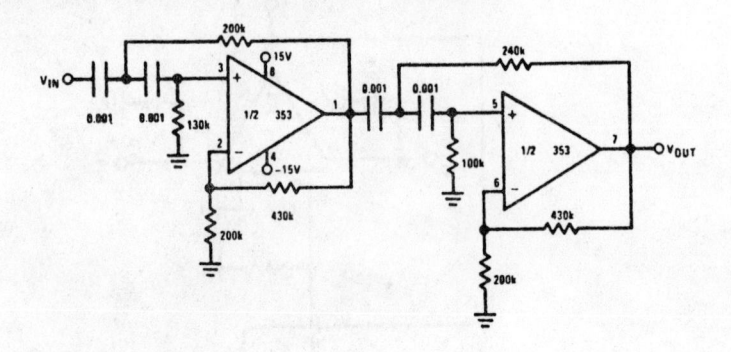

Fourth order high pass Butterworth filter

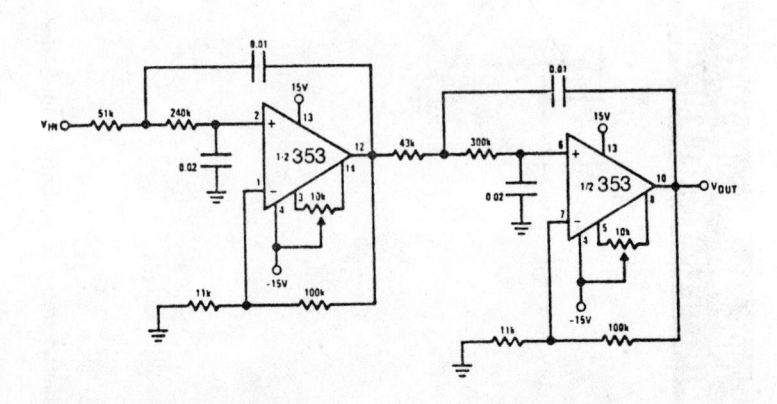

Fourth order low pass Butterworth filter

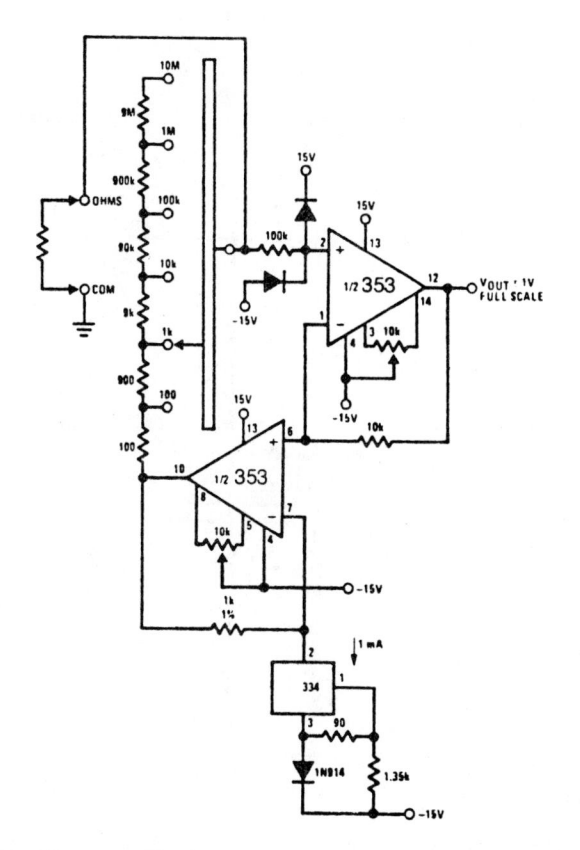

Ohms to volts converter

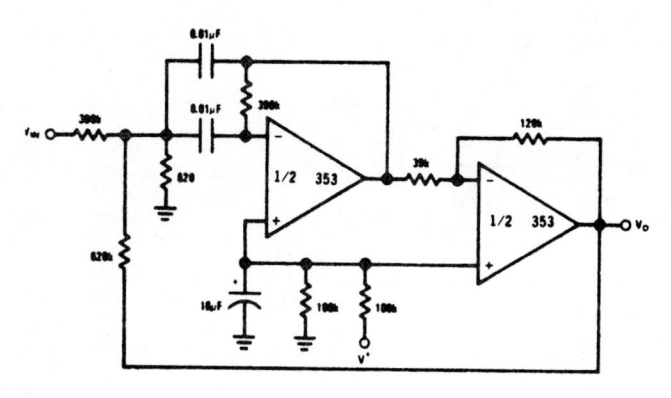

Bandpass active filter

Three-band active tone control

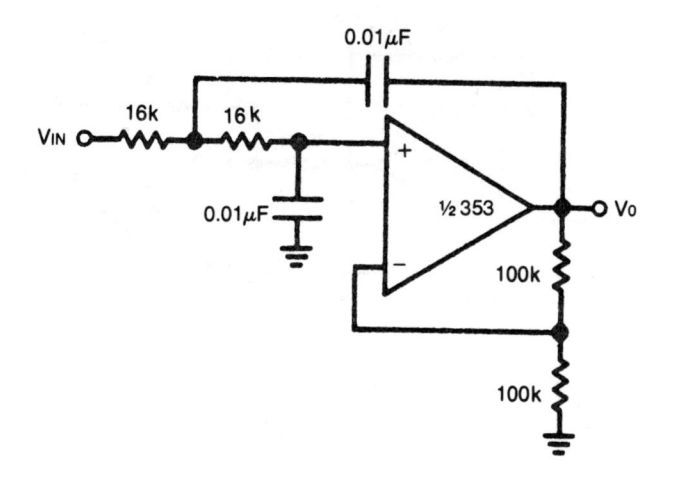

DC coupled low-pass RC active filter

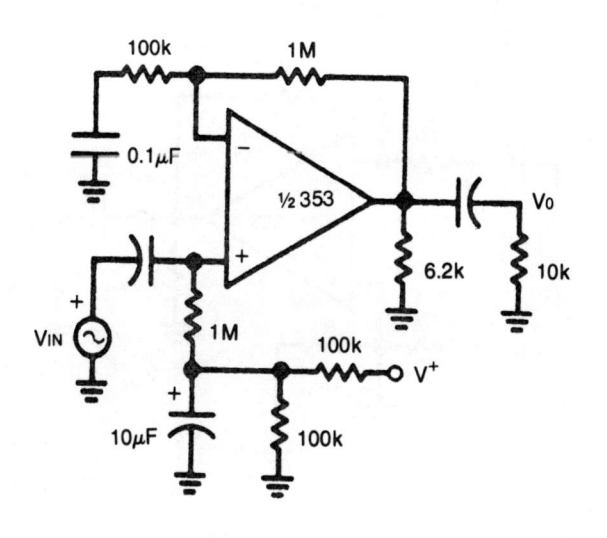

AC coupled noninverting amplifier

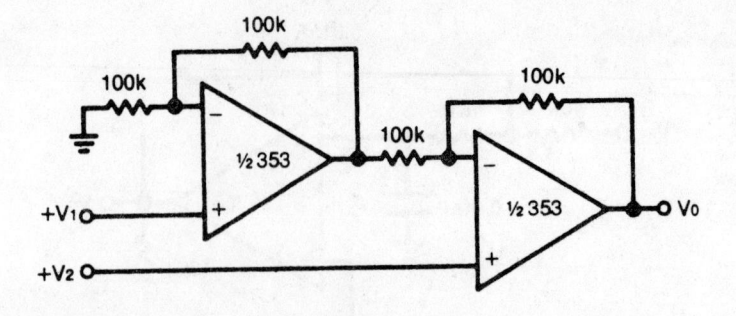

High input Z, dc differential amplifier

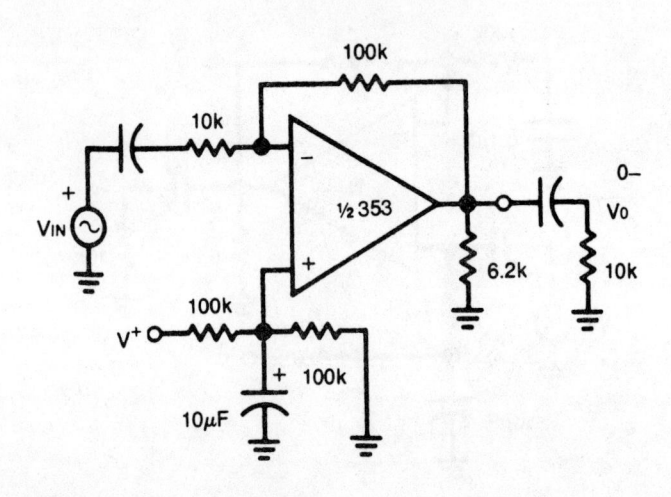

AC coupled inverting amplifier

148

356 — LOW OFFSET MONOLITHIC JFET INPUT OPERTIONAL AMPLIFIER

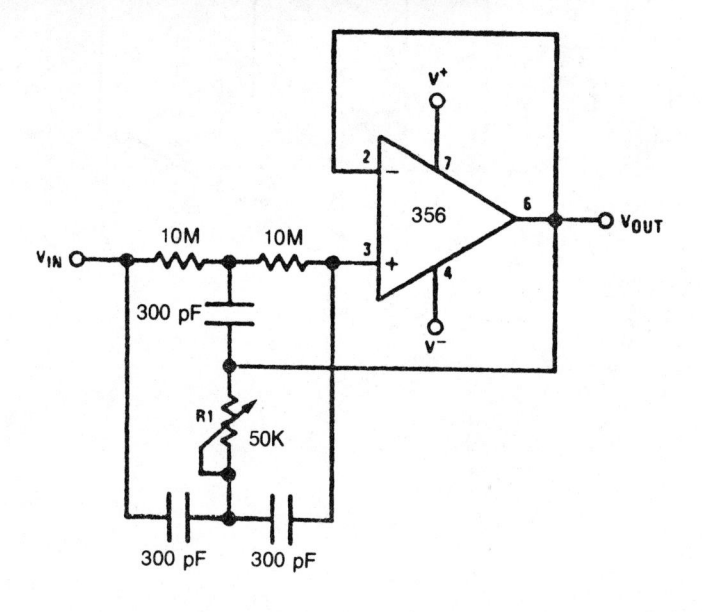

High Q notch filter

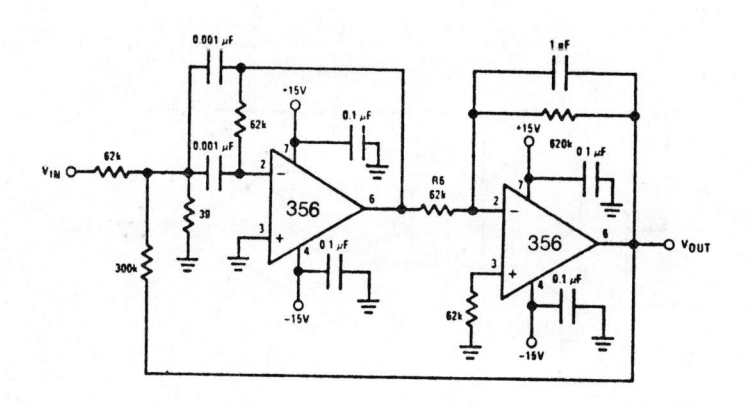

High Q bandpass filter

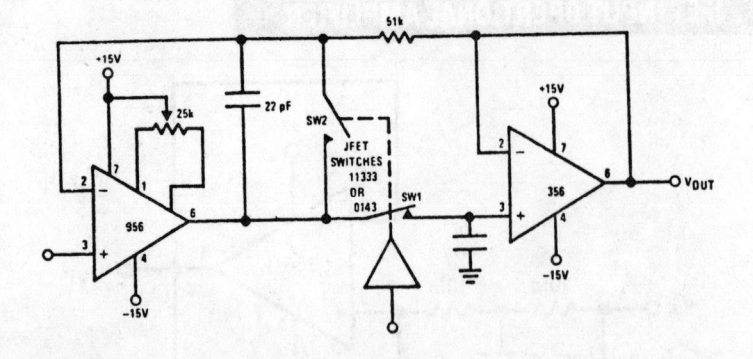

High accuracy sample and hold

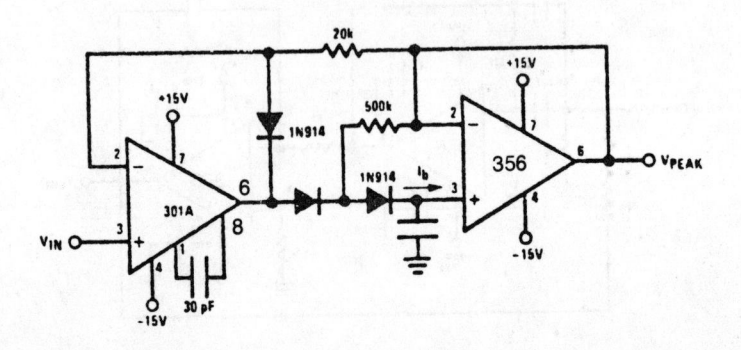

Low drift peak detector

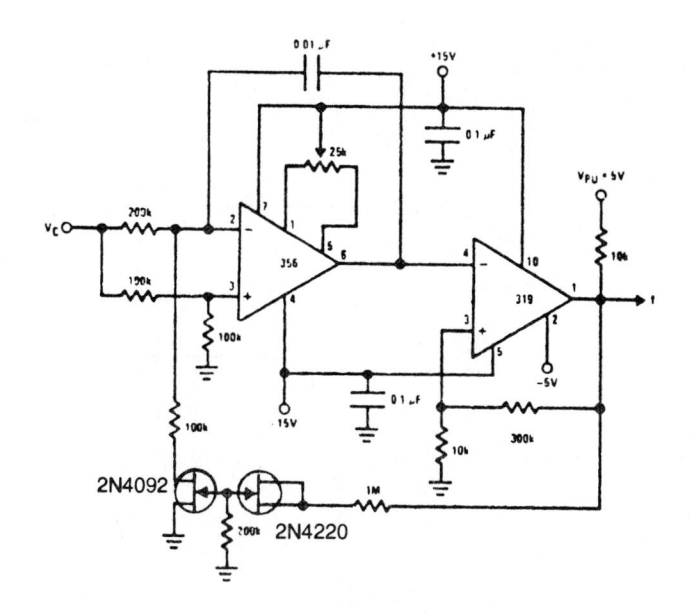

3 decades VCO

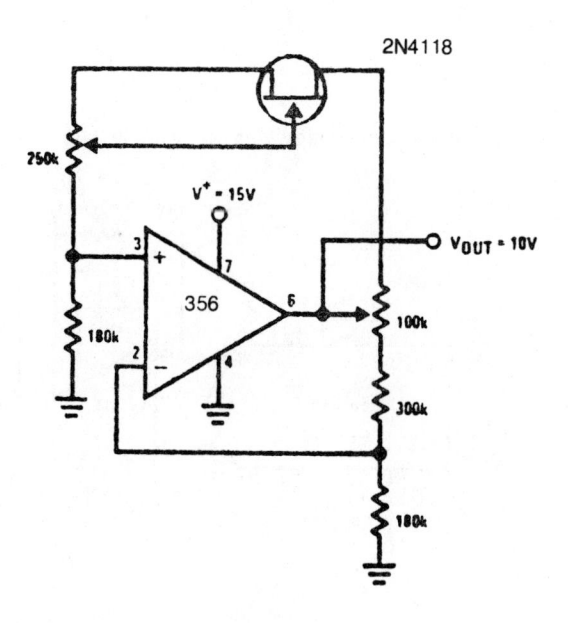

Low drift adjustable voltage reference

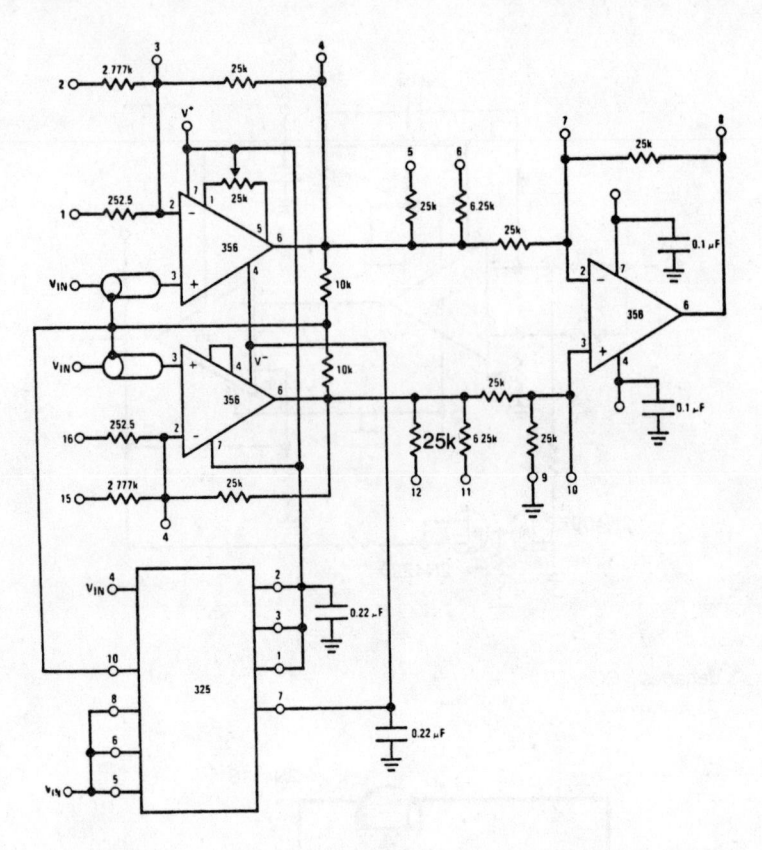

High CMRR, low drift instrumentation amplifier with floating input stage

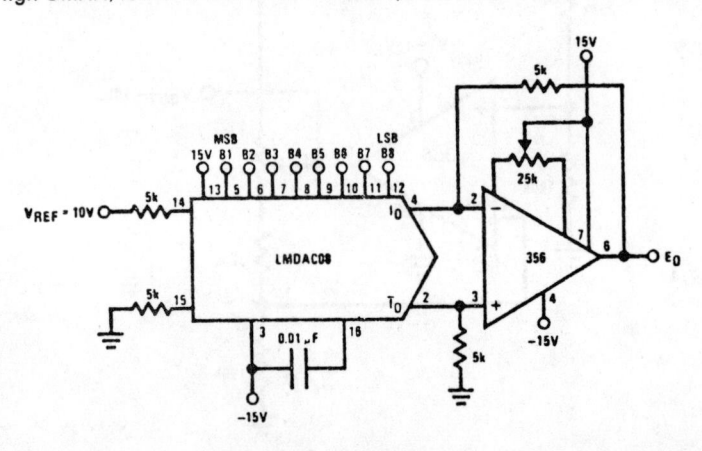

8-bit D/A converter with symmetrical offset binary operation

152

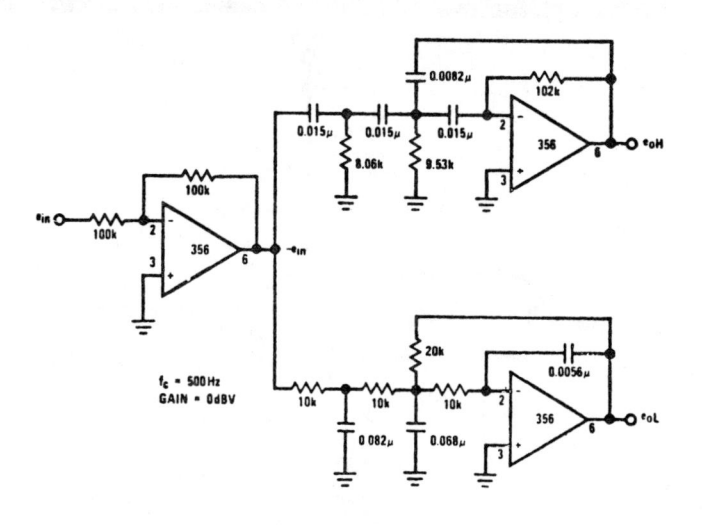

Active crossover network

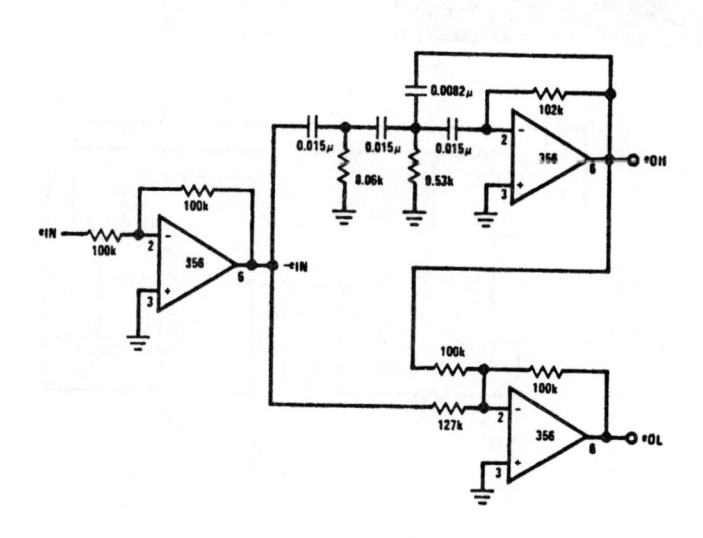

Asymmetrical 3rd order Butterworth active crossover network

153

357 — MONOLITHIC JFET INPUT OPERATIONAL AMPLIFIER

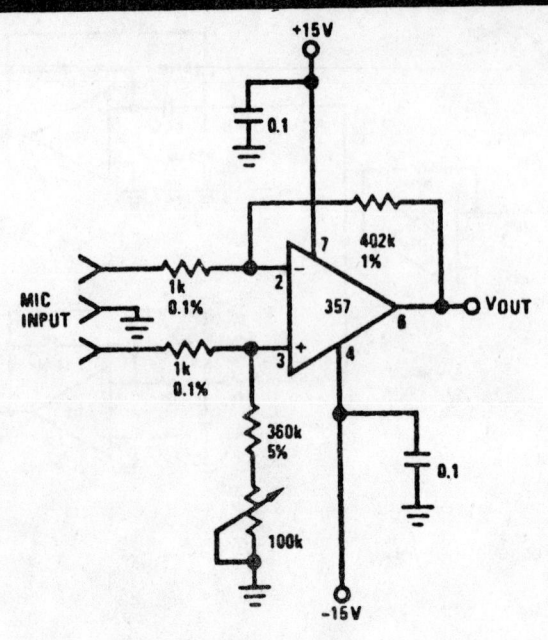

Transformerless mic preamp for balanced inputs

377 — DUAL TWO WATT AUDIO AMPLIFIER

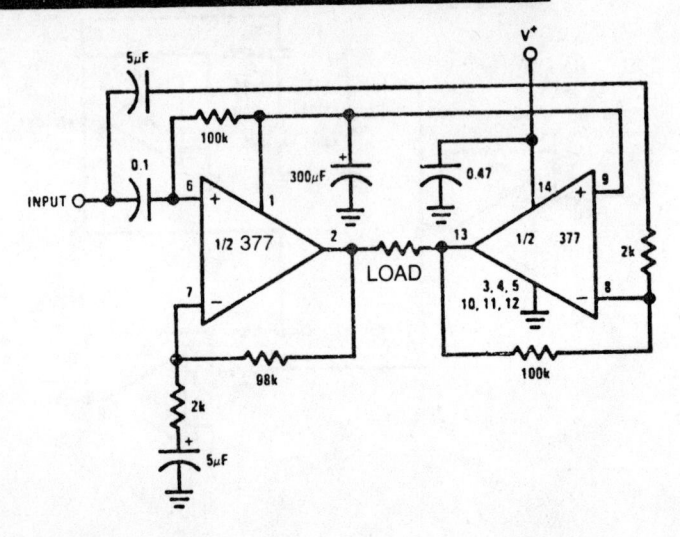

4-watt bridge amplifier

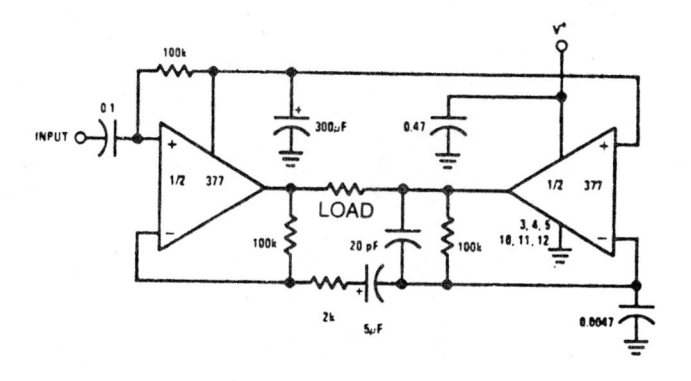

4-watt bridge amplifier with high input impedance

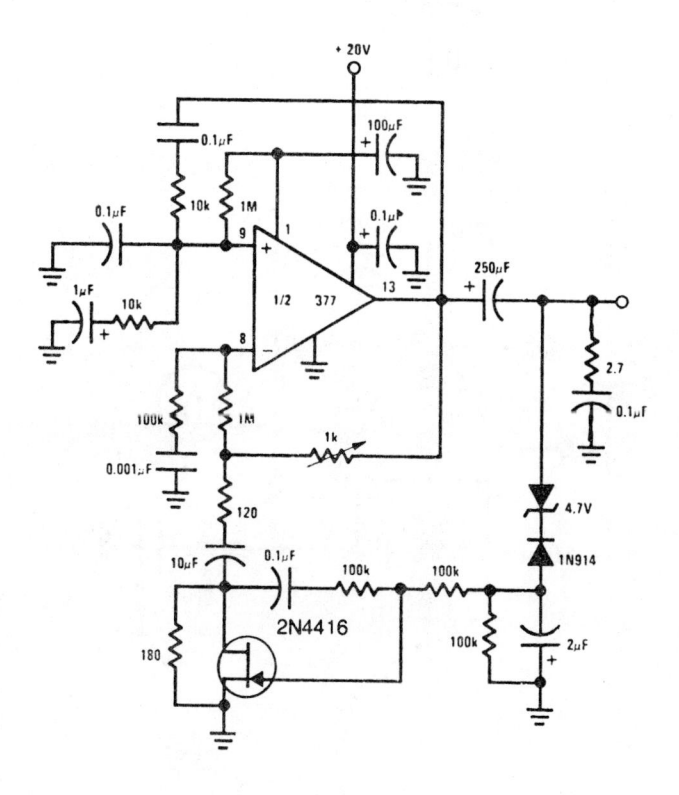

Wien bridge power oscillator

Two-phase motor drive

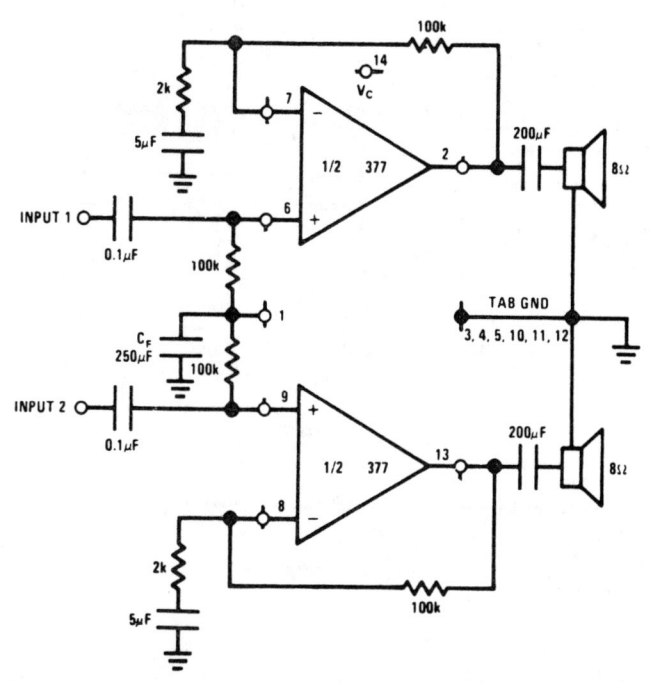

Simple stereo amplifier

378 — DUAL FOUR WATT AUDIO AMPLIFIER

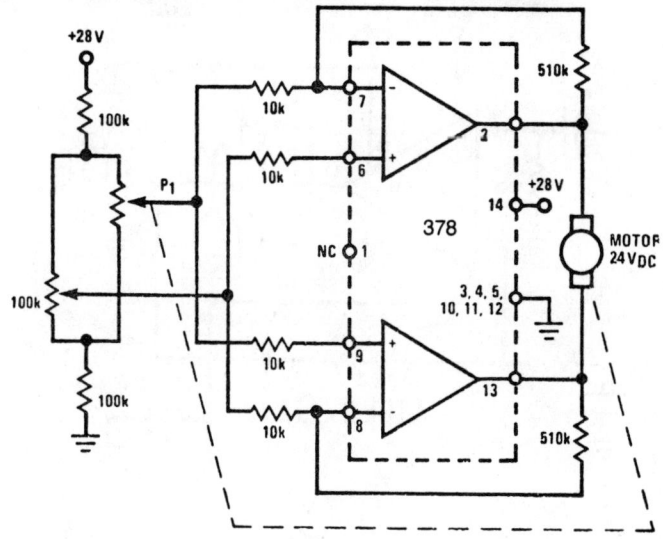

Proportional speed controller

157

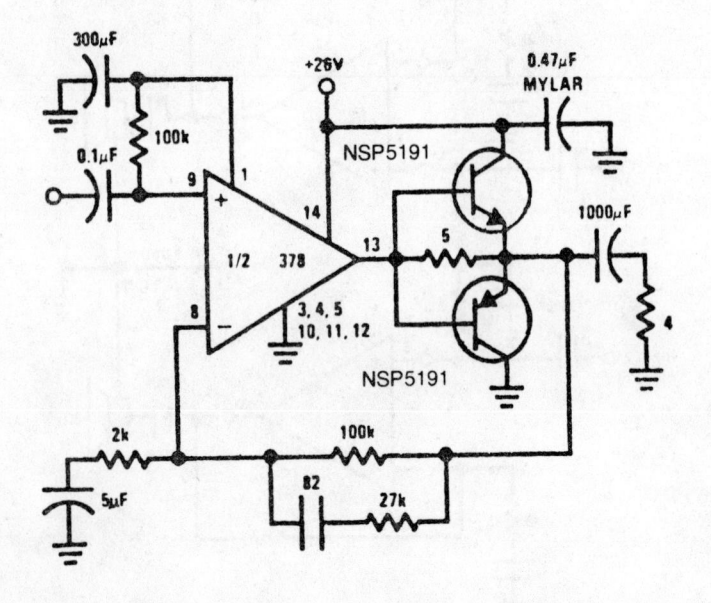

10 Watt power amplifier

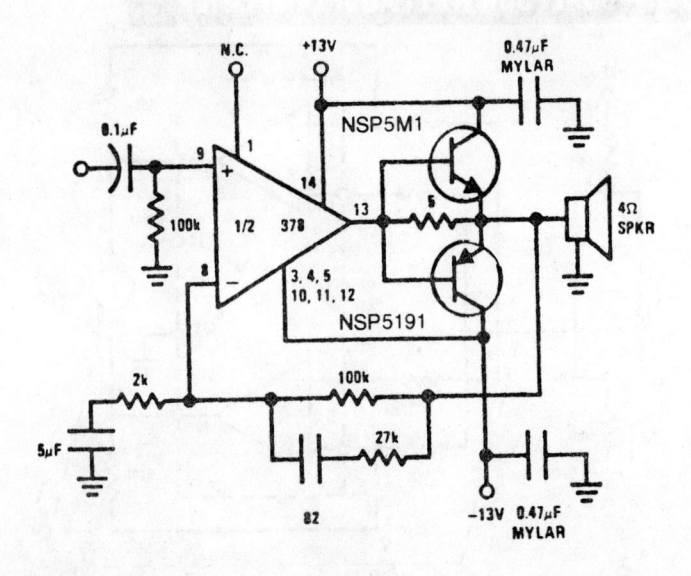

12 watt low-distortion power amplifier

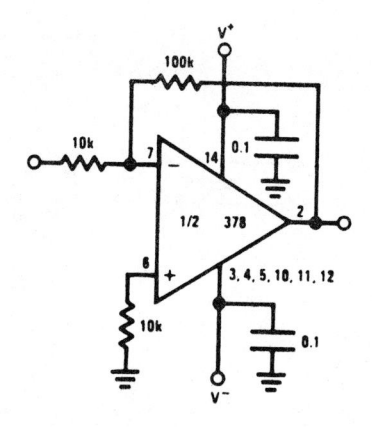

Power op amp (using split supplies)

379—DUAL SIX WATT AUDIO AMPLIFIER

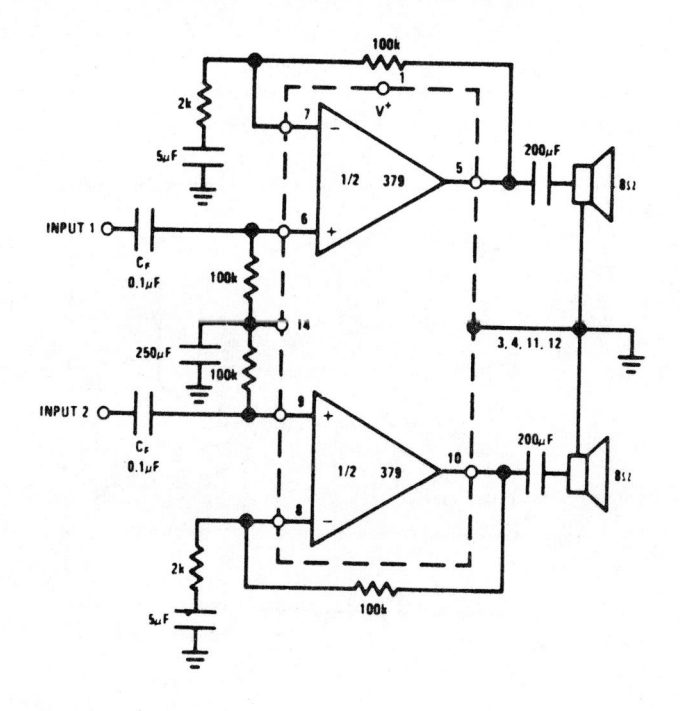

Simple stereo amplifier

159

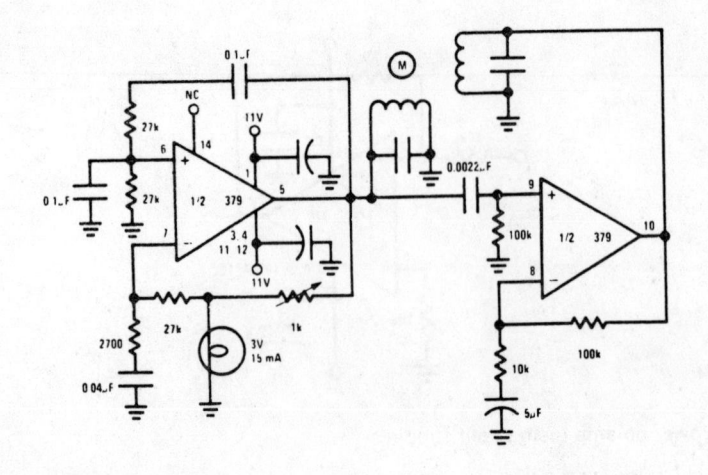

Two-phase motor drive

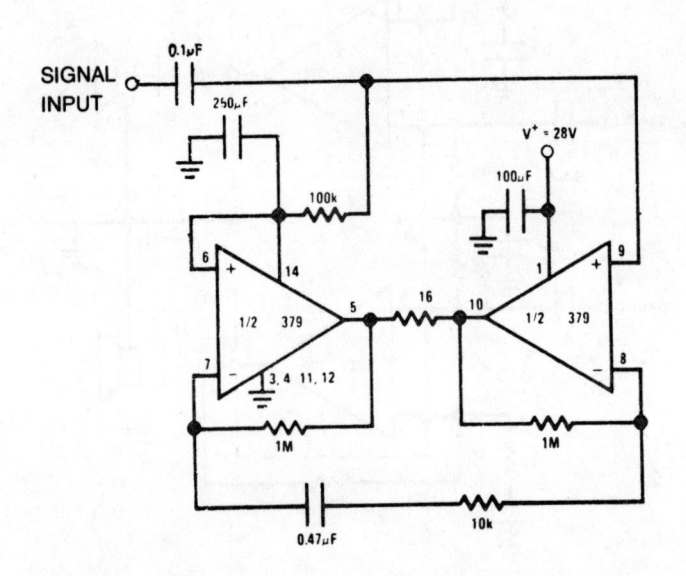

12W bridge amplifier

160

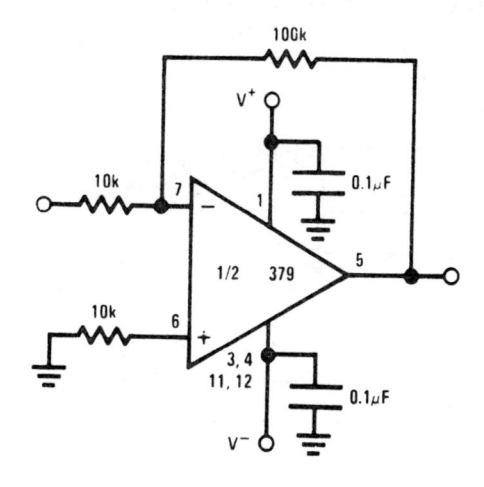

Power op amp (using split supplies)

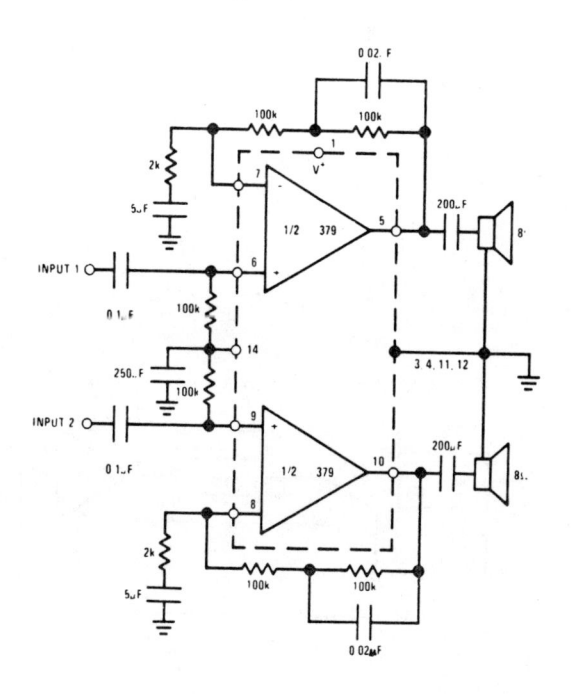

Simple stereo amplifier with bass boost

380—AUDIO POWER AMPLIFIER

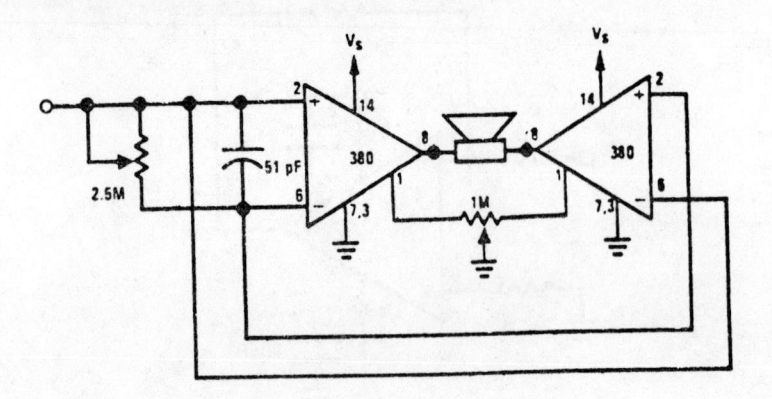

Quiescent balance control

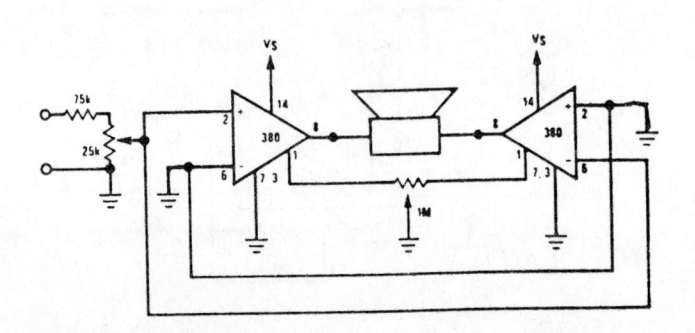

Voltage divider input

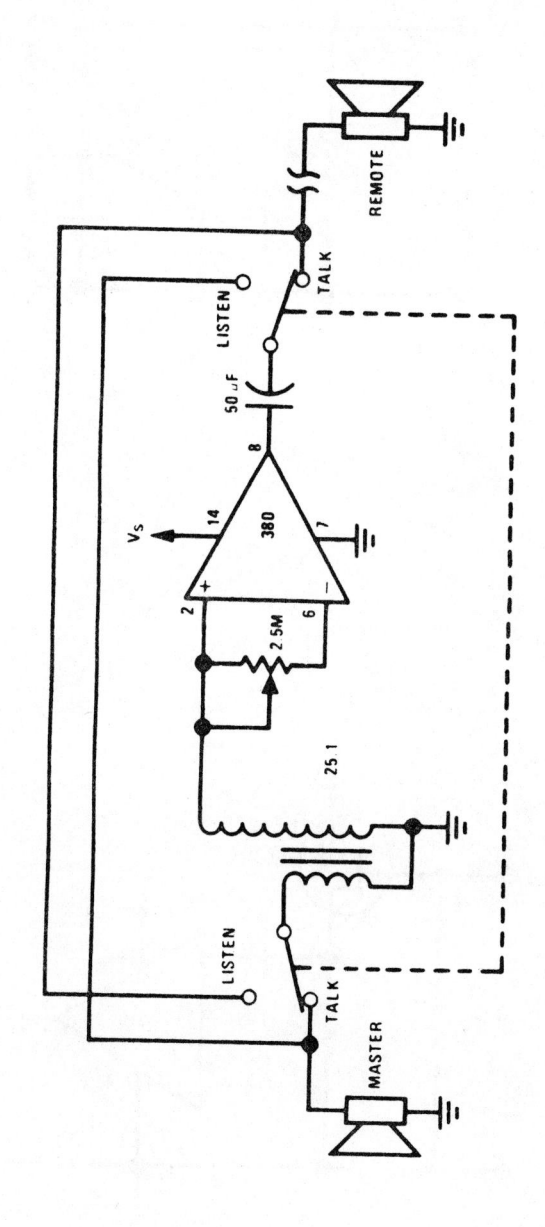

Intercom

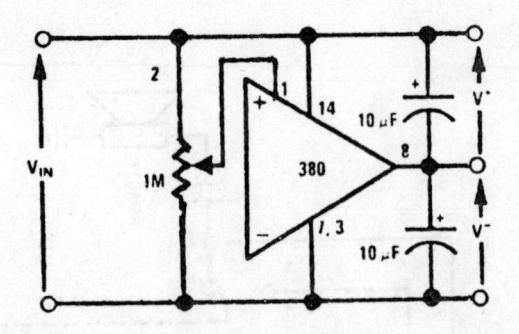

Dual supply

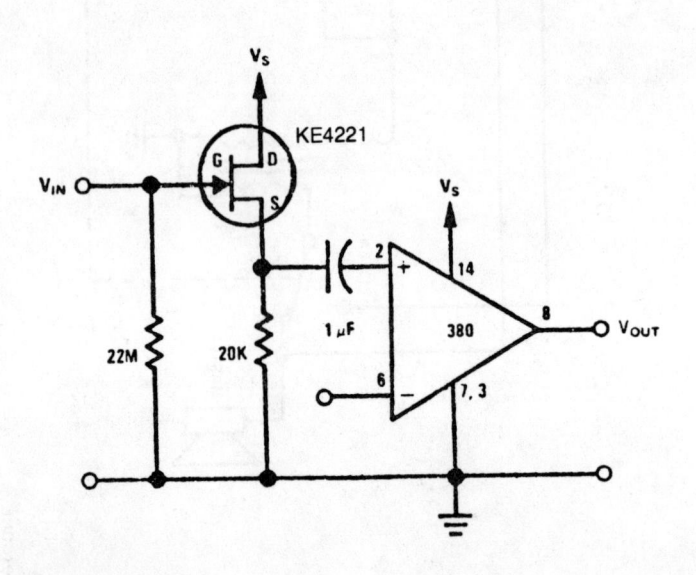

High input impedance

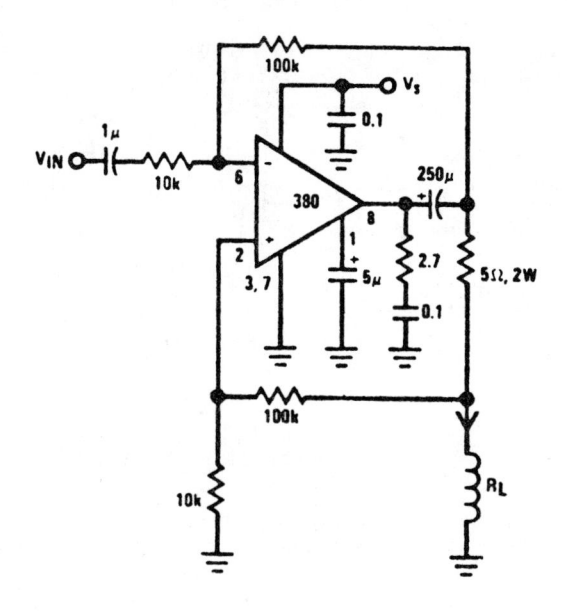

Power voltage-to-current converter

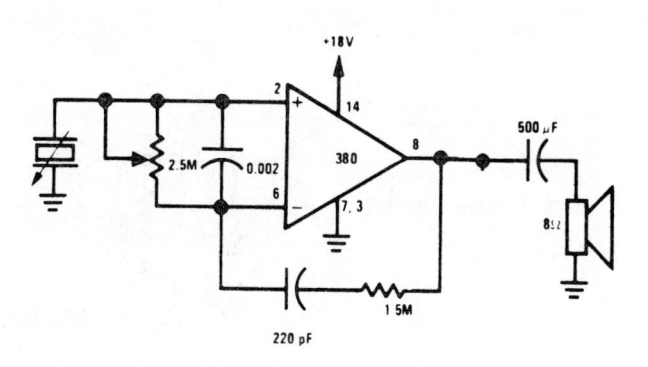

RIAA phono amplifier

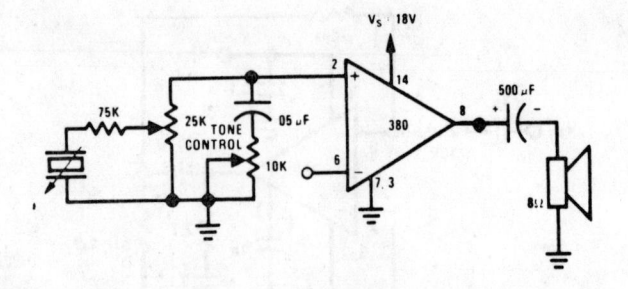

Phono amp

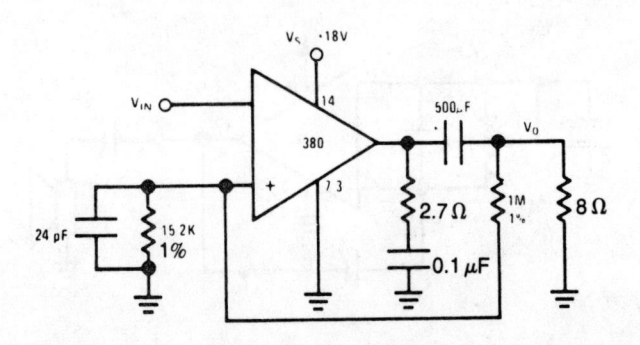

Boosted gain of 200 using positive feedback

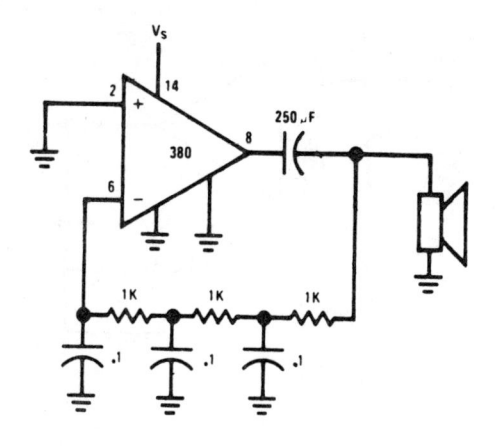

Phase shift oscillator

381—LOW NOISE DUAL PREAMPLIFIER

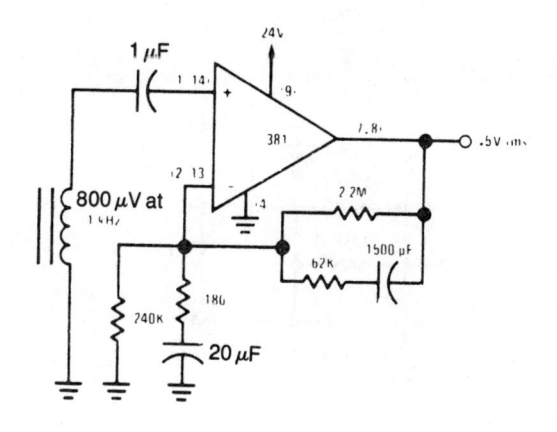

Tape playback amplifier

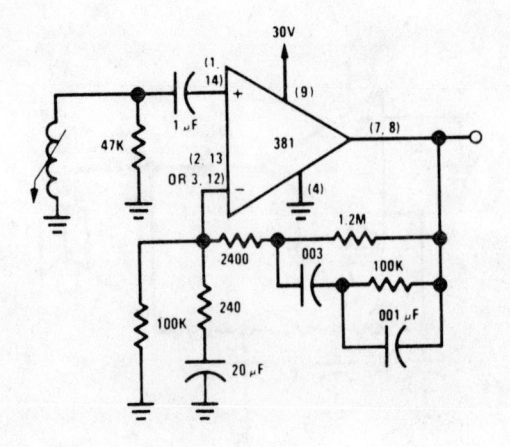

Typical magnetic phono preamp

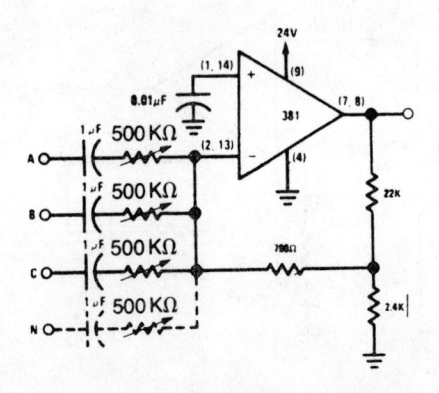

Audio mixer

Single channel of complete phono preamp

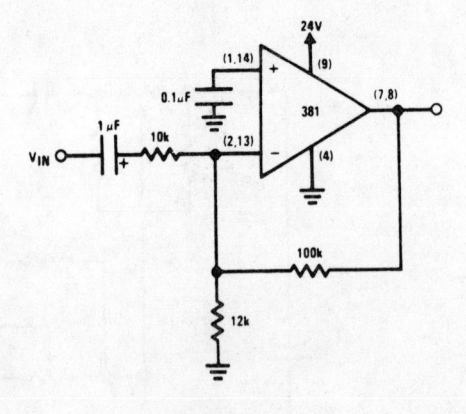

Ultra-low distortion amplifier

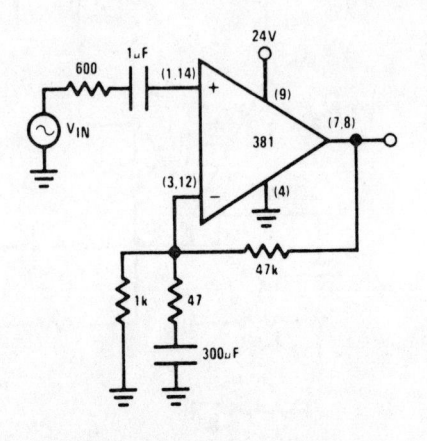

Flat gain circuit

381A—LOW NOISE DUAL PREAMPLIFIER

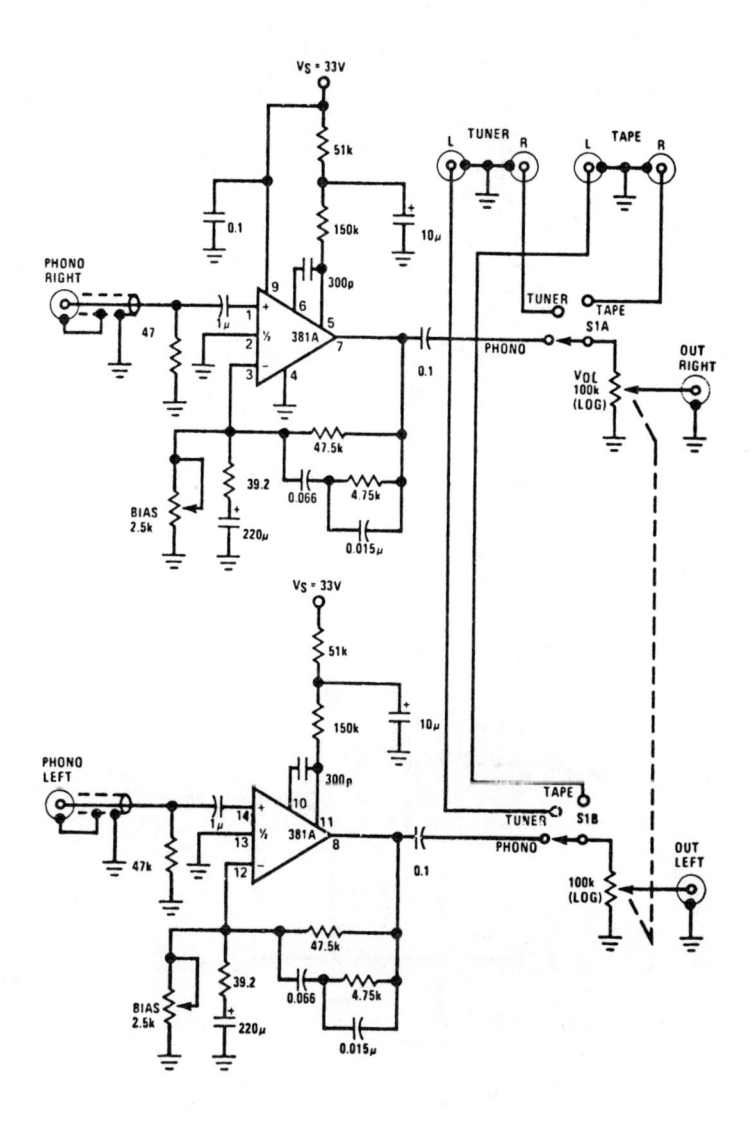

Ultra-low noise mini preamp (RIAA)

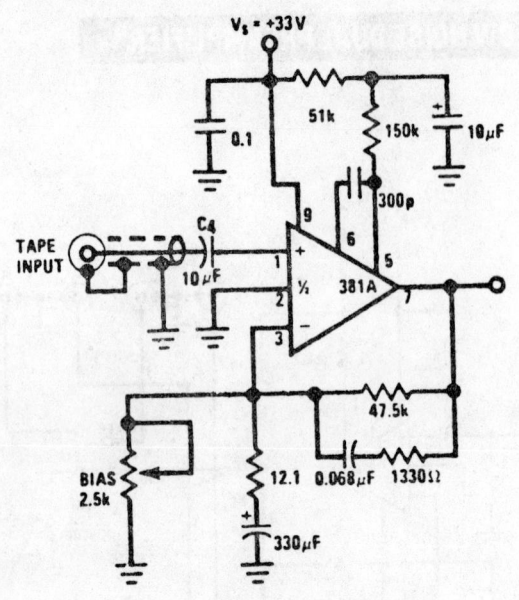

Ultra-low noise tape preamp (NAB, 1-7/8 & 3-3/4 IPS)

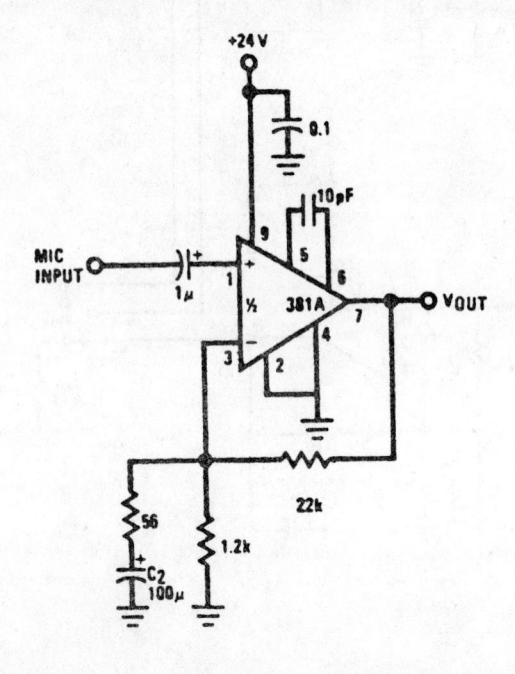

Unbalanced input mic preamp

Balanced input mic preamp

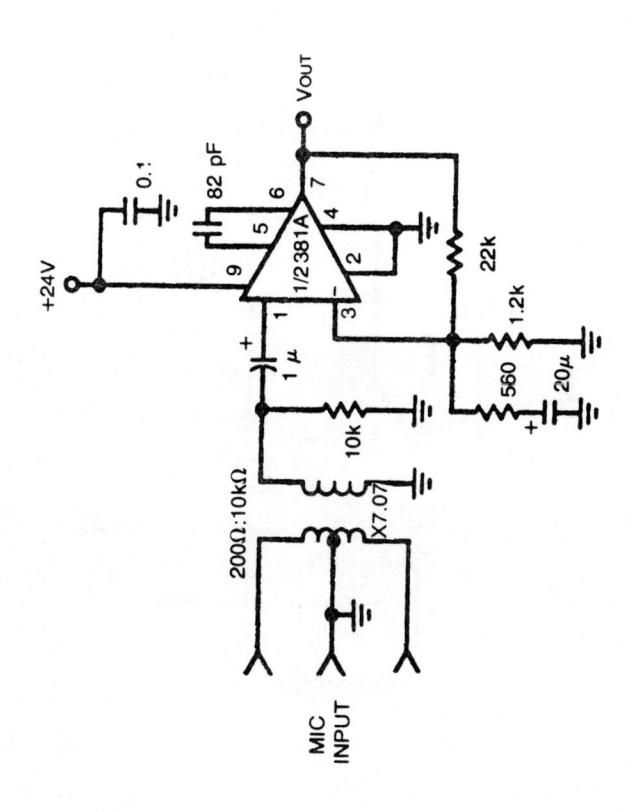

Single channel of complete phono preamp

382—LOW NOISE DUAL PREAMPLIFIER

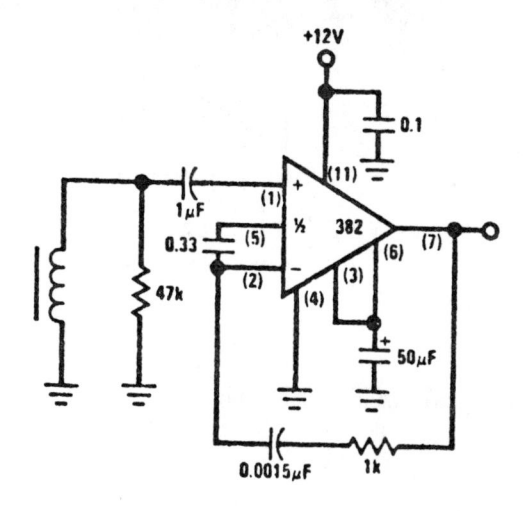

Phono preamp (RIAA)

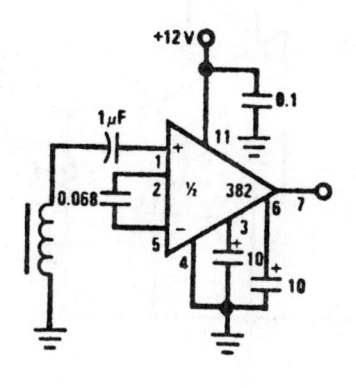

Tape preamp (NAB, 1-7/8 & 3-3/4 IPS)

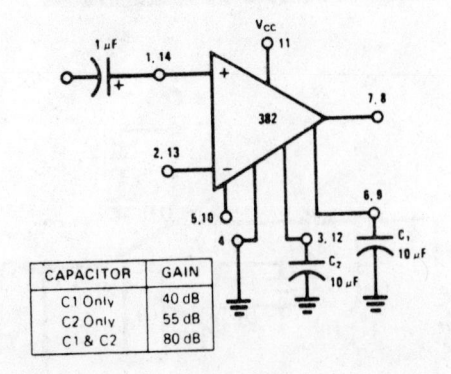

CAPACITOR	GAIN
C1 Only	40 dB
C2 Only	55 dB
C1 & C2	80 dB

Flat response — fixed gain configuration

384—FIVE WATT AUDIO POWER AMPLIFIER

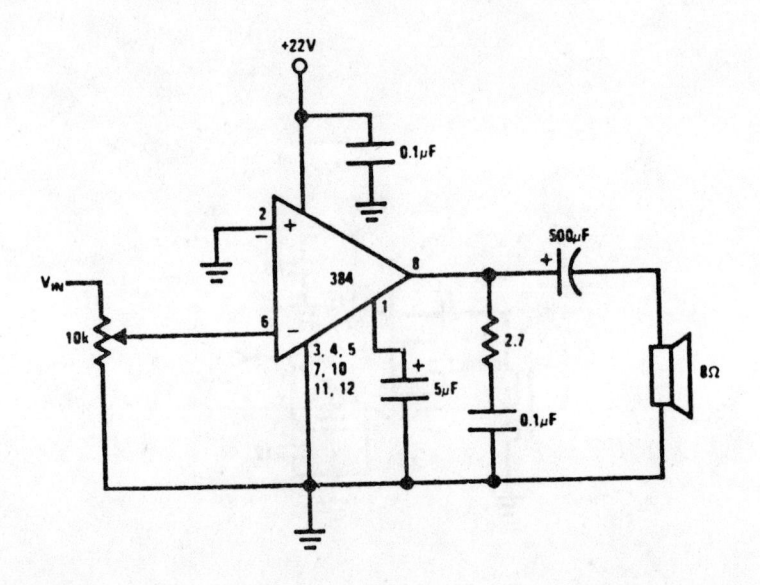

Typical 5W amplifier

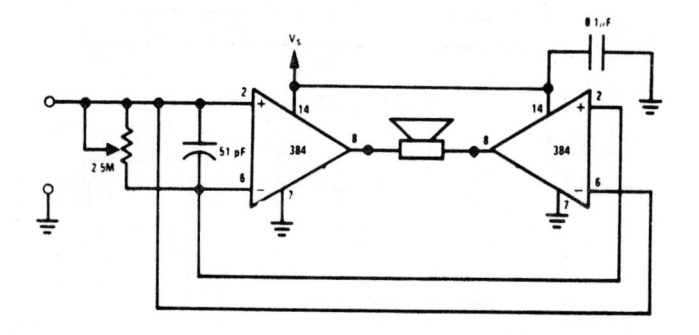

Bridge amplifier

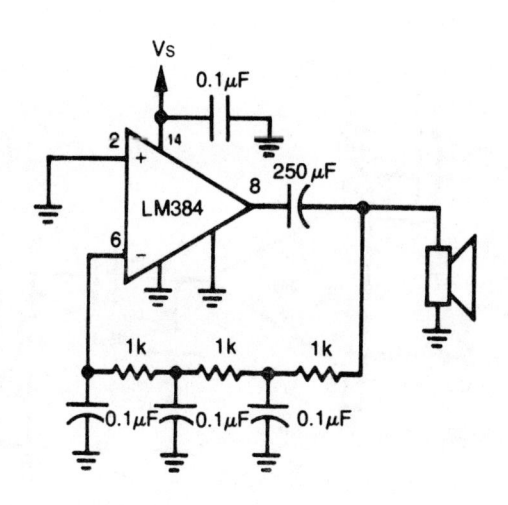

Phase shift oscillator

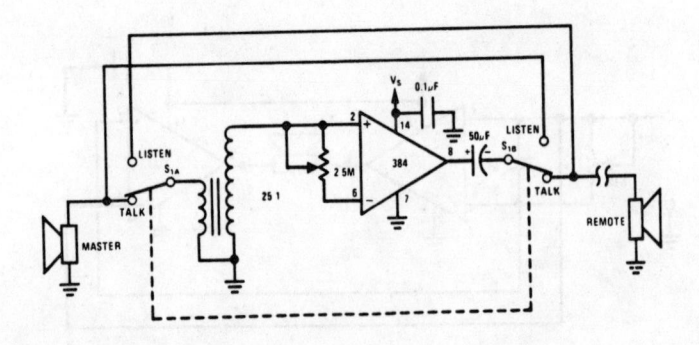

Intercom

386—LOW VOLTAGE AUDIO POWER AMPLIFIER

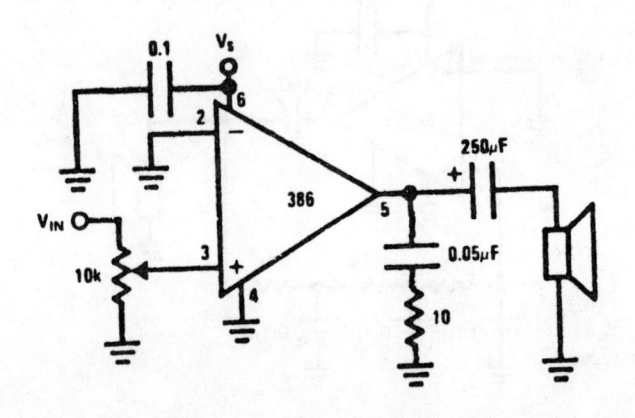

Amplifier with gain = 20 V/V (26 dB)

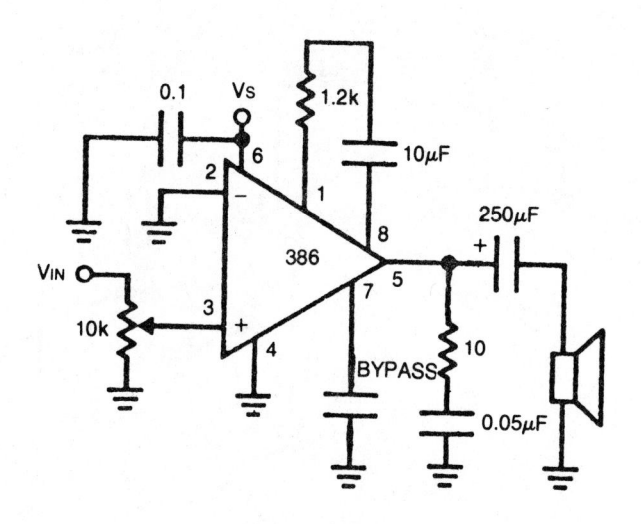

Amplifier with gain = 50 V/V (34 dB)

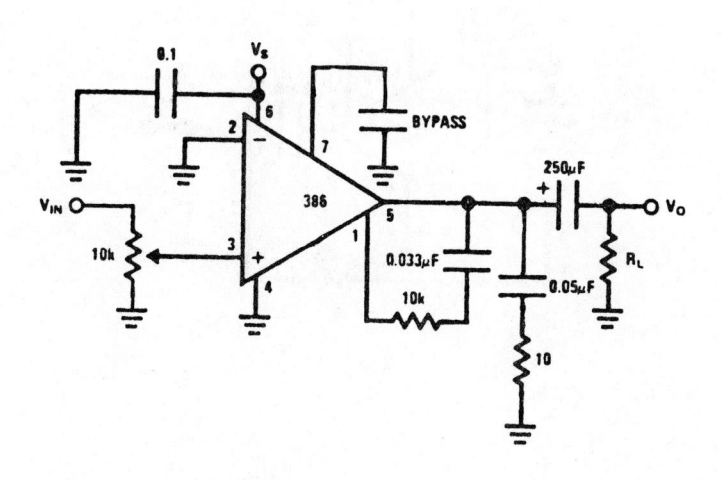

Amplifier with bass boost

179

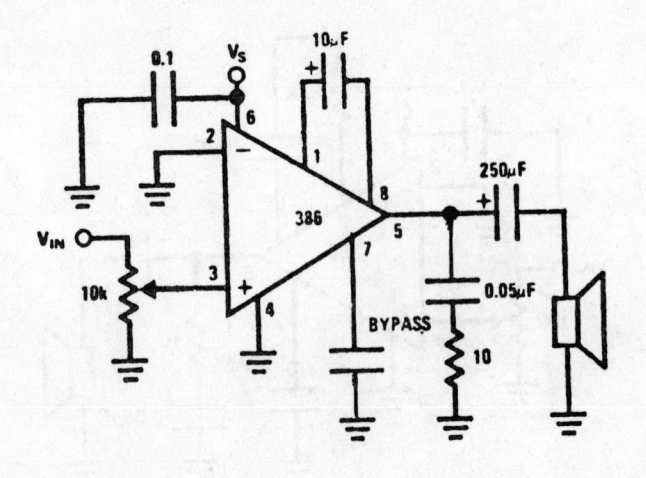

Amplifier with gain = 200 V/V (46 dB)

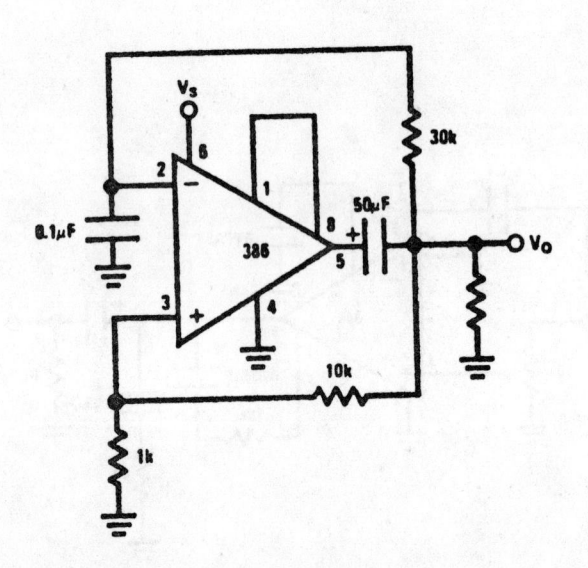

Square wave oscillator

387—LOW NOISE DUAL PREAMPLIFIER

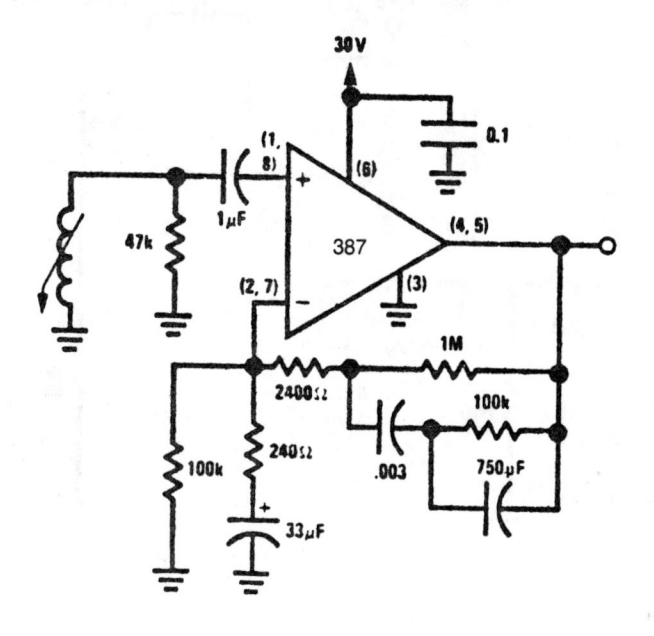

Phono preamp

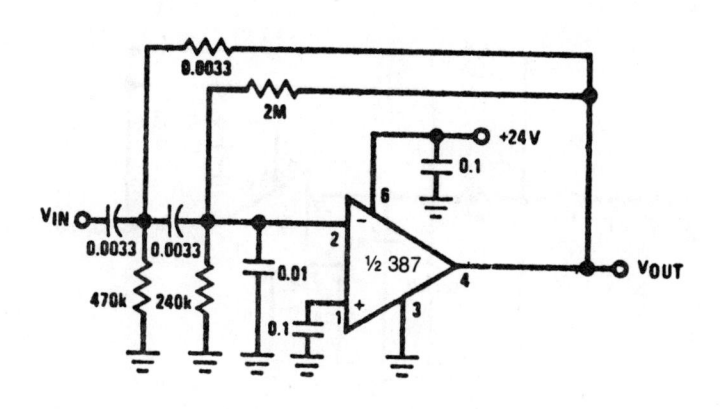

Rumble filter

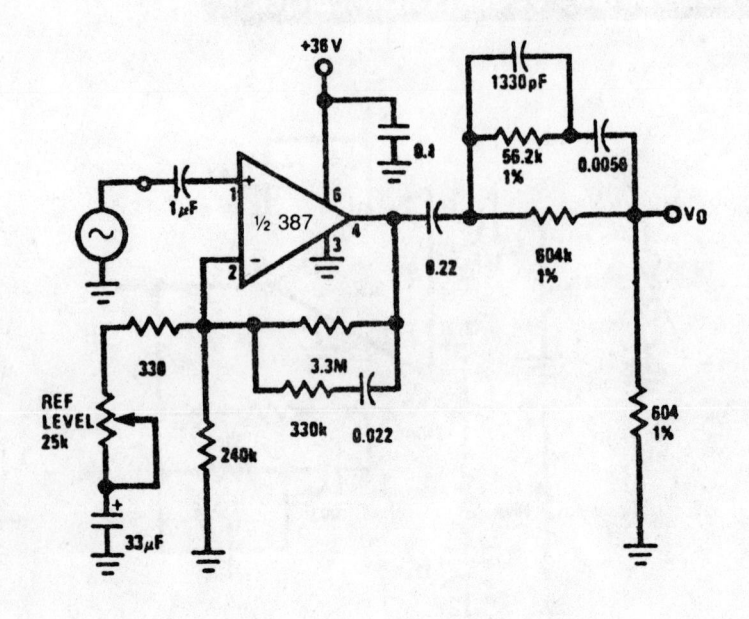

Inverse RIAA response generator

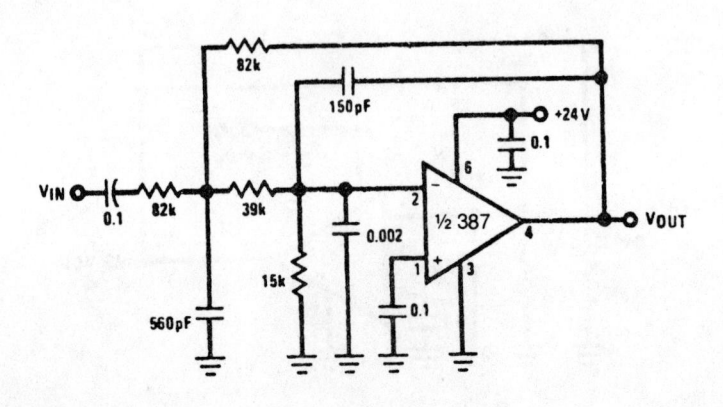

Scratch filter

182

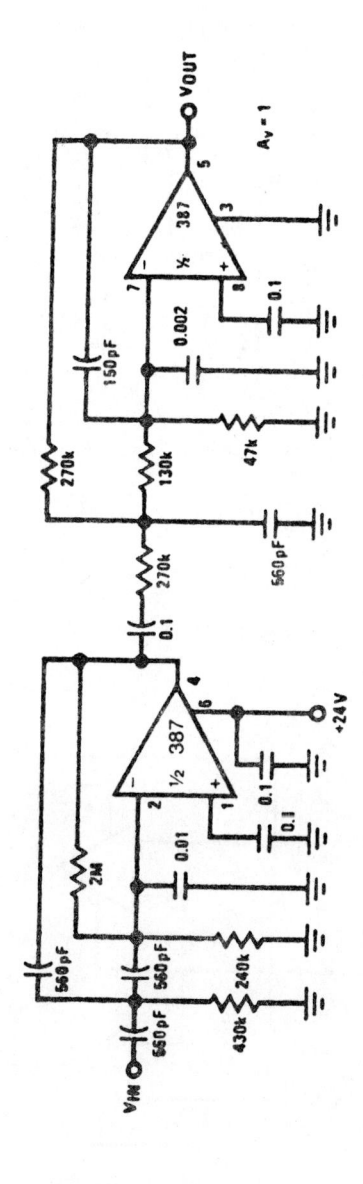

Speech filter (300 Hz-3 kHz bandpass)

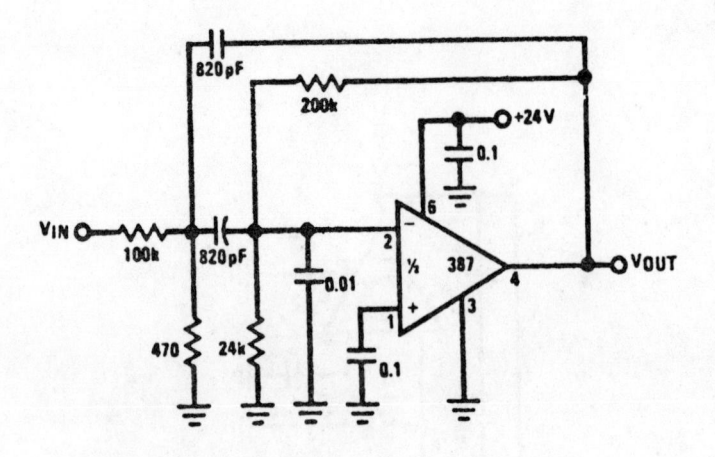

20kHz bandpass active filter

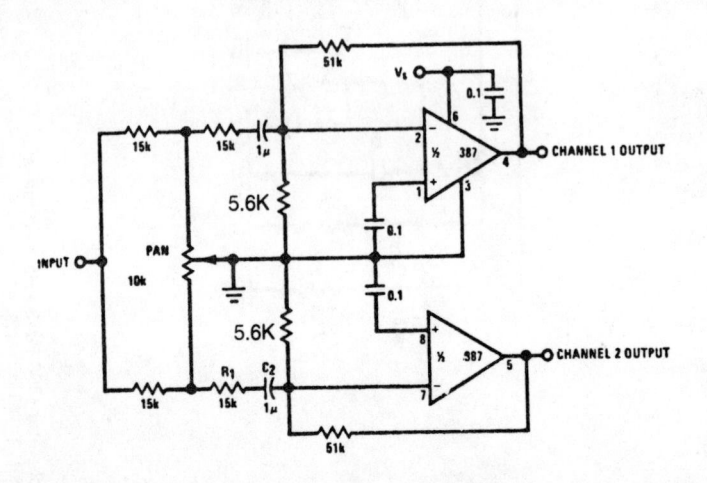

Two channel panning circuit

Acoustic pickup preamp

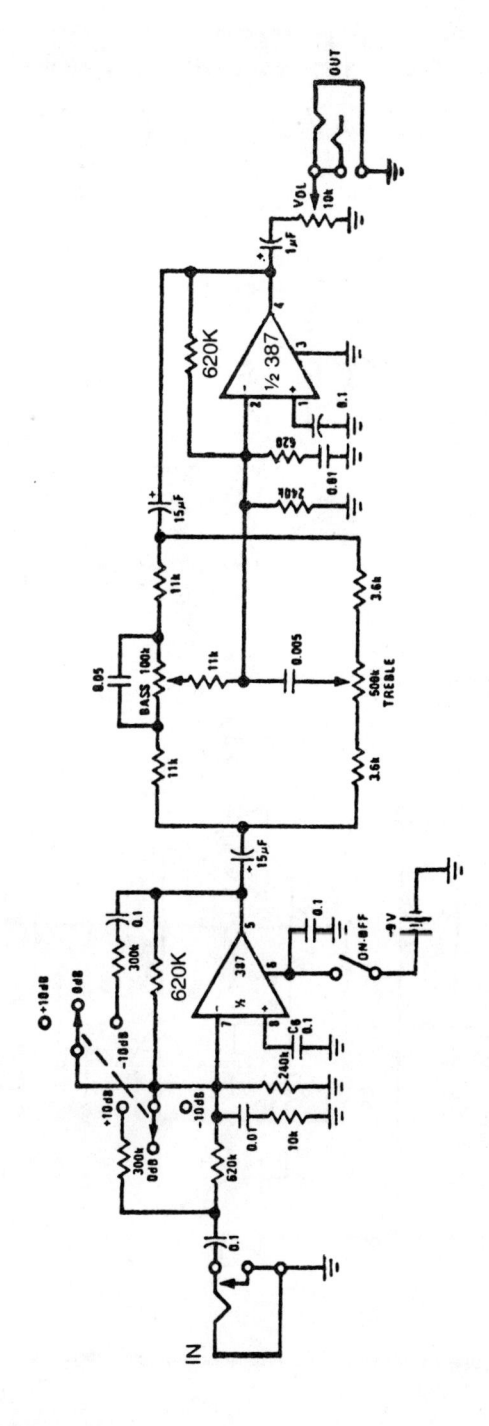

387A—LOW NOISE DUAL PREAMPLIFIER

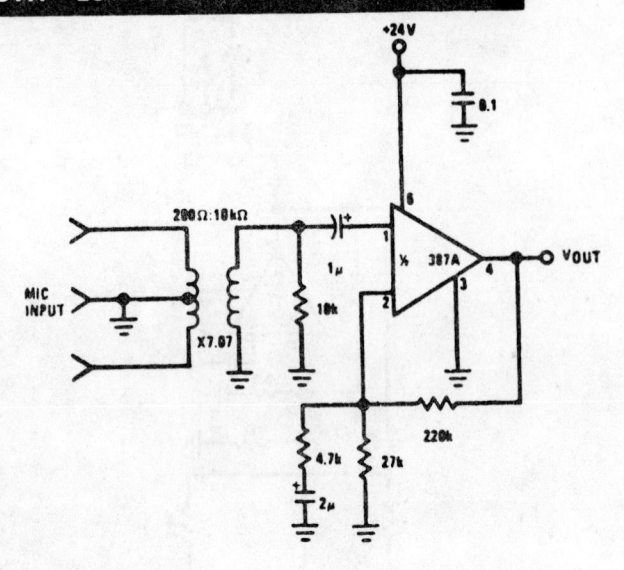

Balanced input mic preamp

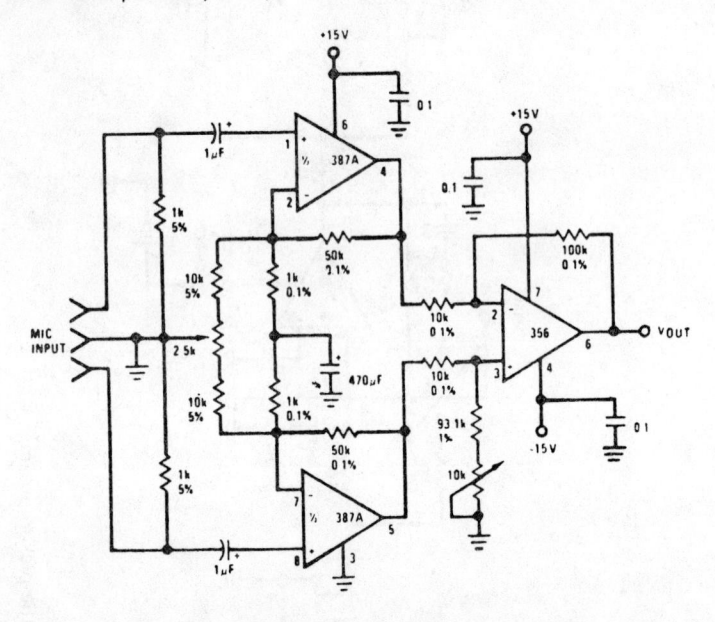

Low noise transformerless balanced mic preamp

388 — 1.5 WATT AUDIO POWER AMPLIFIER

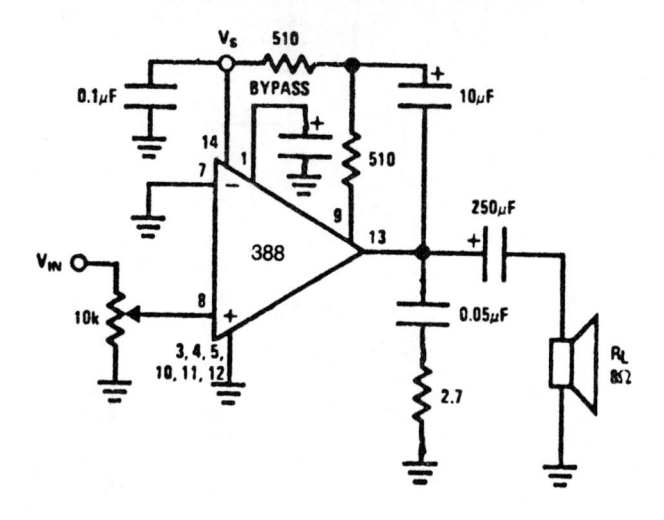

Load returned to ground amplifier with gain = 20

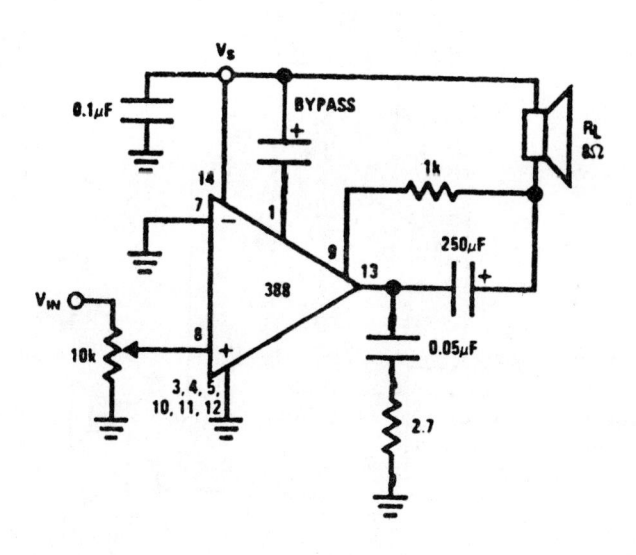

Load returned to Vs amplifier with gain = 20

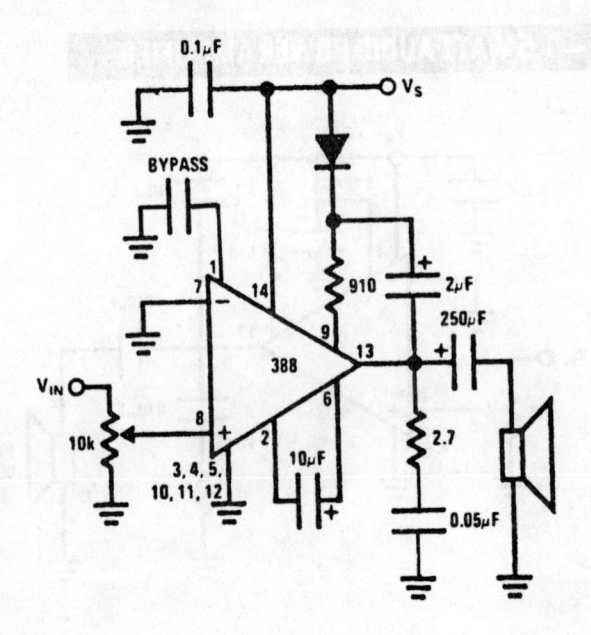

Amplifier with gain = 200

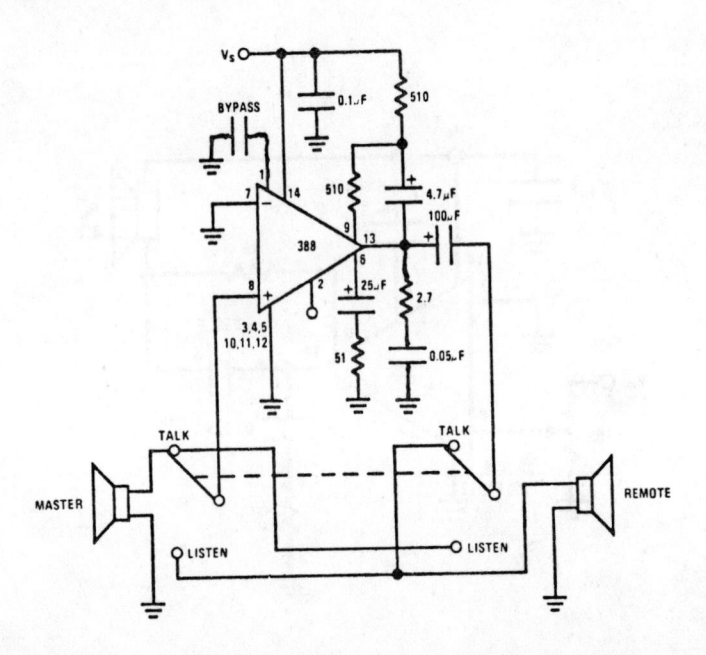

Intercom

Bridge amp

389 — LOW VOLTAGE AUDIO POWER AMPLIFIER WITH NPN TRANSISTOR ARRAY

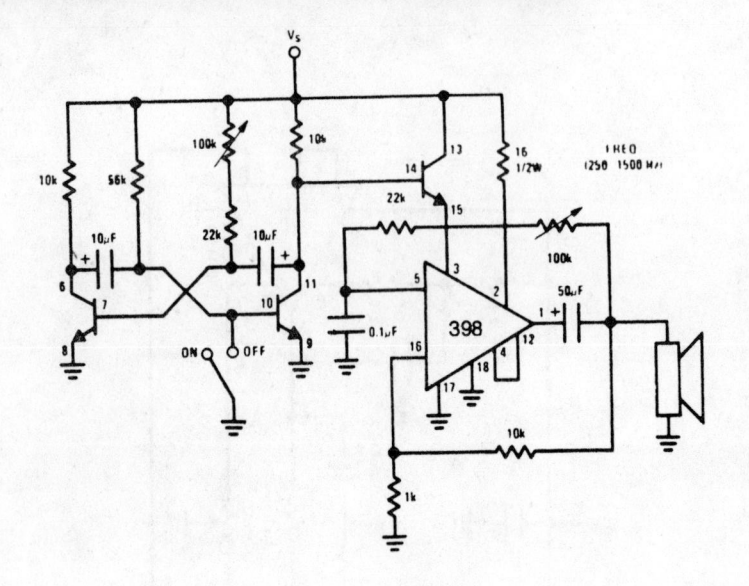

Siren

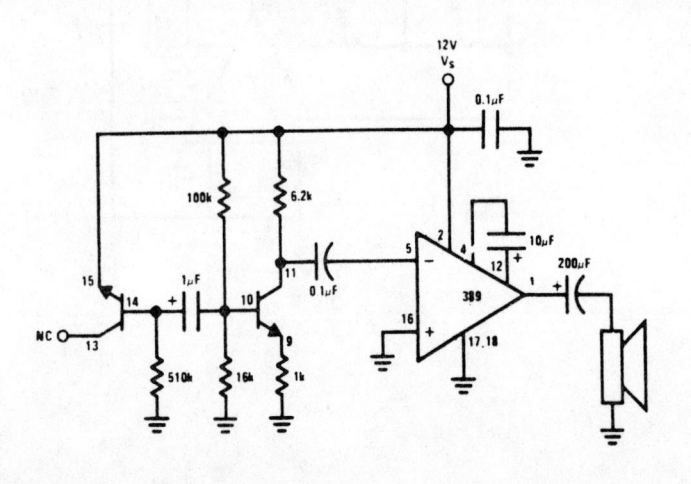

Noise generator using zener diode

190

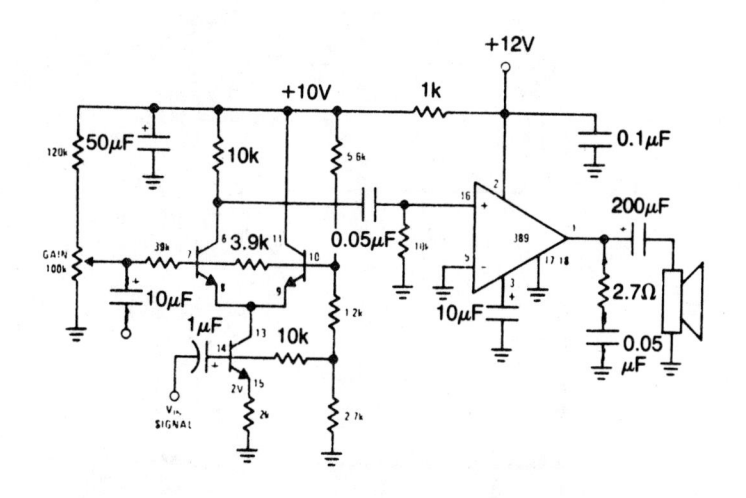

Voltage-controlled amplifier or tremolo circuit

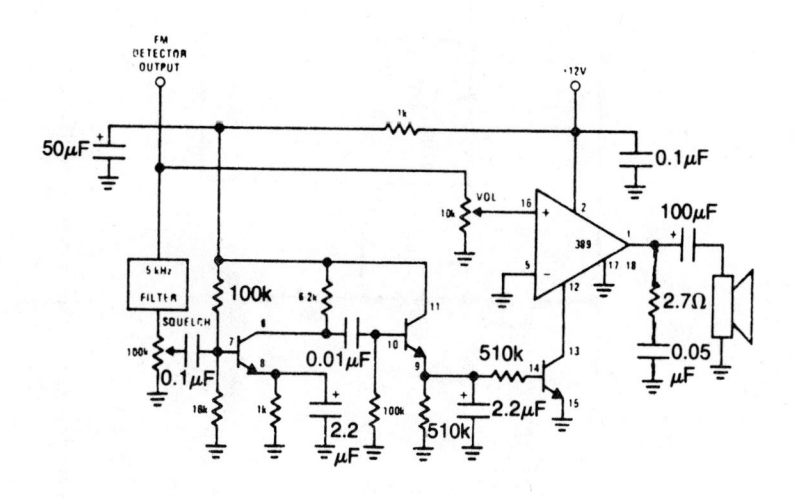

FM scanner noise squelch circuit

Ceramic phono amplifier with tone controls

390 — ONE WATT LOW VOLTAGE AUDIO AMPLIFIER

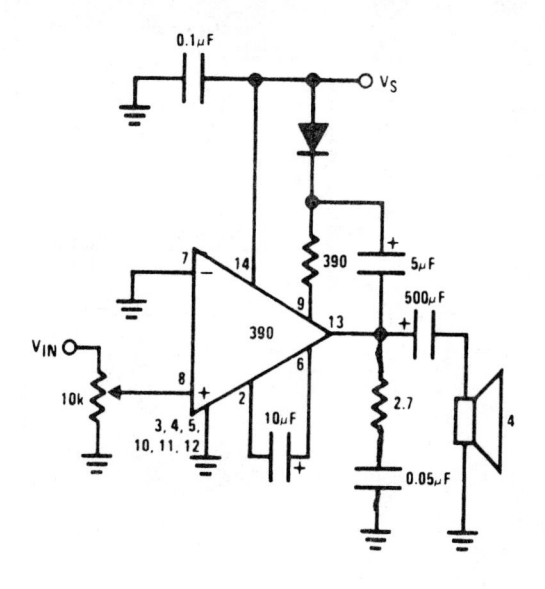

Amplifier with gain = 200 and minimum C_B

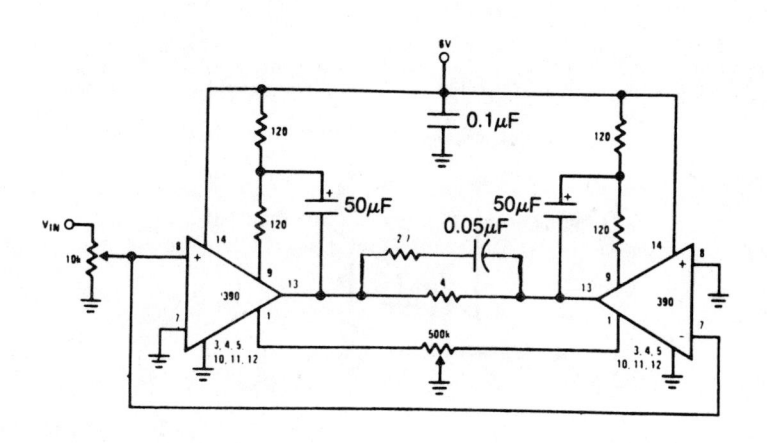

2.5W bridge amplifier

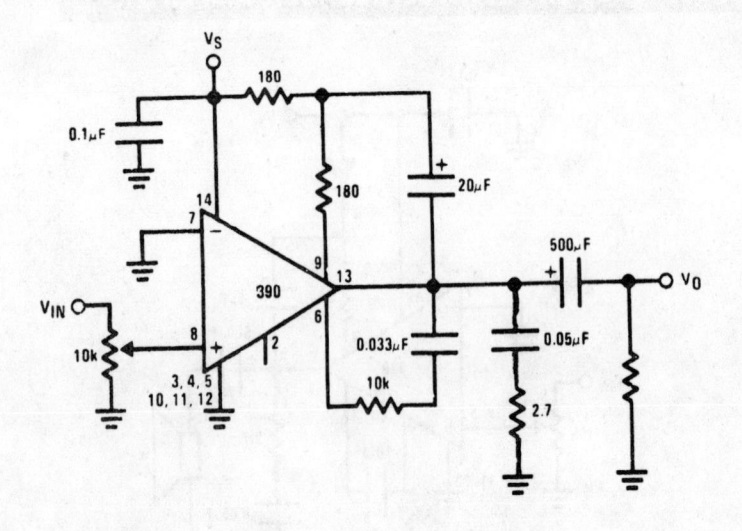

Amplifier with bass boost

398 — MONOLITHIC SAMPLE AND HOLD CIRCUIT

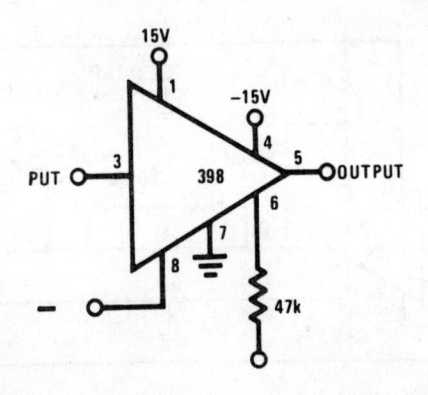

2-channel switch

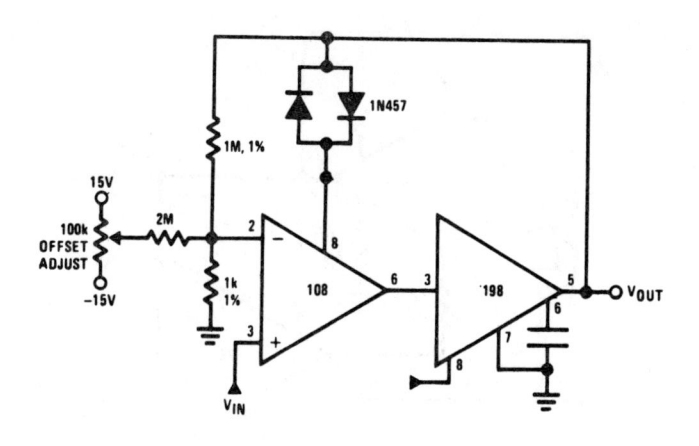

X1000 sample & hold

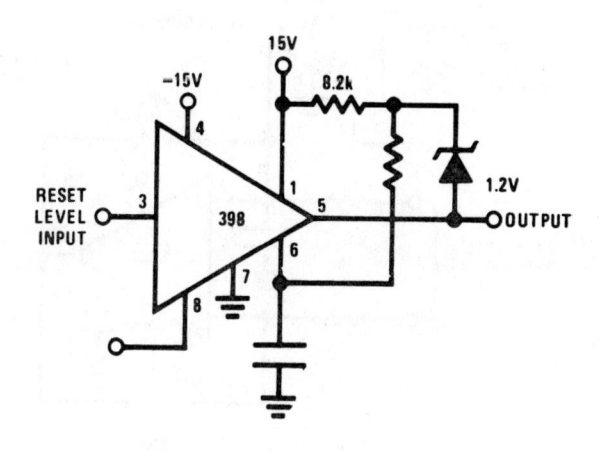

Ramp generator with variable reset level

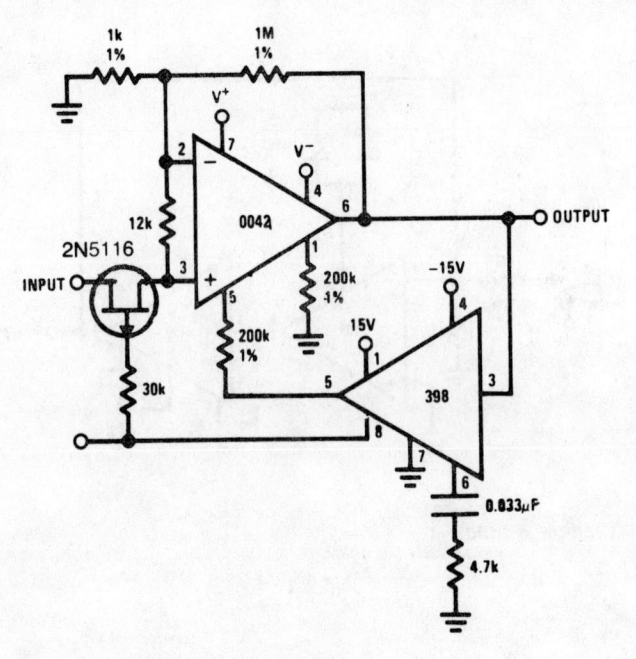

Reset stabilized amplifier (gain of 1000)

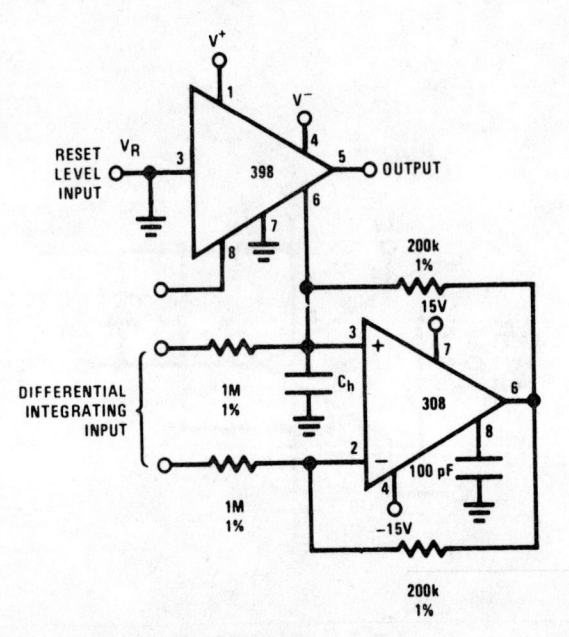

Integrator with programmable reset level

196

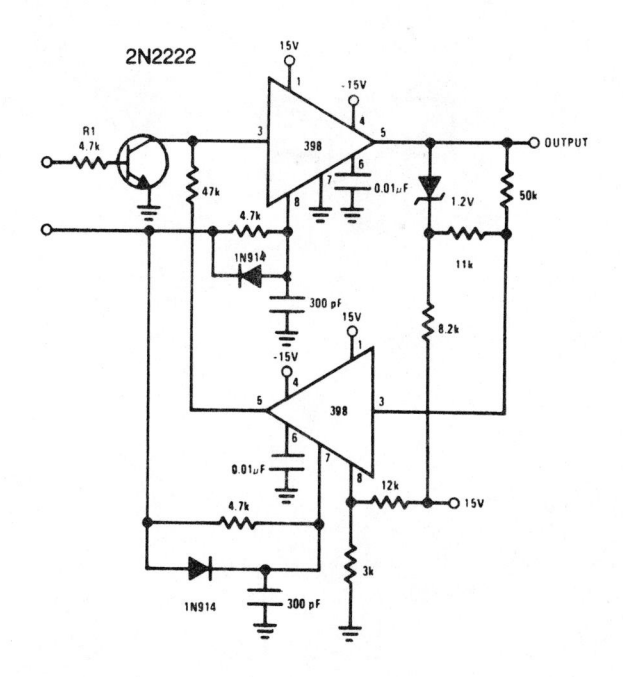

Staircase generator

531—HIGH SLEW RATE OPERATIONAL AMPLIFIER

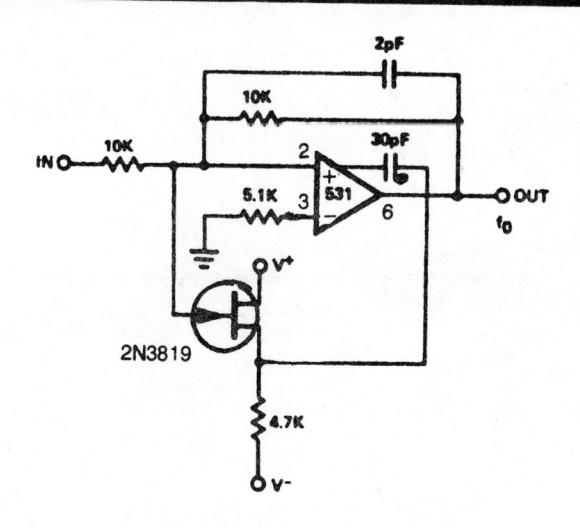

High speed inverter (10MHz bandwidth)

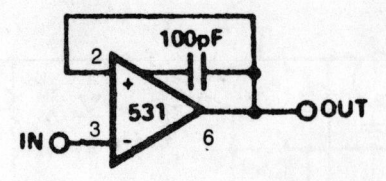

Fast settling voltage follower

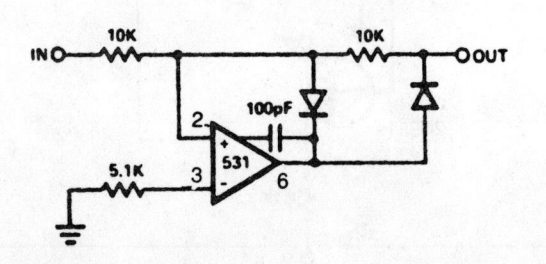

Half-wave rectifier

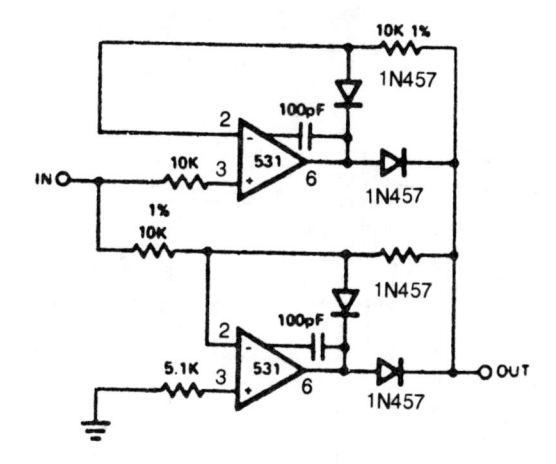

Full-wave rectifier

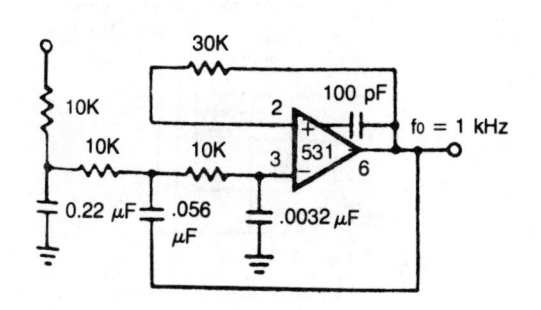

3 pole active low pass filter

540 — INTEGRATED POWER DRIVER

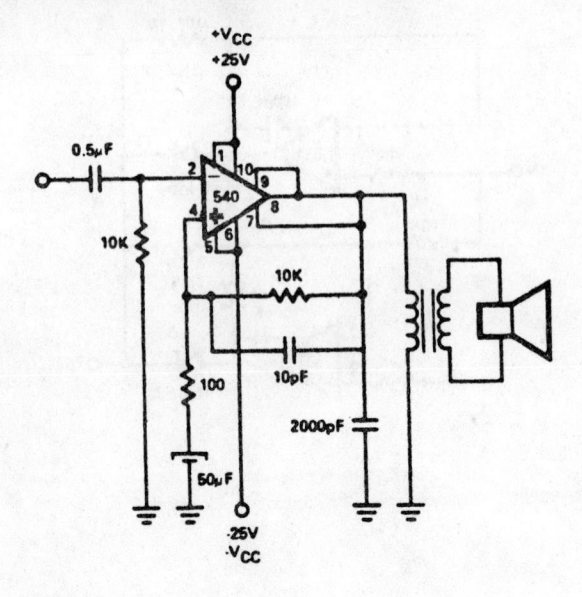

1 watt power amplifier

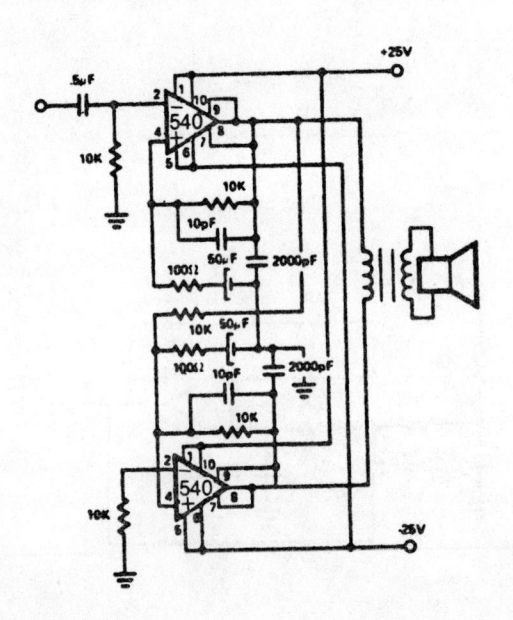

3 watt power amplifier

555—INTEGRATED TIMER

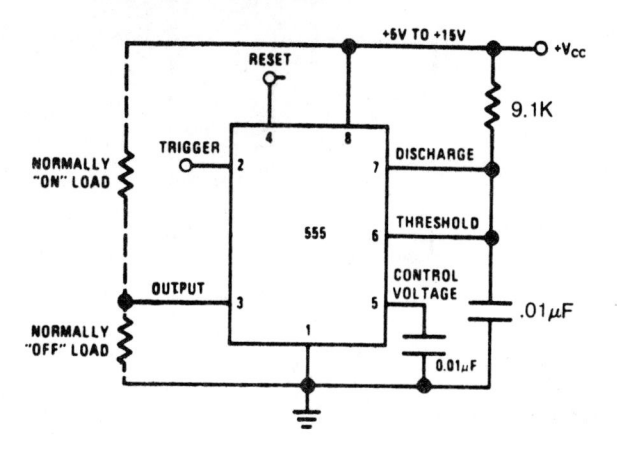

Monostable multivibrator

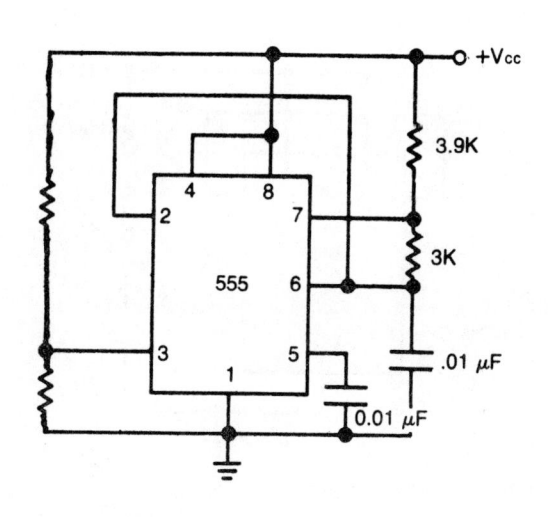

Astable multivibrator

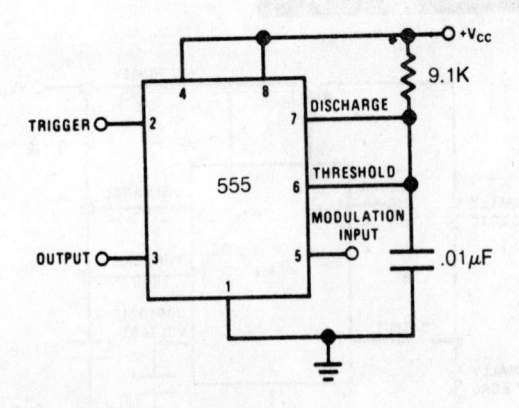

Pulse width modulator

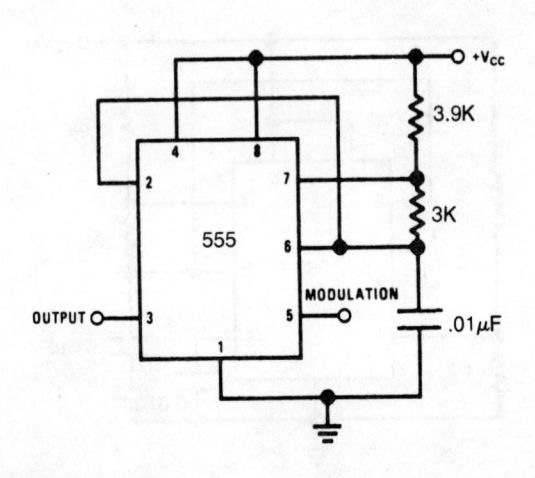

Pulse position modulator

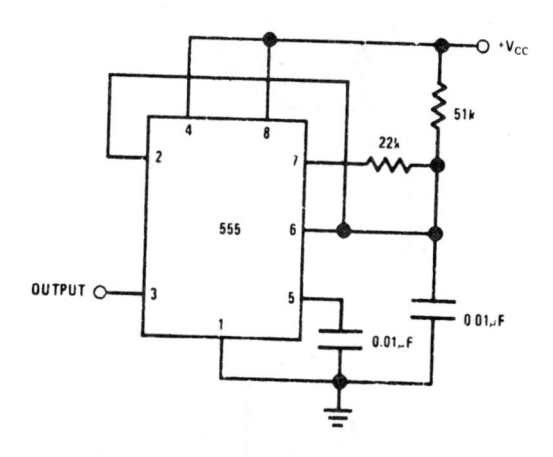

50% duty cycle oscillator

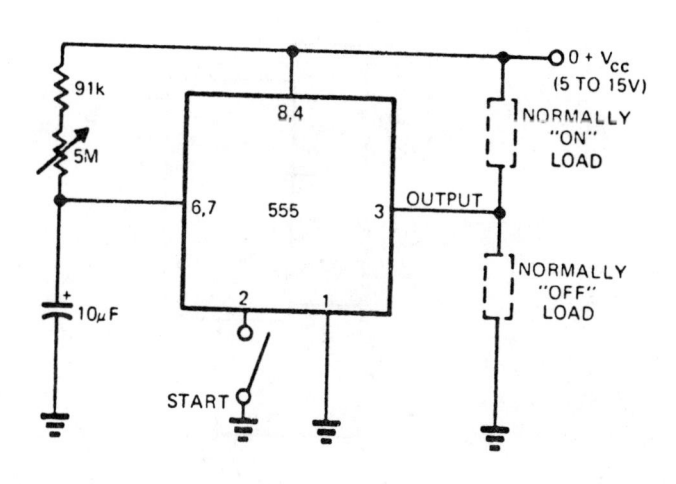

On or off control

203

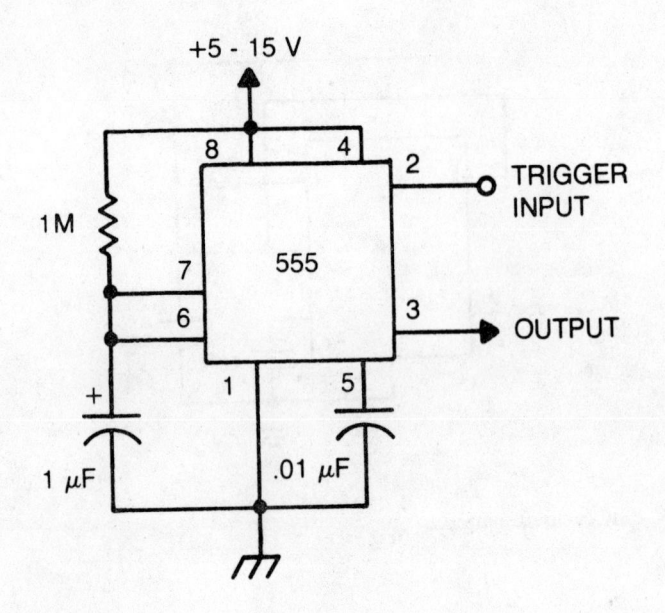

One-shot timer

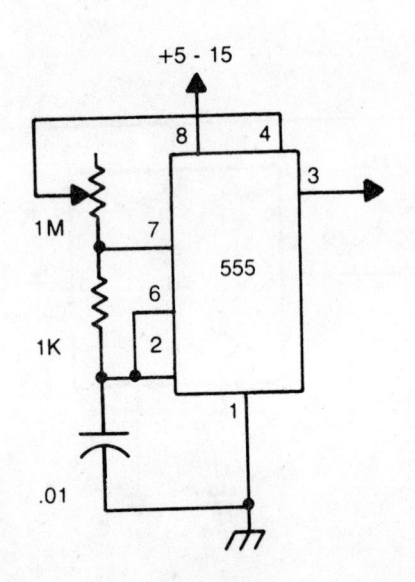

Pulse generator

204

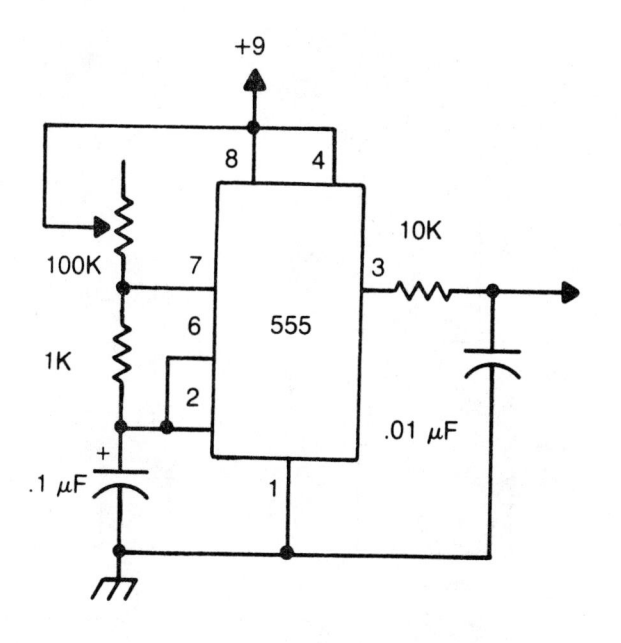

Triangle wave generator

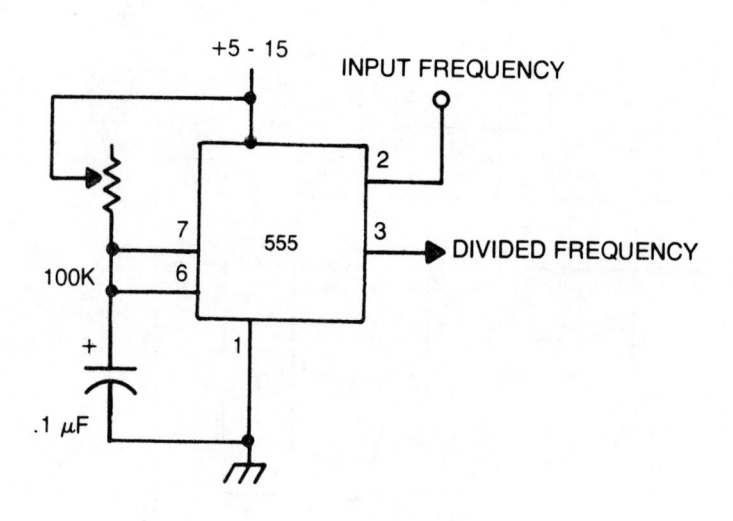

Frequency divider

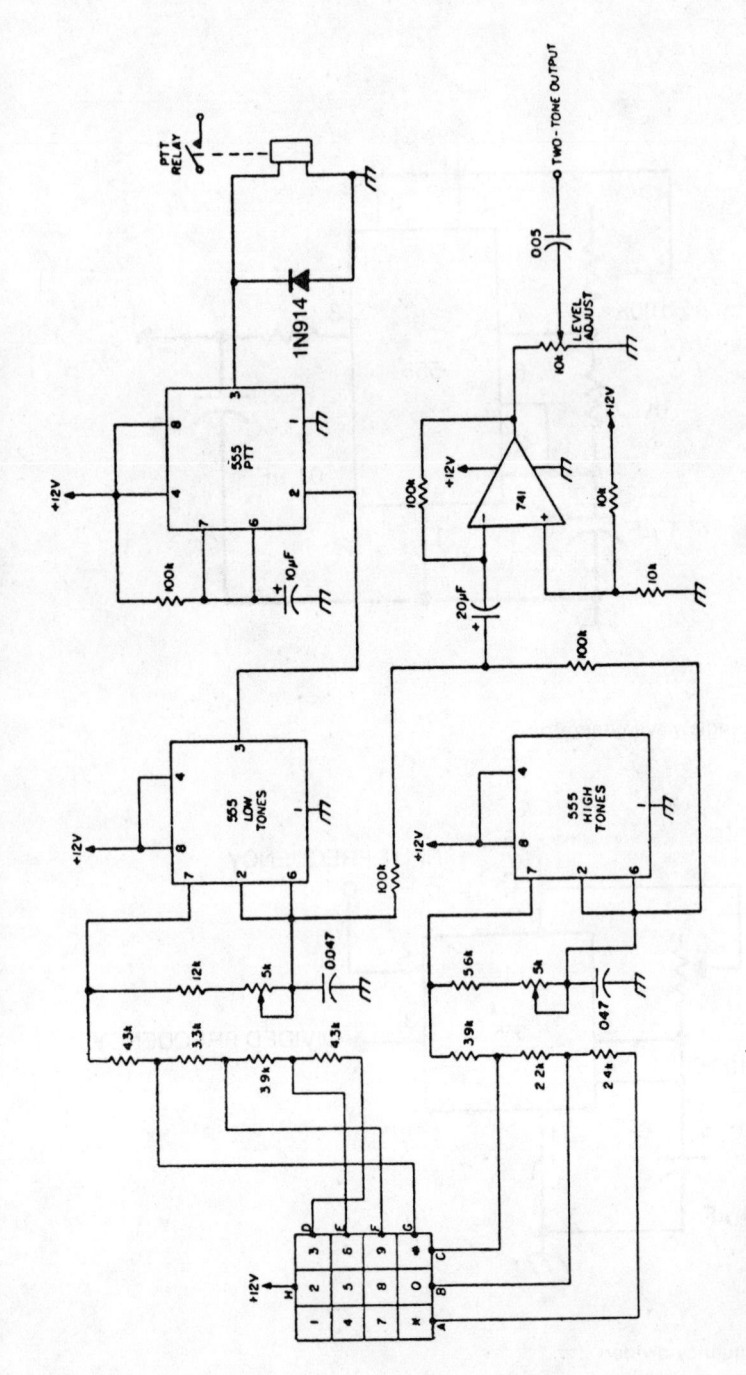

Telephone dialing tone encoder

706 — FIVE WATT AUDIO AMPLIFIER

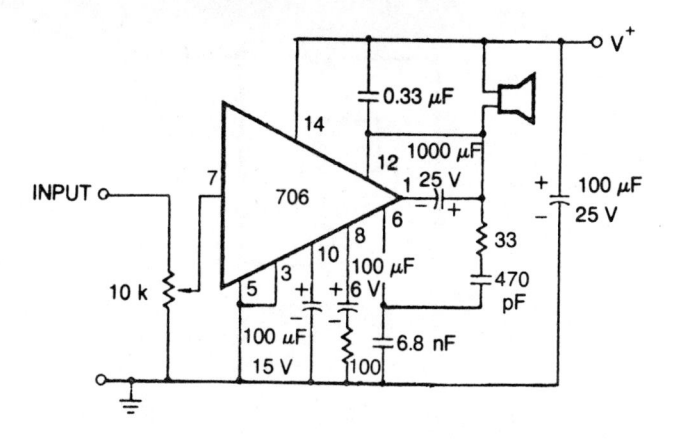

5 watt audio amplifier

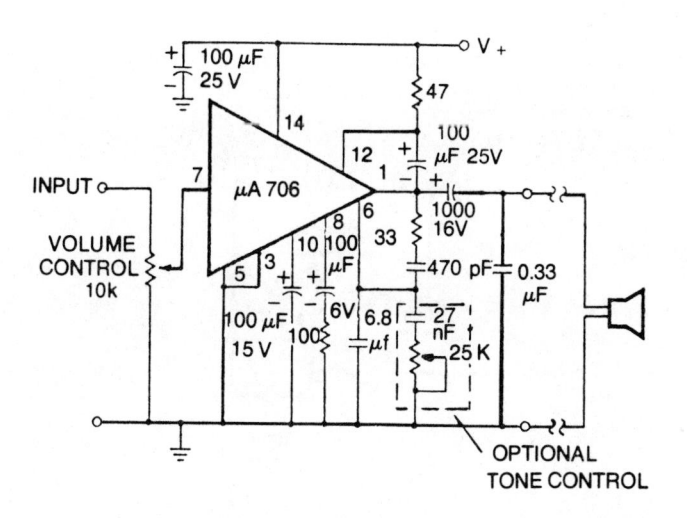

5 watt audio amplifier with load connected to ground

709 — MONOLITHIC OPERATIONAL AMPLIFIER

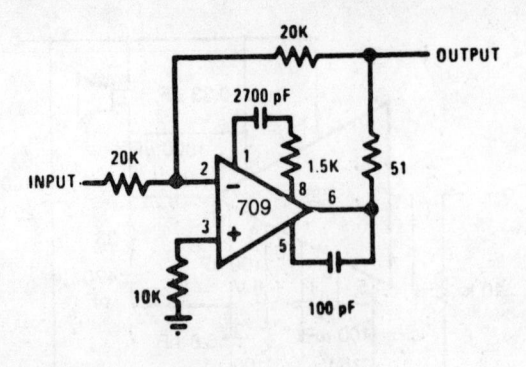

Unity gain inverting amplifier

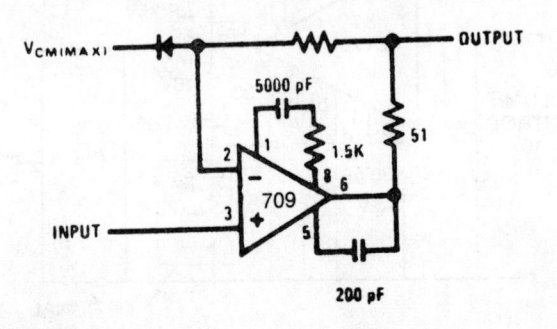

Voltage follower

715—HIGH SPEED OPERATIONAL AMPLIFIER

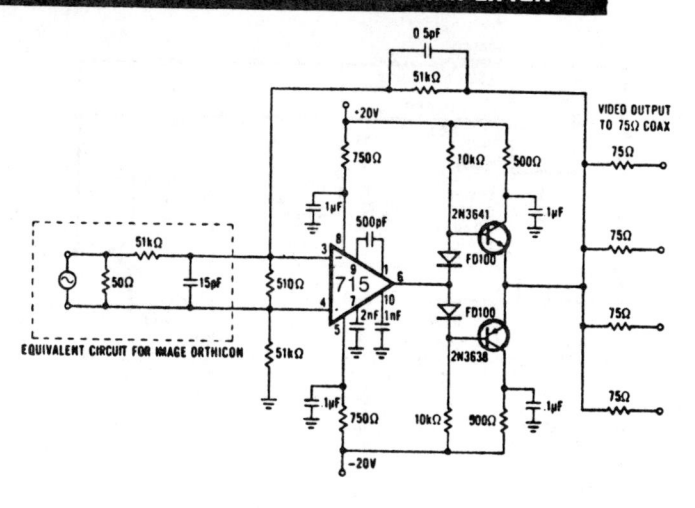

Wide band video amplifier with 75 Ω coax cable drive capability

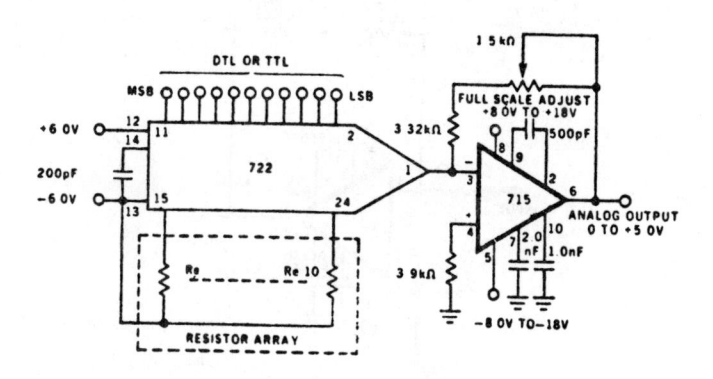

High speed 10-bit digital to analog converter

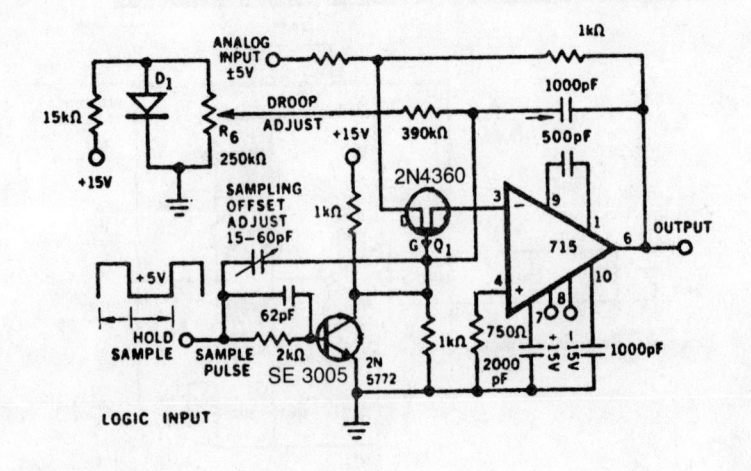

High speed sample and hold

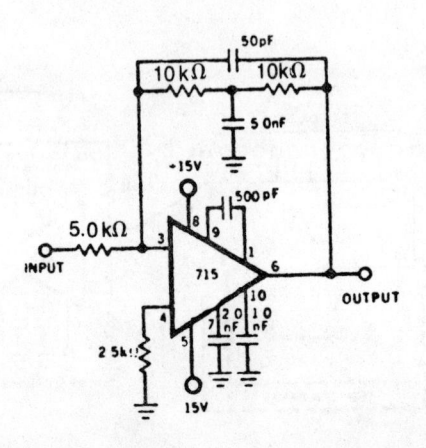

High speed integrator

725 — INSTRUMENTATION OPERATIONAL AMPLIFIER

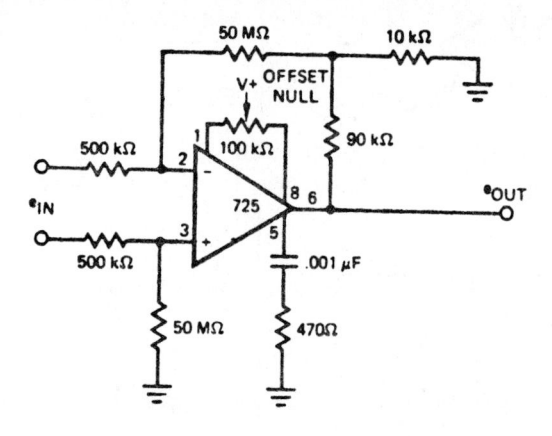

Precision amplifier

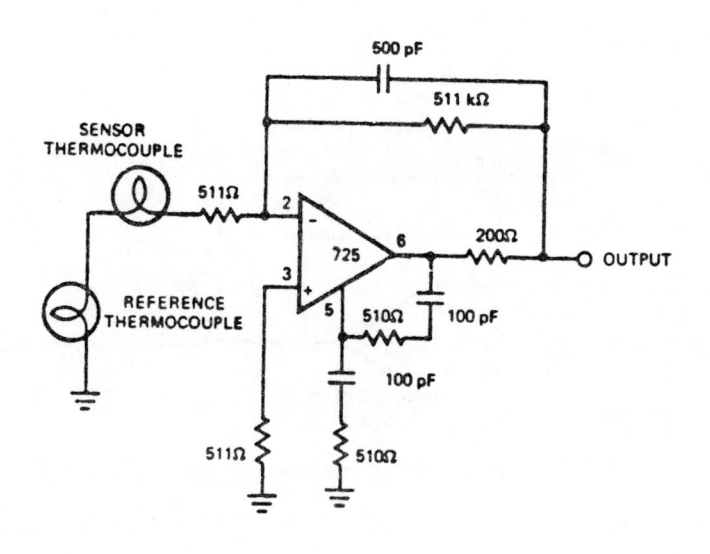

Thermocouple amplifier

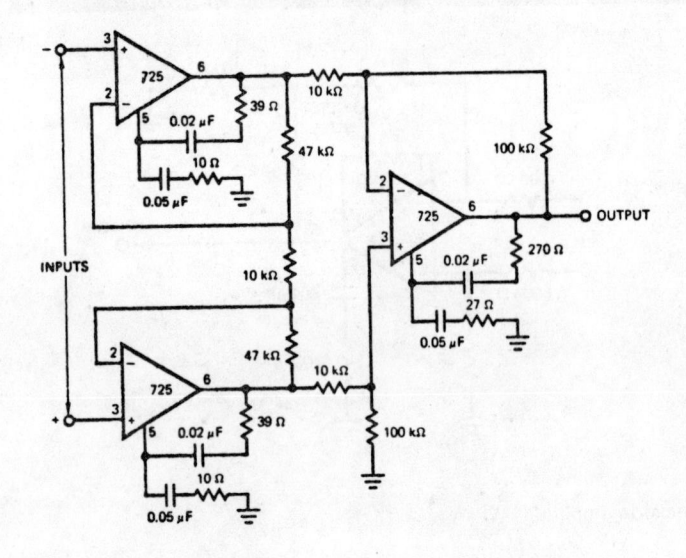

Instrumentation amplifier with high common mode rejection

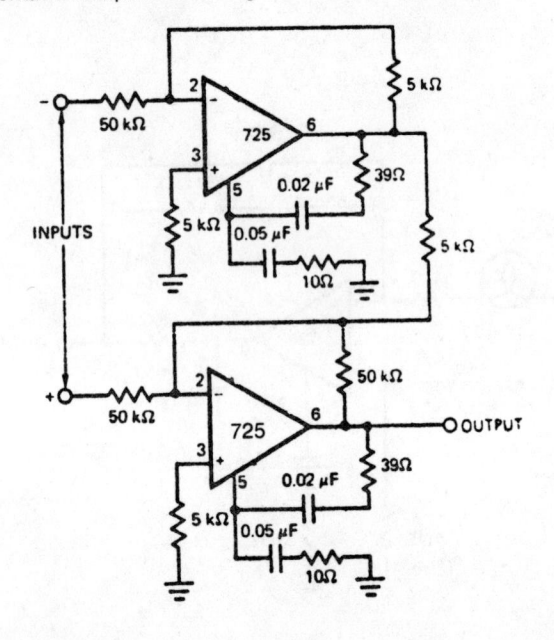

± 100 V common mode range differential amplifier

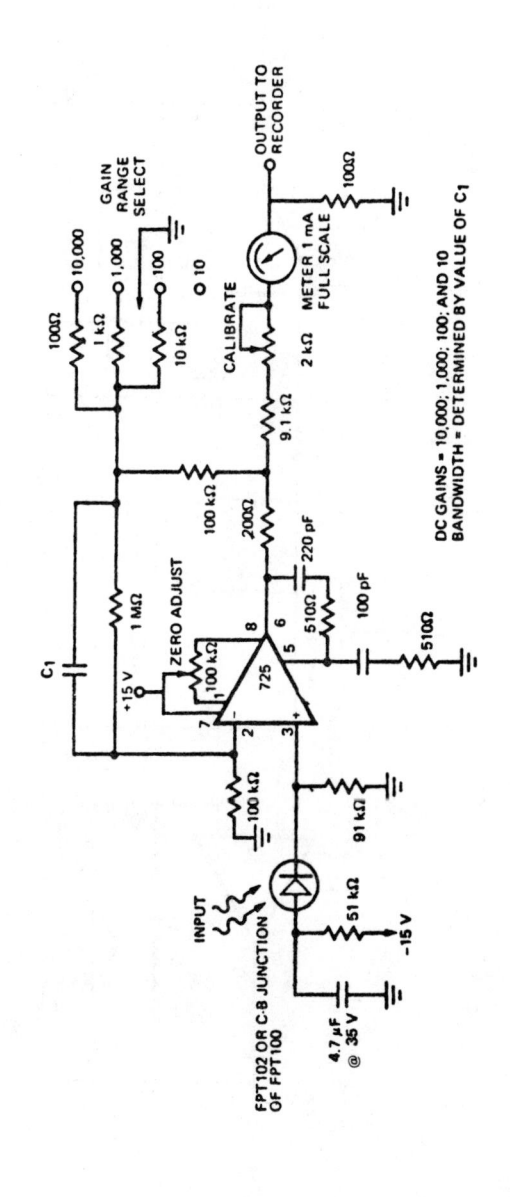

Photodiode amplifier

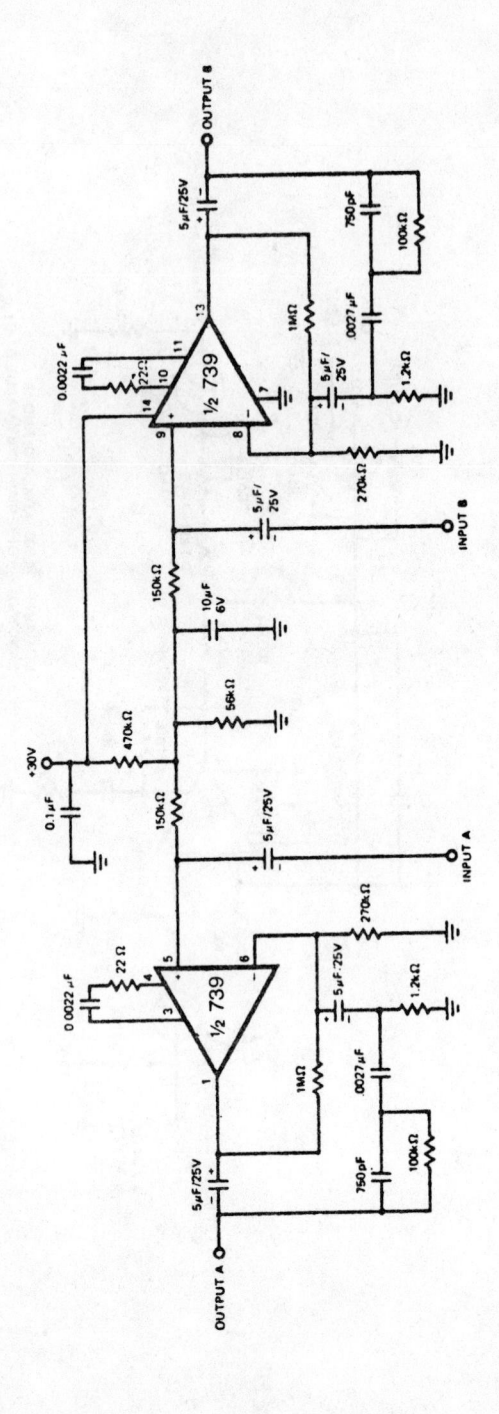

739 — DUAL LOW NOISE AUDIO AMPLIFIER/OPERATIONAL AMPLIFIER

Stereo phono preamplifier — RIAA equalized

741—FREQUENCY COMPEN-SATED OPERATIONAL AMPLIFIER

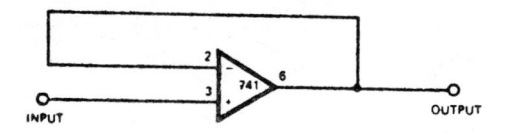

Unity-gain voltage follower

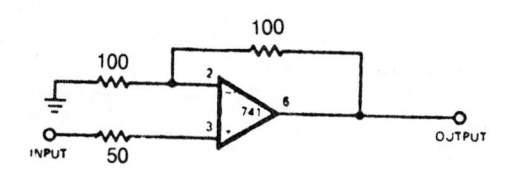

Non-inverting amplifier

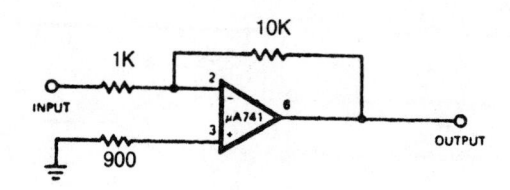

Inverting amplifier

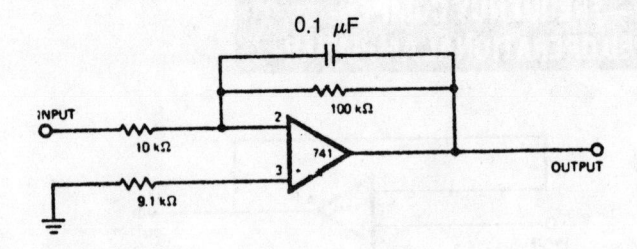

Simple integrator

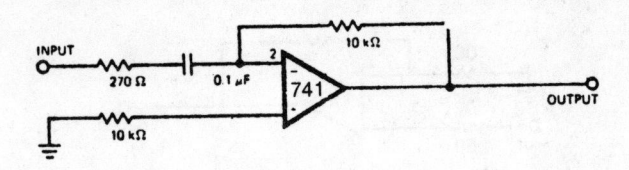

Simple differentiator

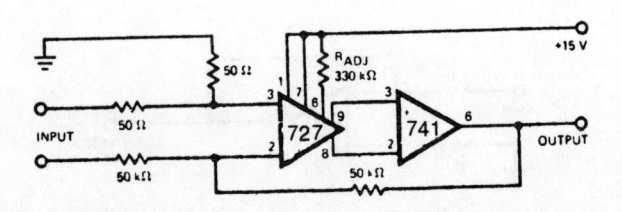

Low drift low noise amplifier

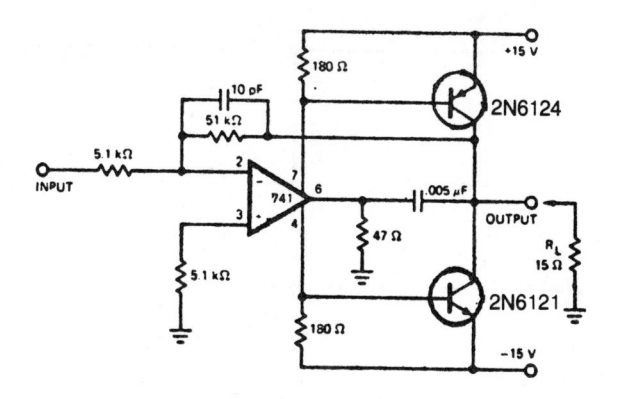

High slew rate power amplifier

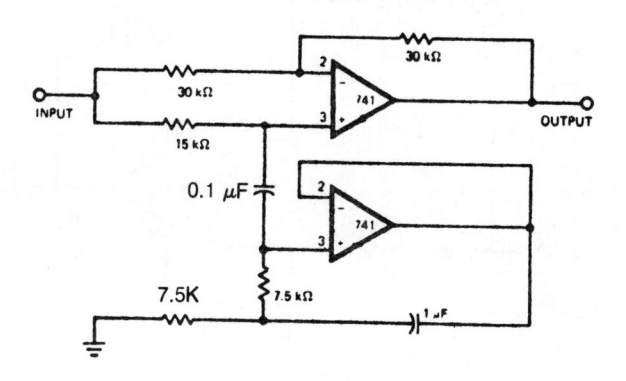

Notch filter

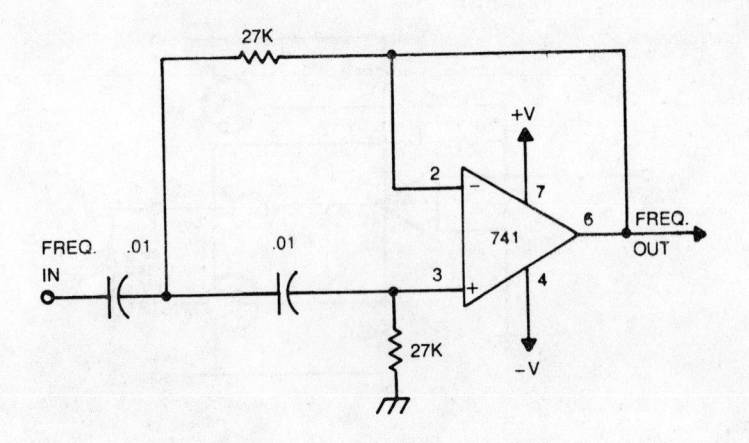

High pass active filter

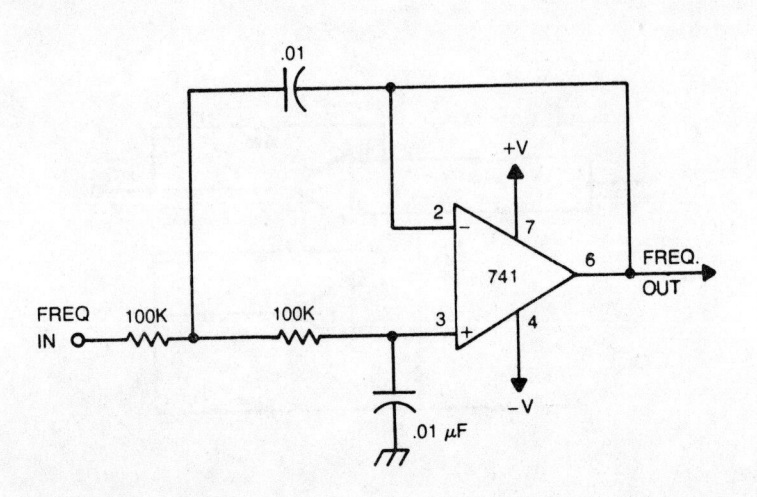

Low pass active filter

747 — DUAL FREQUENCY COMPENSATED OPERATIONAL AMPLIFIER

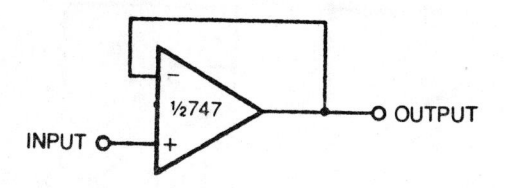

Unity-gain voltage follower

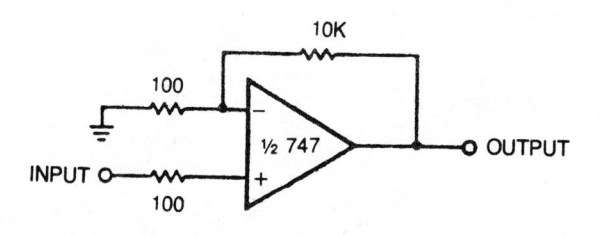

Noninverting amplifier

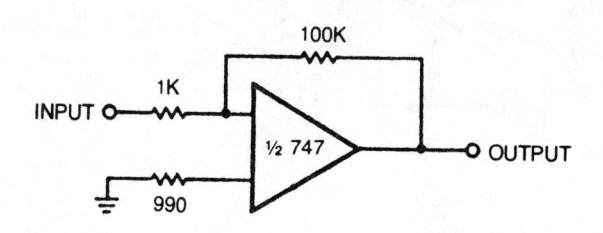

Inverting amplifier

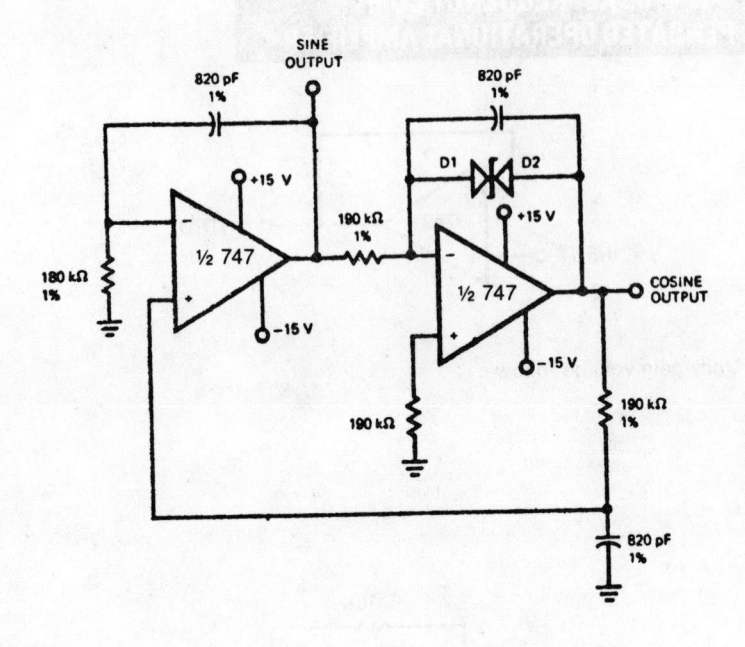

Quadrature oscillator

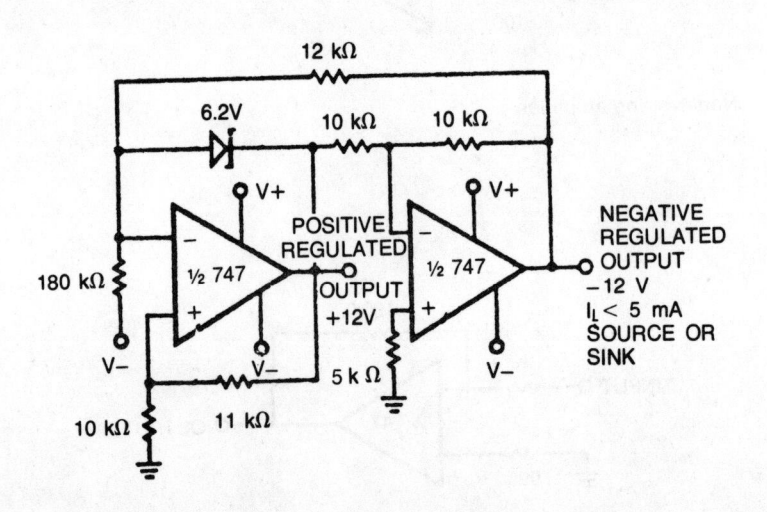

Tracking positive and negative voltage references

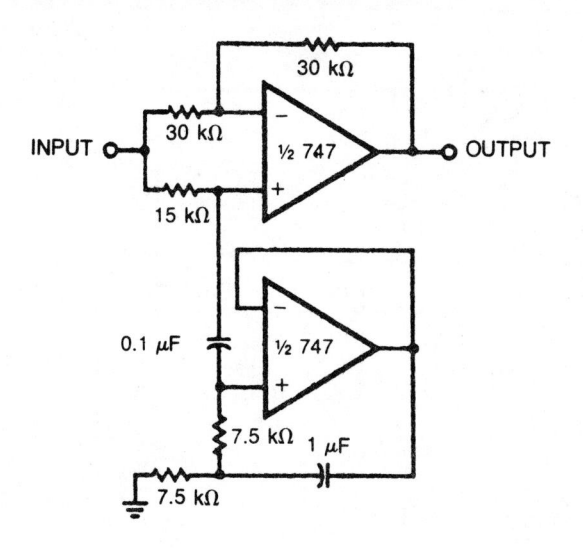

Notch filter

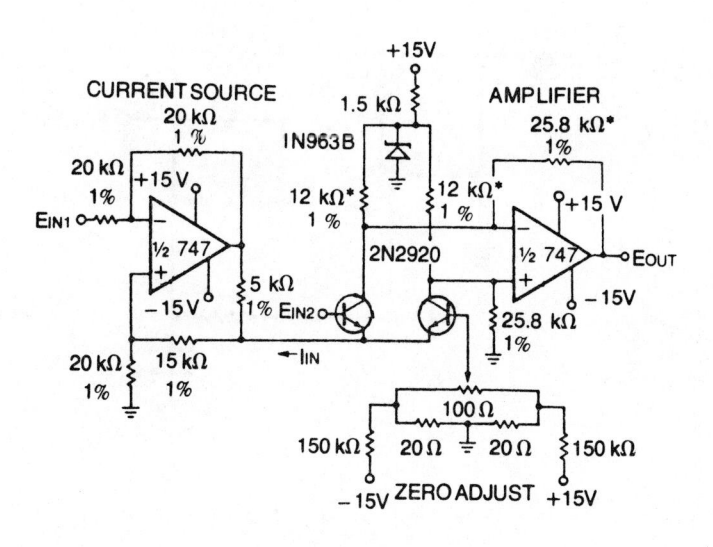

Analog multiplier

748 — HIGH PERFORMANCE OPERATIONAL AMPLIFIER

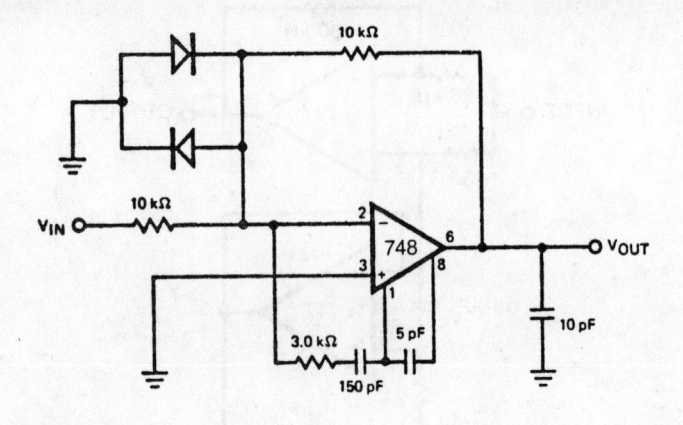

Feed forward compensation

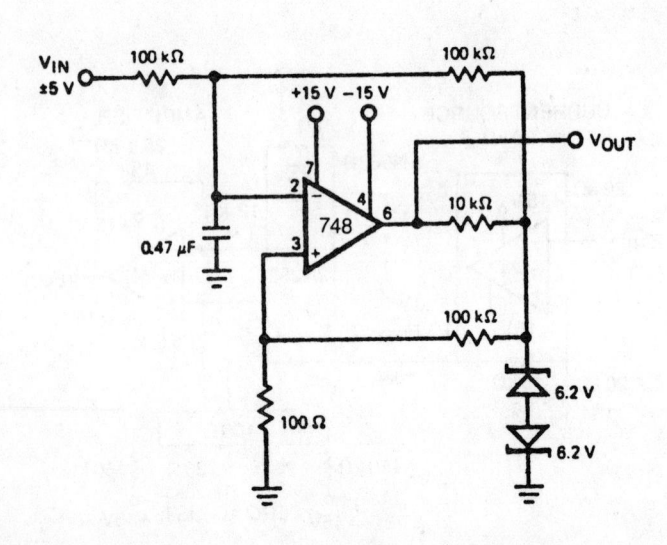

Pulse width modulator

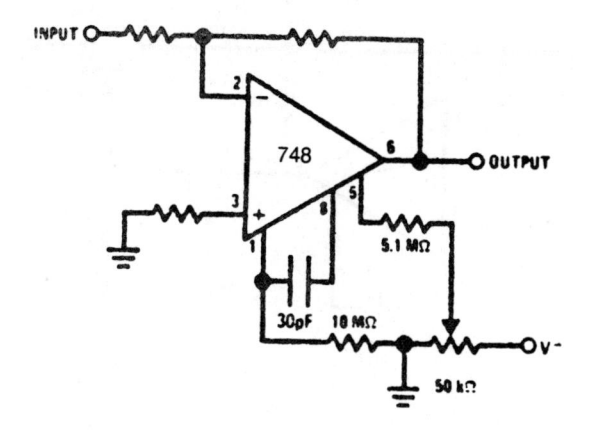

Inverting amplifier with balancing circuit

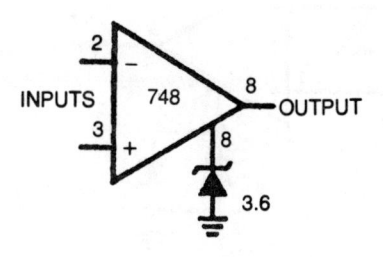

Voltage comparator for driving DTL or TTL integrated circuits

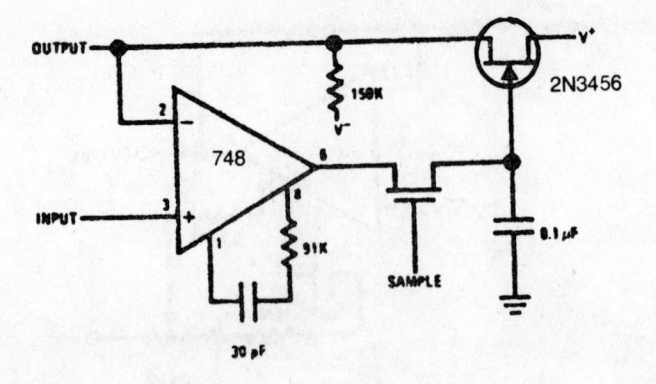

Low drift sample and hold

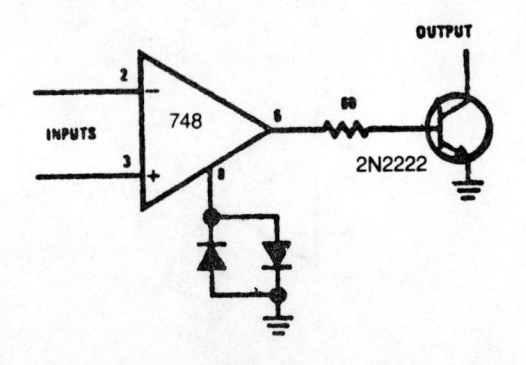

Voltage comparator for driving RTL logic or high current driver

749 — DUAL AUDIO OPERATIONAL AMPLIFIER/PREAMPLIFIER

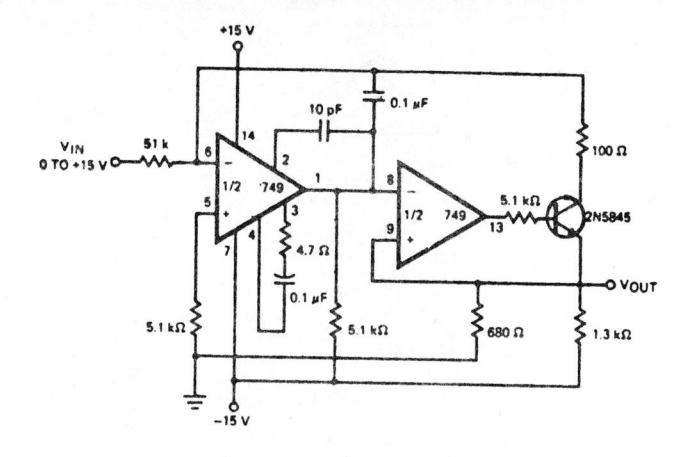

Voltage to frequency converter

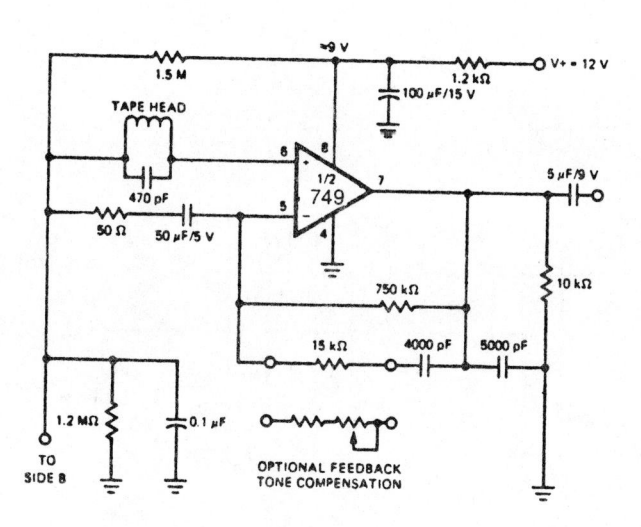

Stereo tape preamplifier

760—HIGH SPEED DIFFERENTIAL COMPARATOR

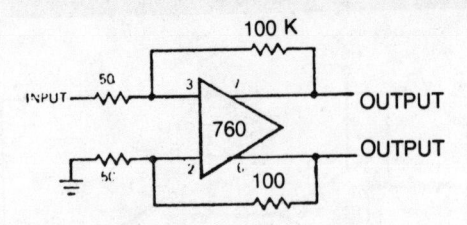

Level detector with hysteresis

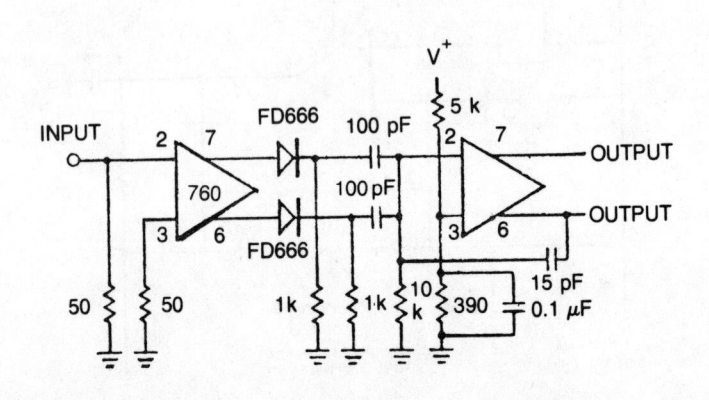

Zero crossing detector

775 — QUAD COMPARATOR

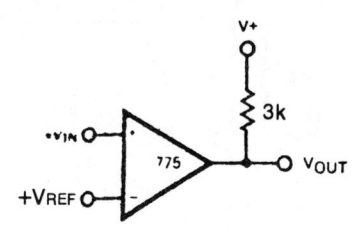

Basic comparator

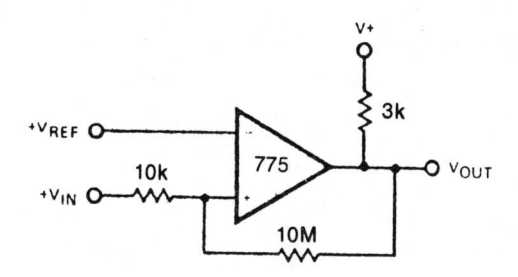

Noninverting comparator with hysteresis

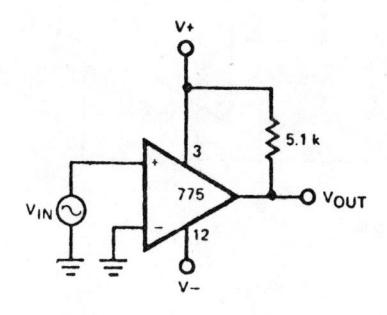

Zero crossing detector (dual supply)

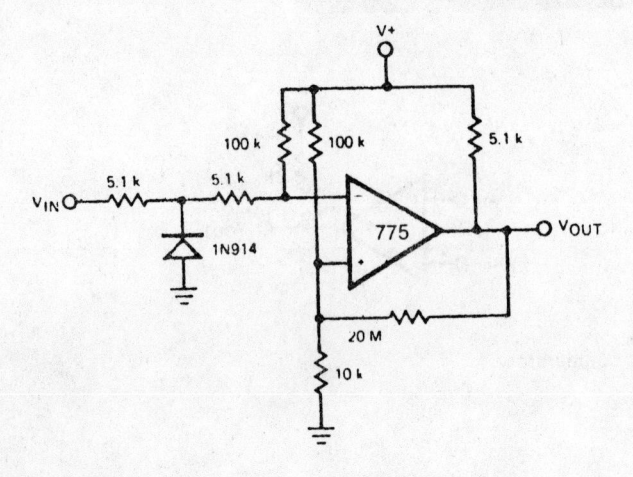

Zero crossing detector (single power supply)

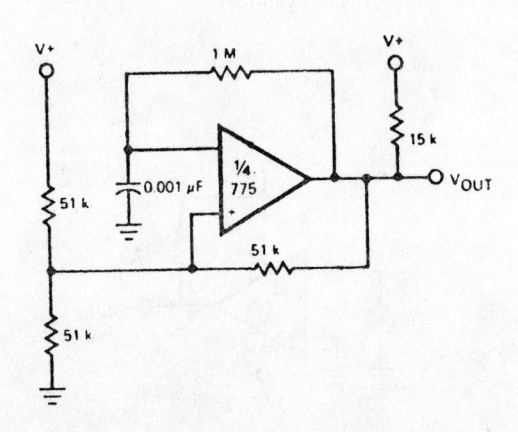

Free running square wave oscillator

228

776 — PROGRAMMABLE OPERATIONAL AMPLIFIER

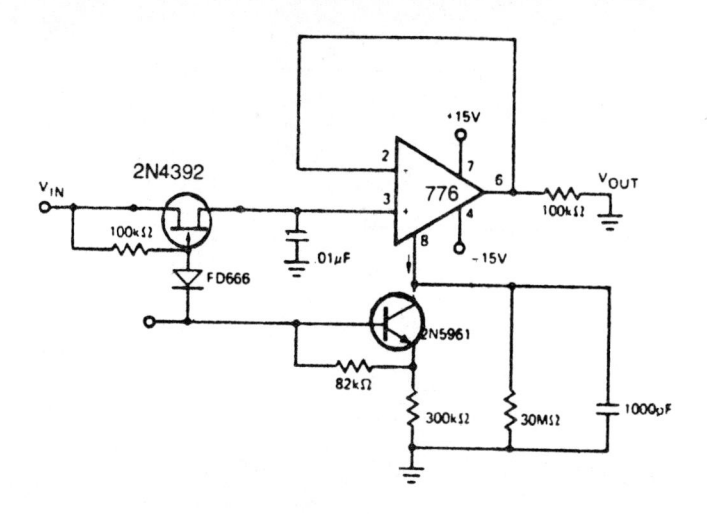

High accuracy sample and hold

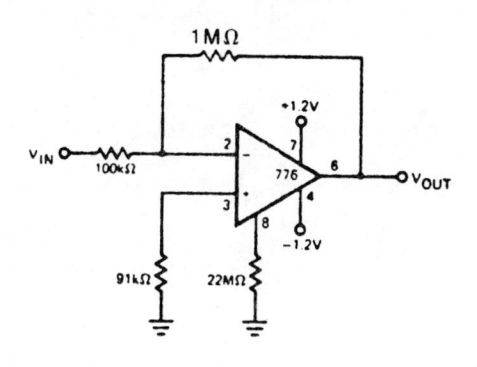

Nano-watt amplifier

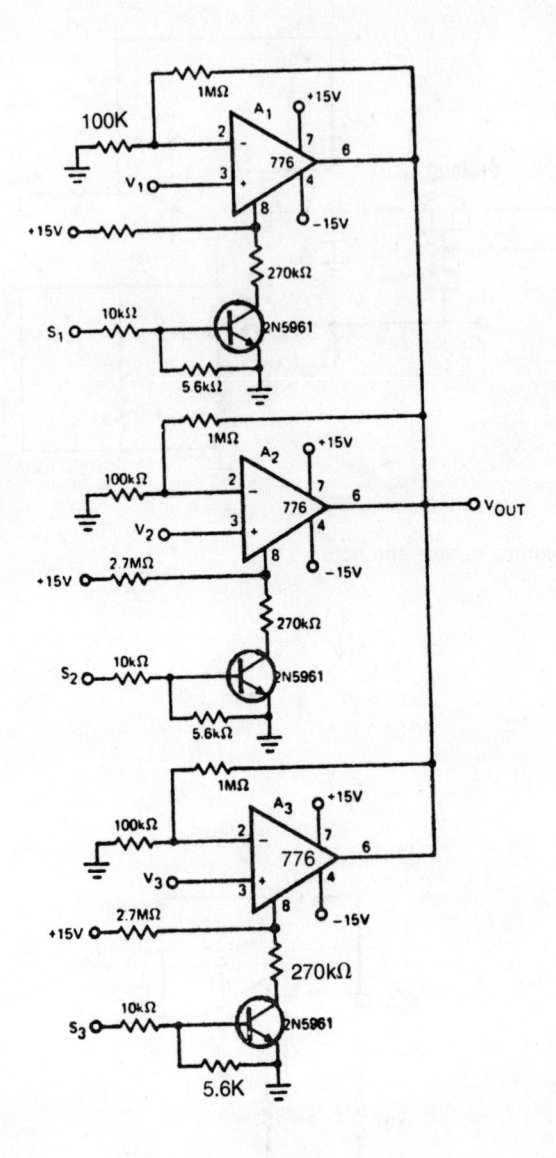

Multiplexing and signal conditioning without FETs

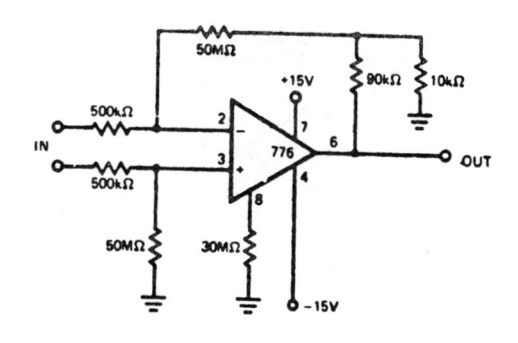

High input impedance amplifier

777—PRECISION OPERATIONAL AMPLIFIER

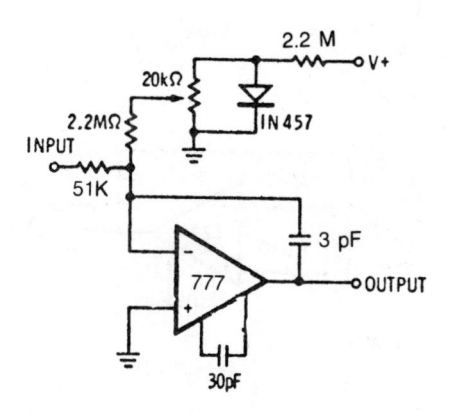

Bias compensated long time integrator

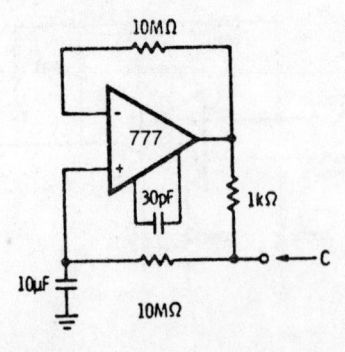

Capacitance multiplier

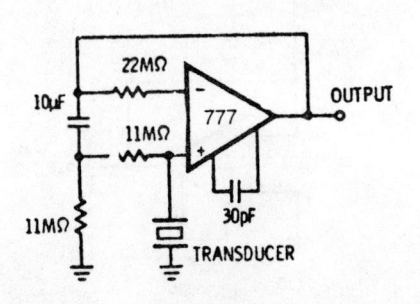

Amplifier for capacitance transducers

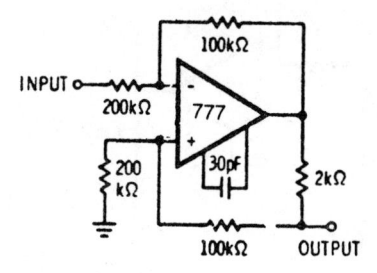

Bilateral current source

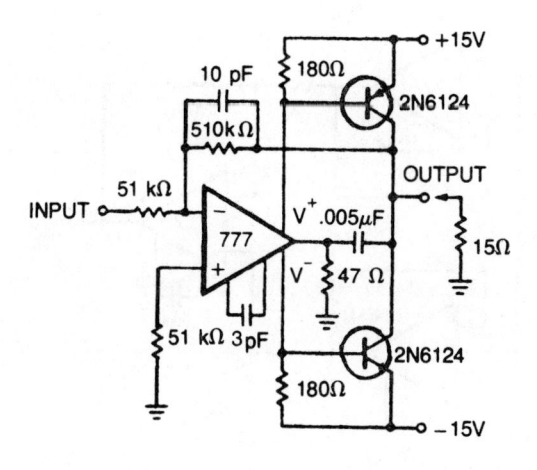

High slew rate power amplifier

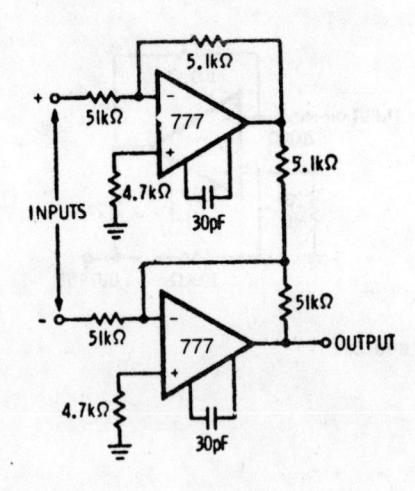

±100 V common mode range instrumentation amplifier

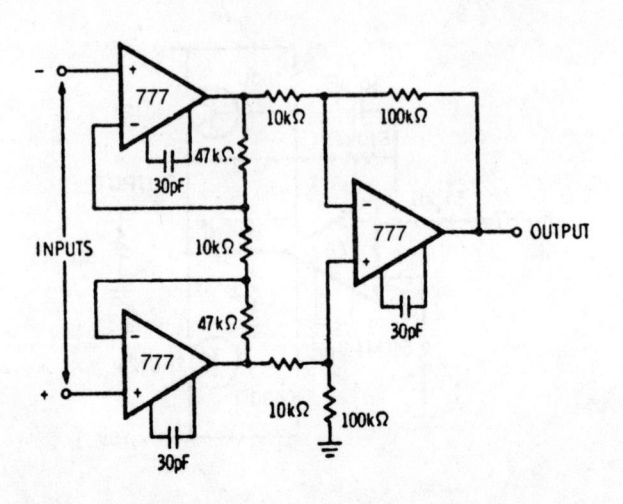

Instrumentation amplifier with high common mode rejection

234

791—POWER OPERATIONAL AMPLIFIER

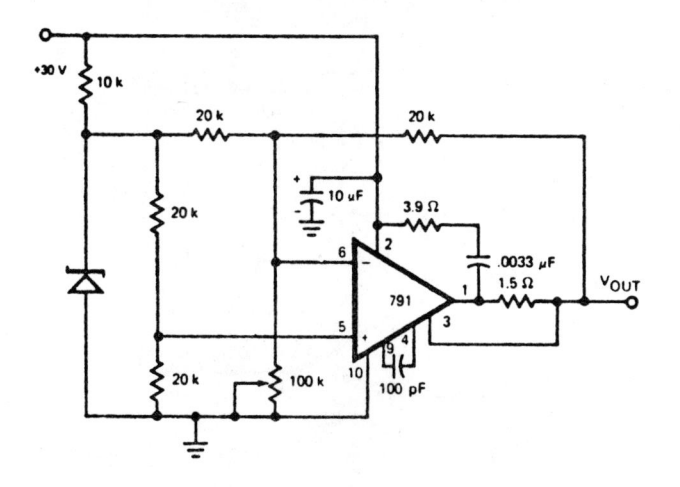

Positive voltage regulator

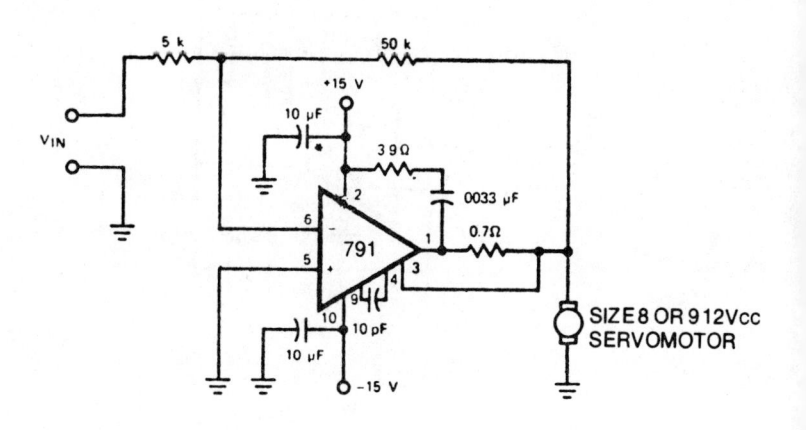

DC servo amplifier

235

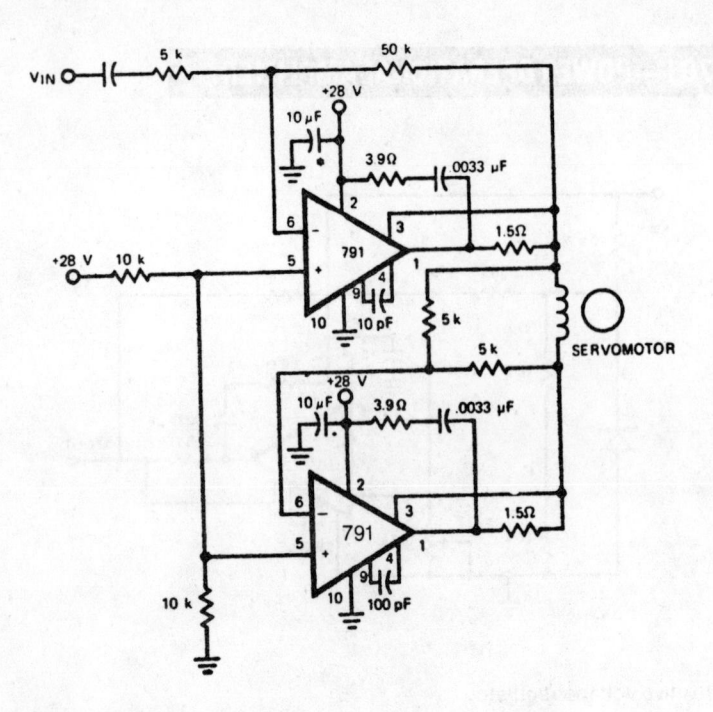

AC servo amplifier bridge type

798 — DUAL OPERATIONAL AMPLIFIER

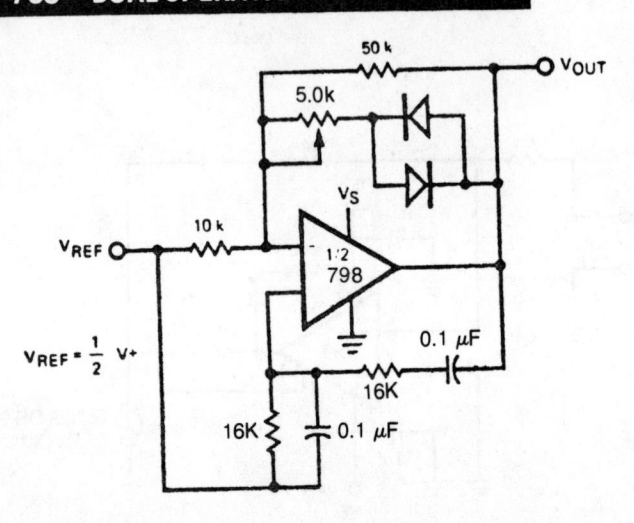

$V_{REF} = \frac{1}{2} V^+$

Wien bridge oscillator

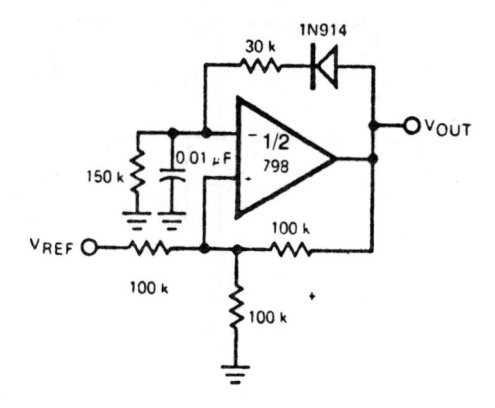

Pulse generator

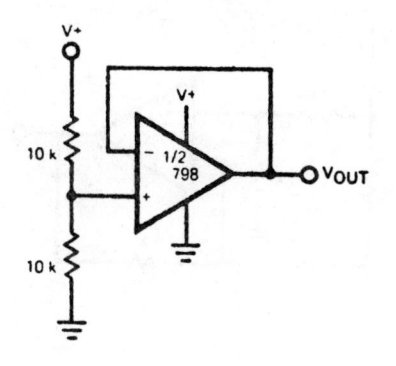

Voltage reference

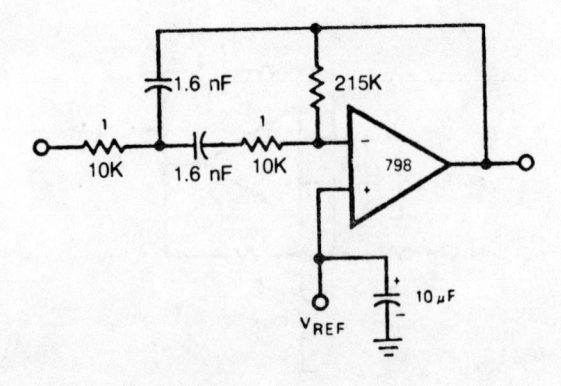

Multiple feedback bandpass filter

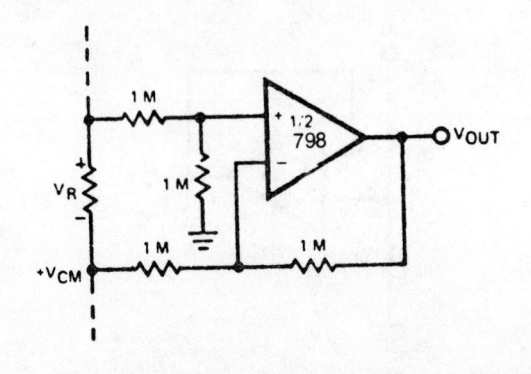

Ground referencing a differential input signal

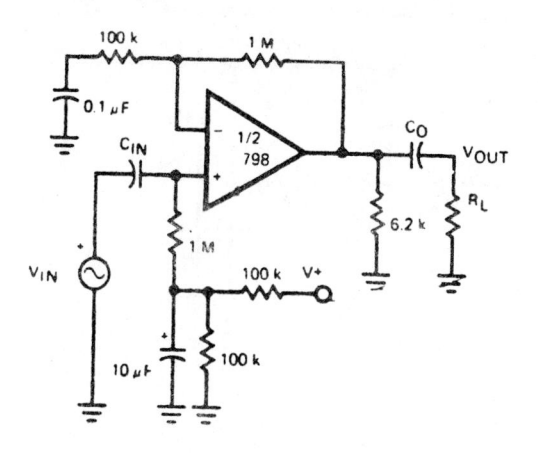

AC coupled noninverting amplifier

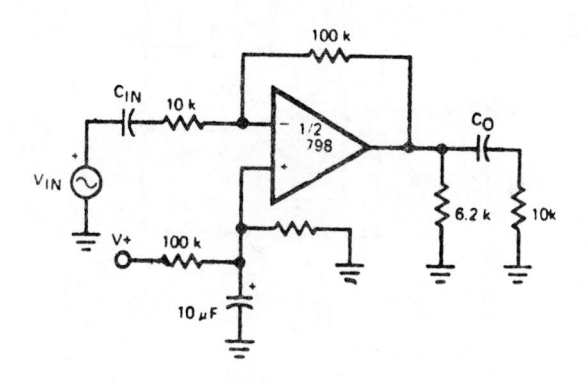

AC coupled inverting amplifer

799—HIGH GAIN OPERATIONAL AMPLIFIER

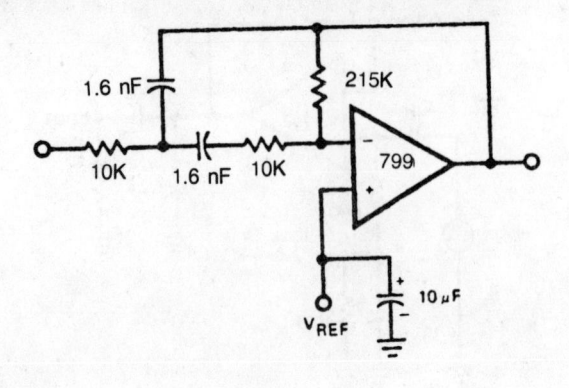

Multiple feedback bandpass filter

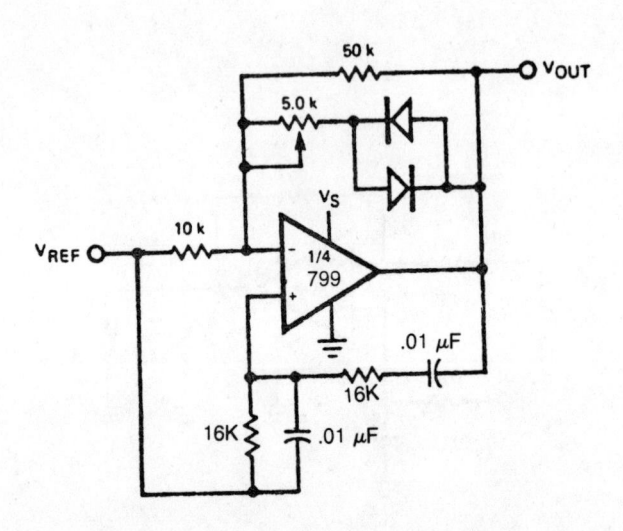

Wien bridge oscillator

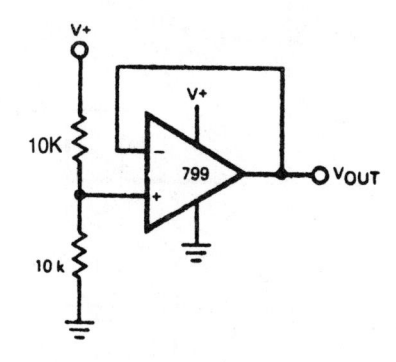

Voltage reference

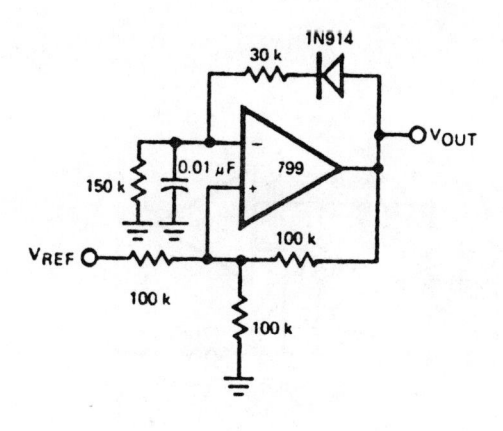

Pulse generator

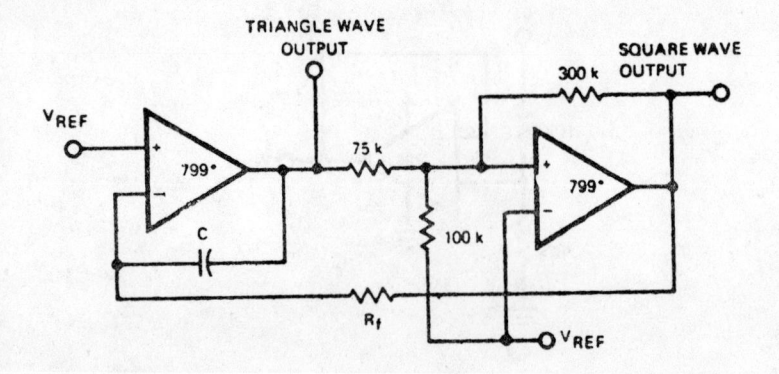

Function generator

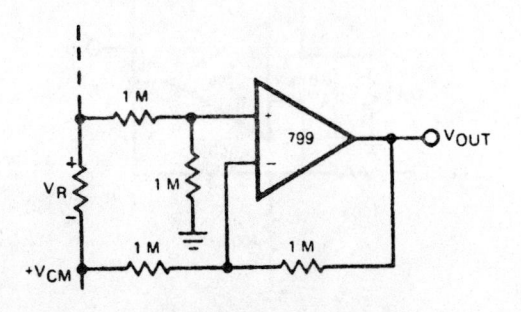

Ground referencing a differential input signal

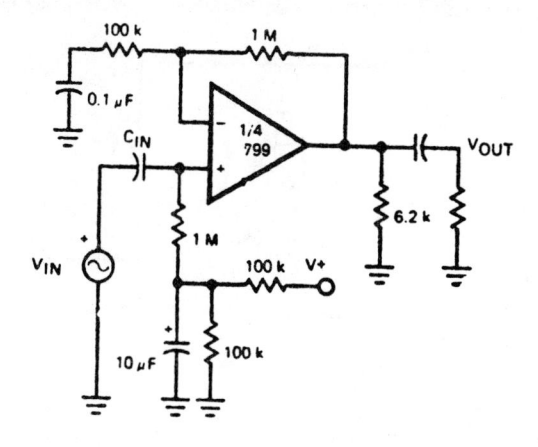

AC couple noninverting amplifier

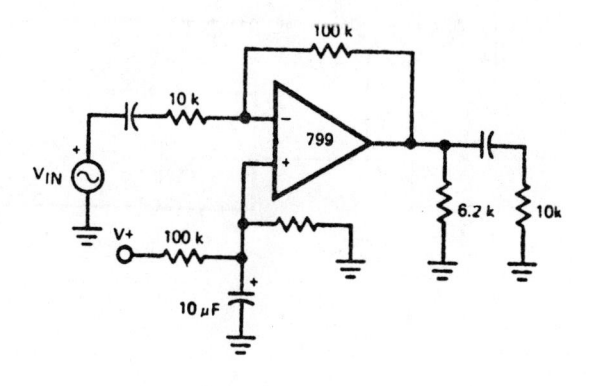

AC coupled inverting amplifier

0001—LOW POWER OPERATIONAL AMPLIFIER

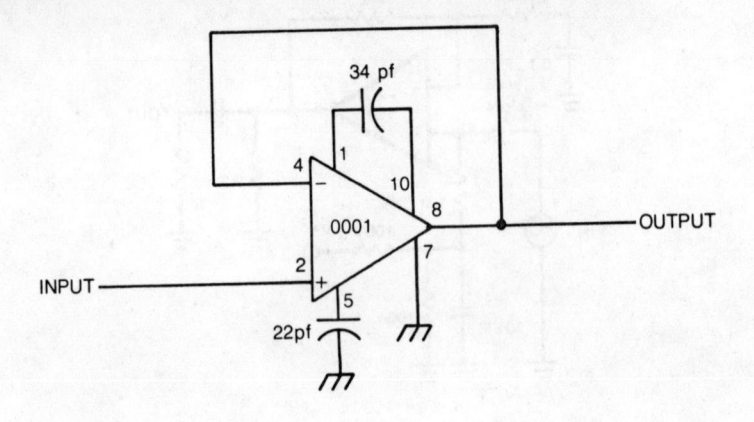

Voltage follower

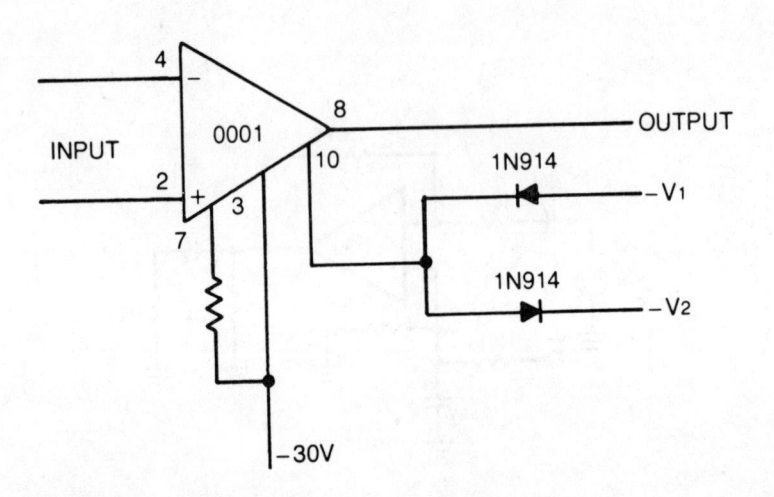

Voltage comparator and MOS driver

0003 — WIDE BANDWIDTH OPERATIONAL AMPLIFIER

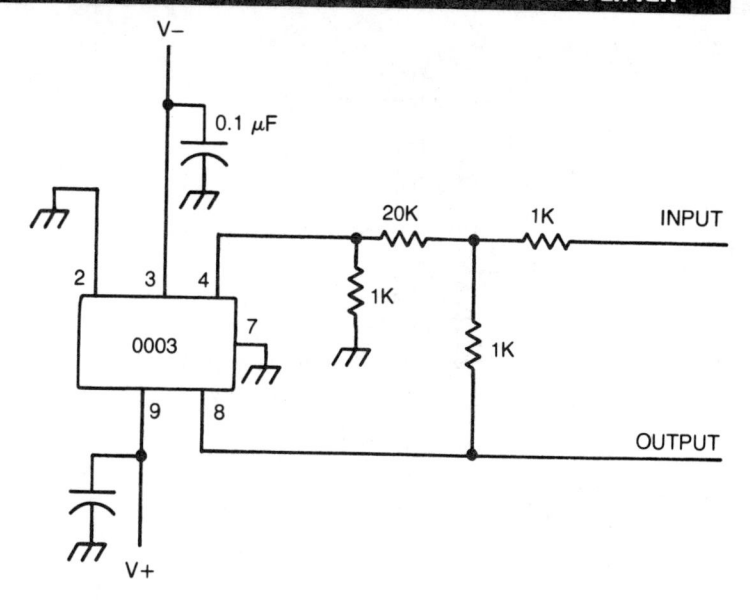

Unity gain inverting amplifier

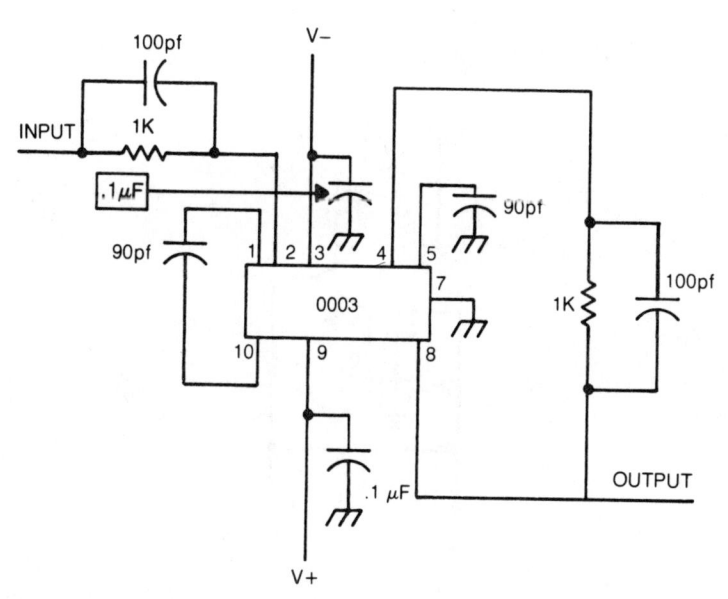

Unity gain follower

0004 — HIGH VOLTAGE OPERATIONAL AMPLIFIER

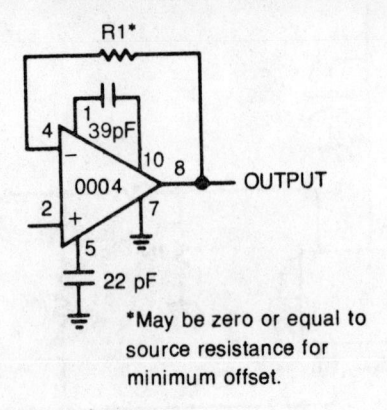

*May be zero or equal to source resistance for minimum offset.

Voltage follower

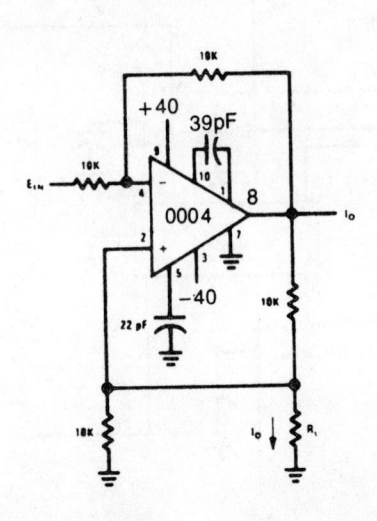

High compliance current source

246

0005—HIGH INPUT RESISTANCE OPERATIONAL AMPLIFIER

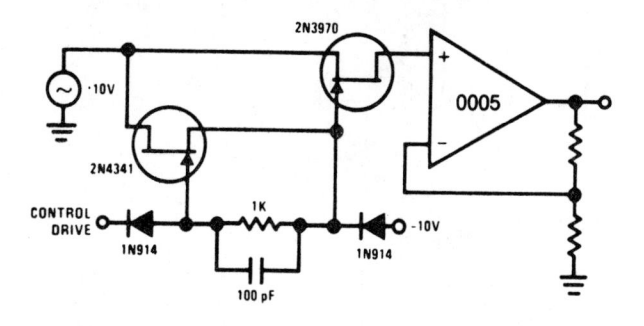

High toggle rate high frequency analog switch

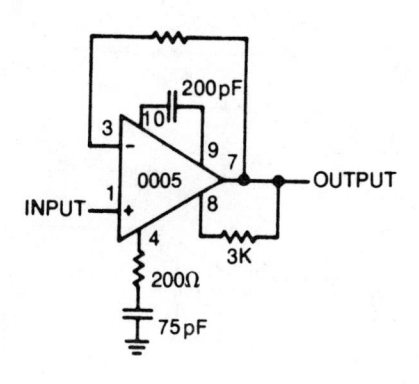

Voltage follower

247

0021 — 1.0 AMP OPERATIONAL AMPLIFIER

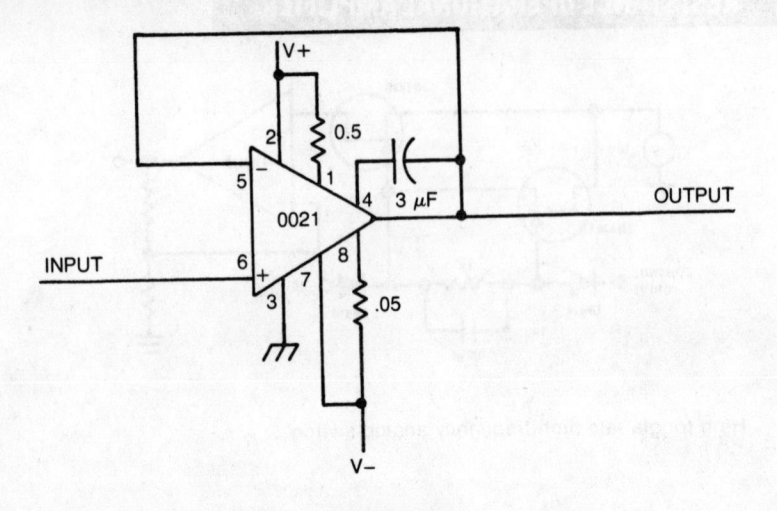

Unity gain amplifier with short circuit limiting

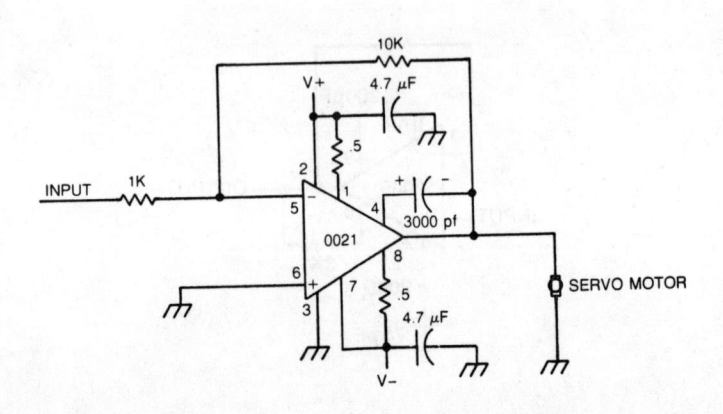

DC servo motor amplifier

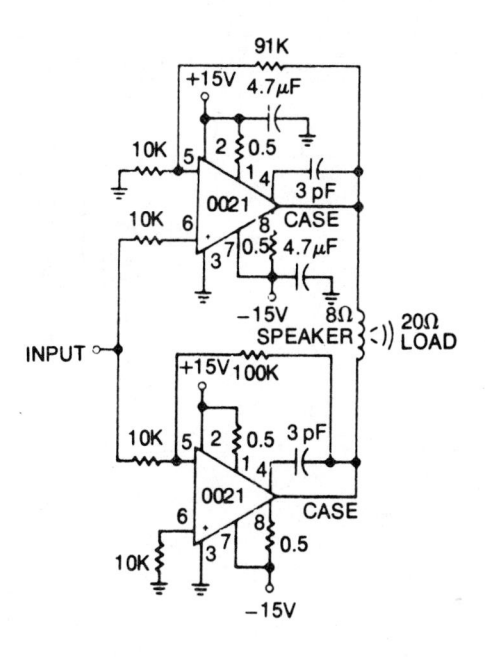

10 watt (rms) audio amplifier

0022—FET INPUT OPERATIONAL AMPLIFIER

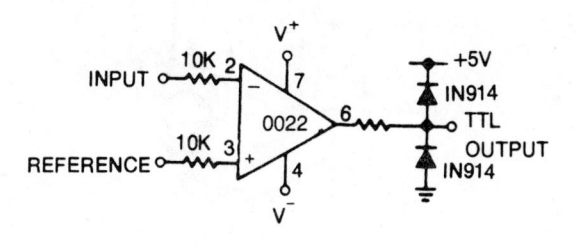

Precision voltage comparator

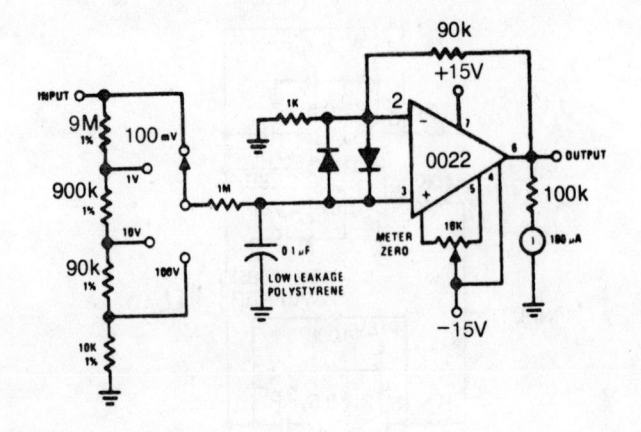

Sensitive low cost "VTVM"

0024—HIGH SLEW RATE OPERATIONAL AMPLIFIER

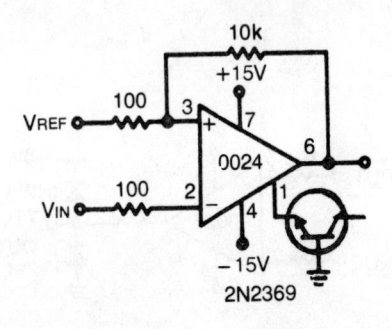

TTL compatible comparator

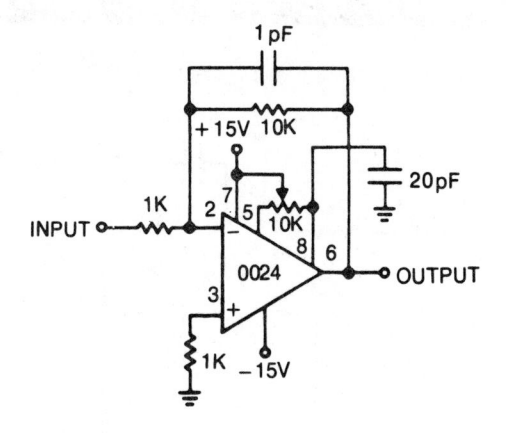

Offset null

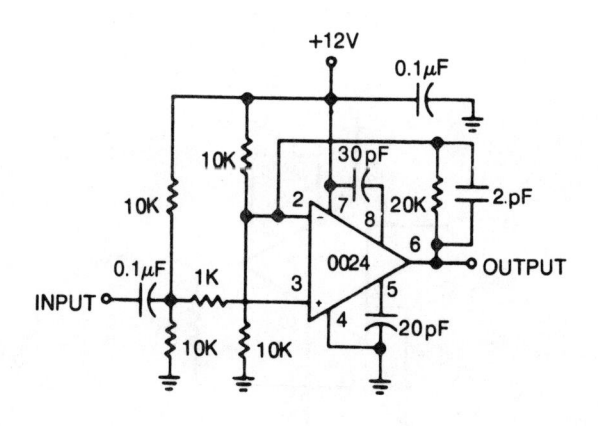

Video amplifier

0032—ULTRA FAST FET OPERATIONAL AMPLIFIER

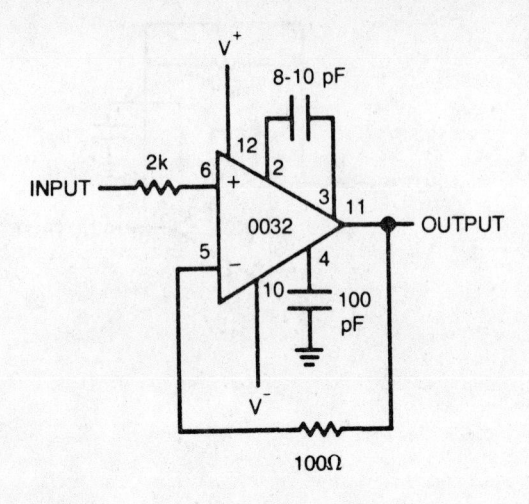

Unity gain amplifier

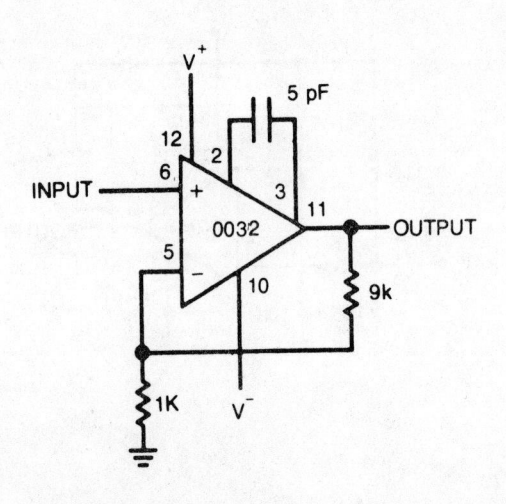

10X buffer amplifier

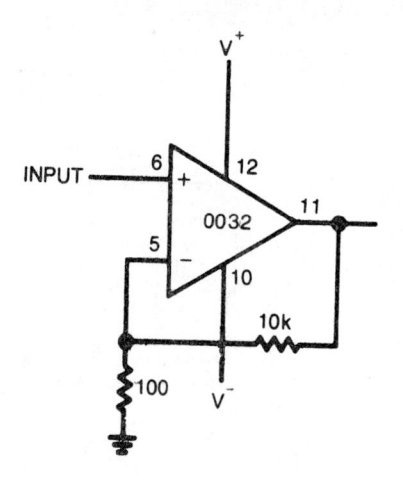

100X buffer amplifier

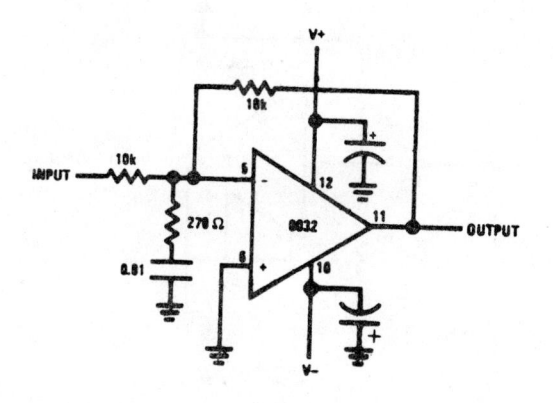

Noncompensated unity gain inverter

0033—FAST BUFFER AMPLIFIER

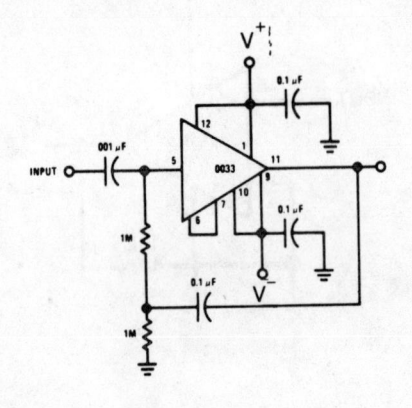

High input impedance ac coupled amplifier

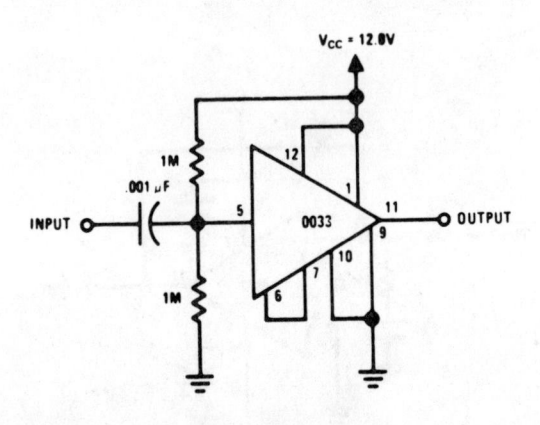

Single supply ac amplifier

254

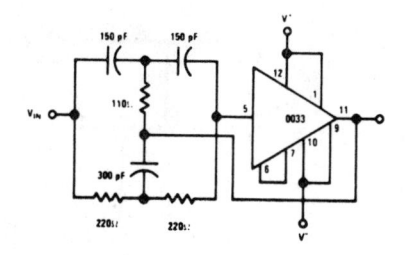

4.5 MHz notch filter

0042—FET INPUT OPERATIONAL AMPLIFIER

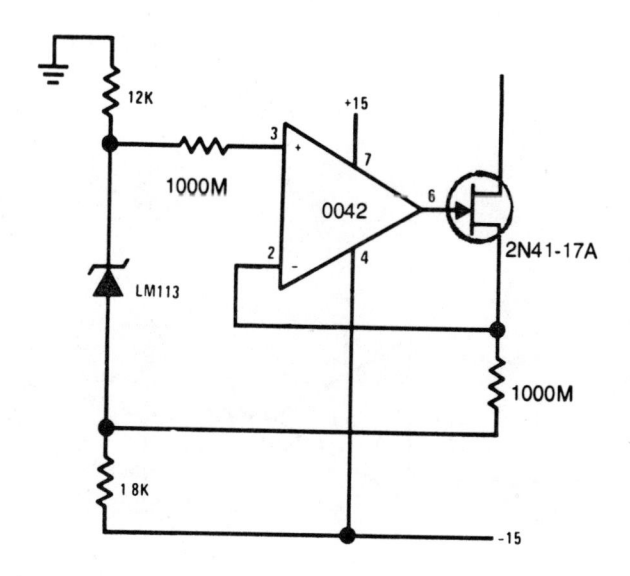

Precision current sink

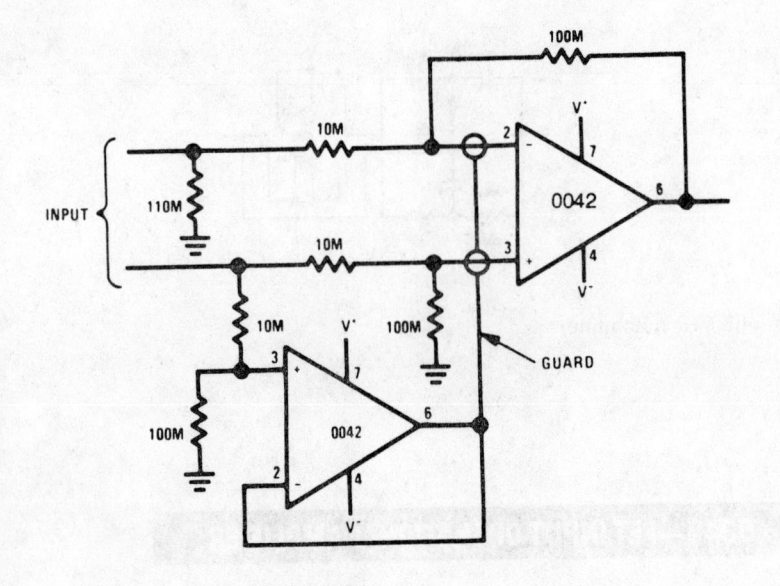

Guarded full differential amplifier

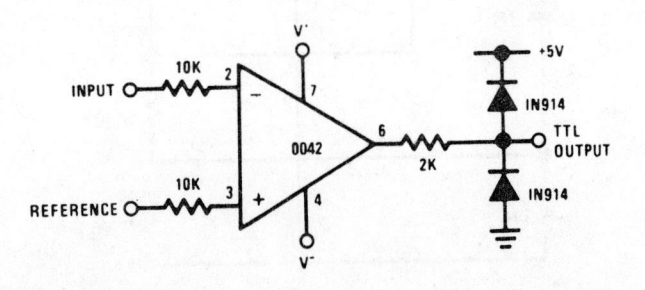

Precision voltage comparator

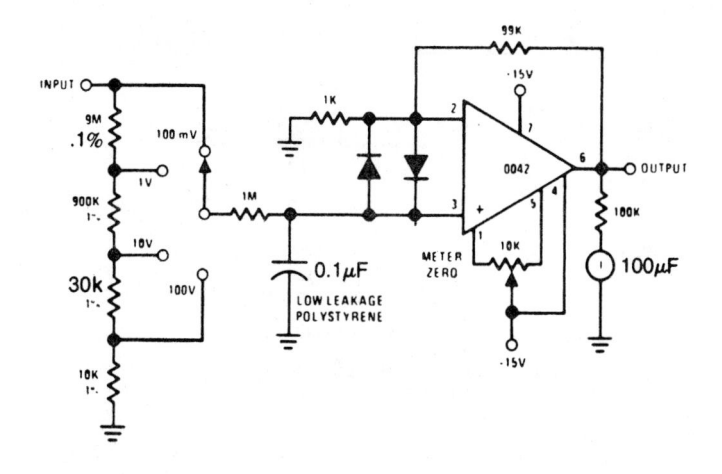

Sensitive low cost "VTVM"

0044—LOW NOISE OPERATIONAL AMPLIFIER

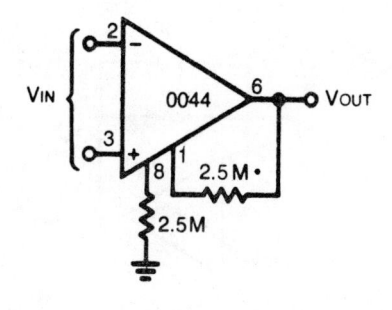

X1000 Instrumentation amp

257

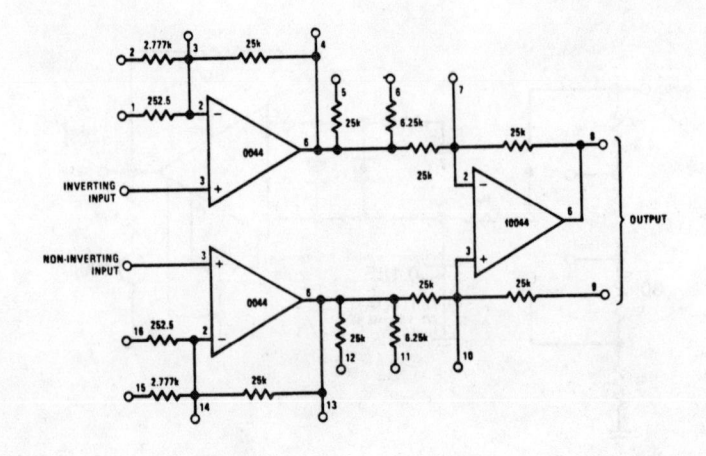

Precision instrumentation amplifier

0062 — HIGH SPEED FET OPERATIONAL AMPLIFIER

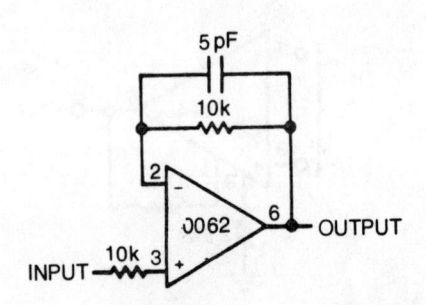

Fast voltage follower

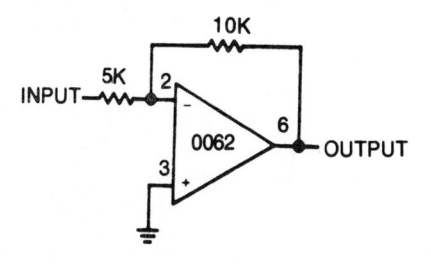

Fast summing amplifier

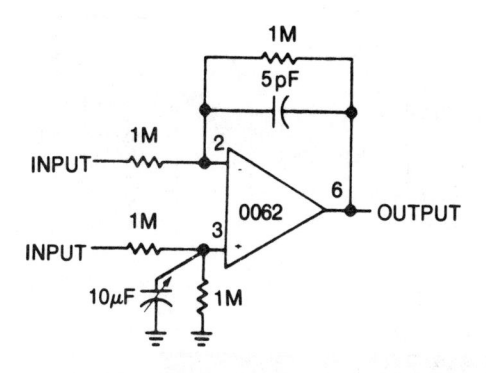

Differential amplifier

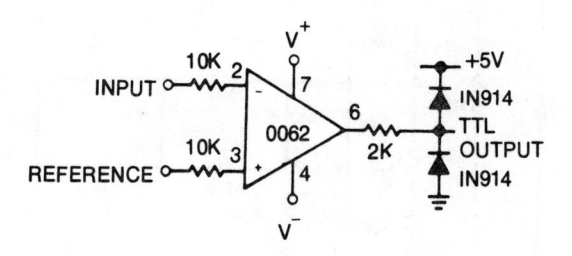

Fast precision voltage comparator

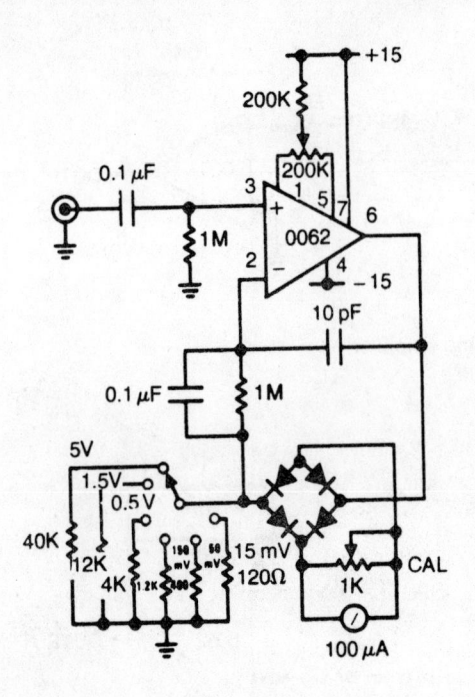

Wide range ac voltmeter

1303—STEREO PREAMPLIFIER

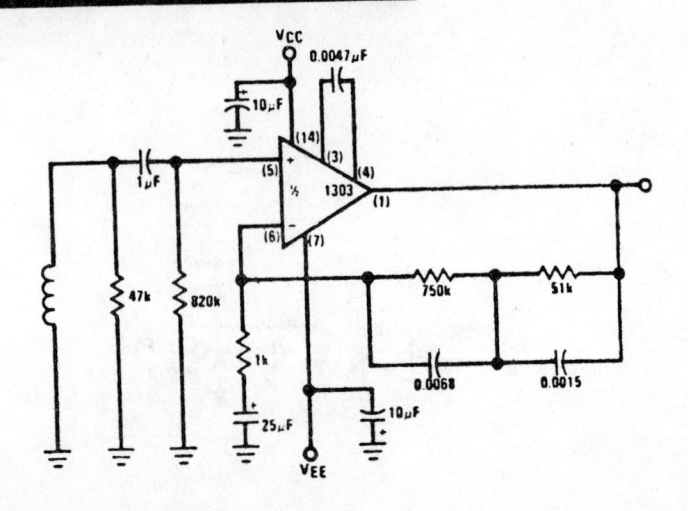

Phono preamp (RIAA)

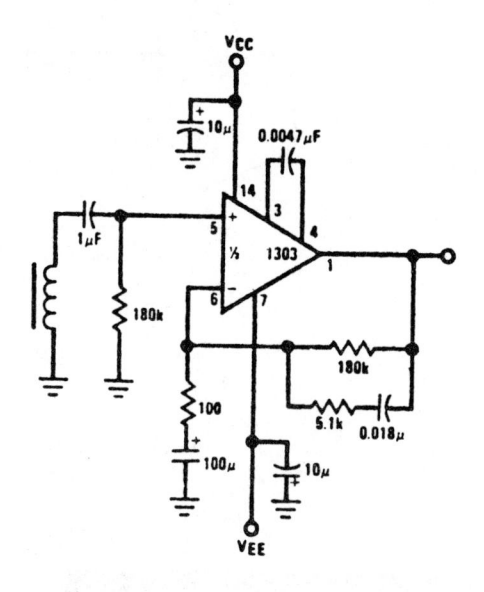

Tape preamp (NAB, 1-7/8 & 3-3/4 IPS)

1306—HALF WATT INTEGRATED AUDIO AMPLIFIER

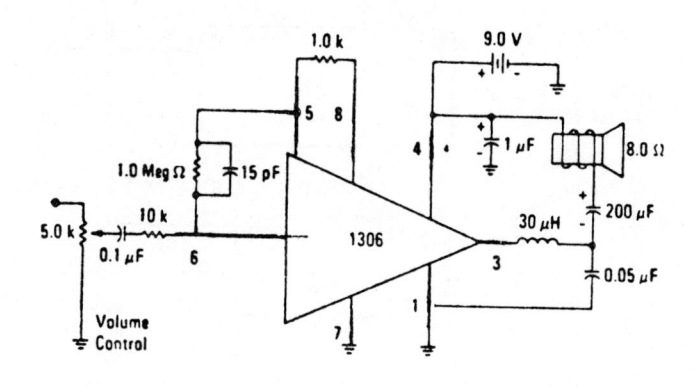

AM-FM radio, audio section

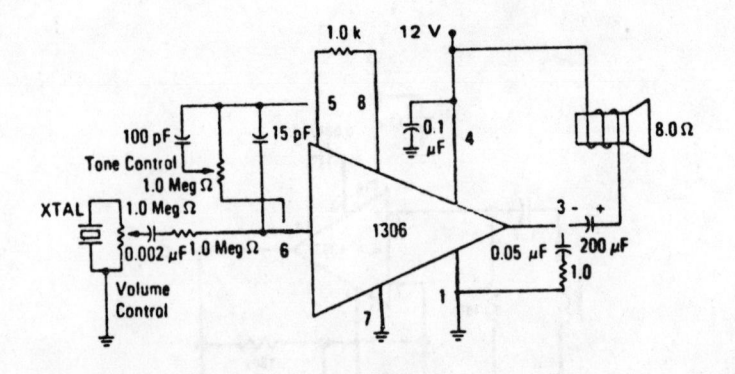

Phonograph amplifier (ceramic cartridge)

1422—MONOLITHIC TIMER WITH EXTERNALLY ADJUSTABLE THRESHOLD

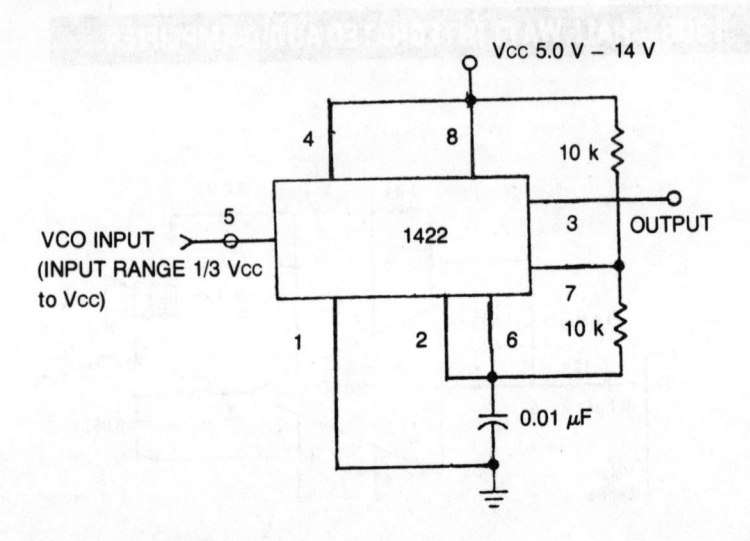

Voltage controlled oscillator

○ Vcc 5.0 V – 14 V

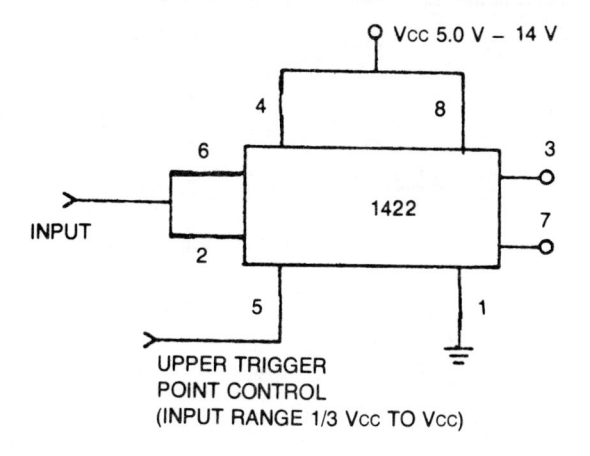

INPUT

UPPER TRIGGER POINT CONTROL
(INPUT RANGE 1/3 Vcc TO Vcc)

Schmitt trigger

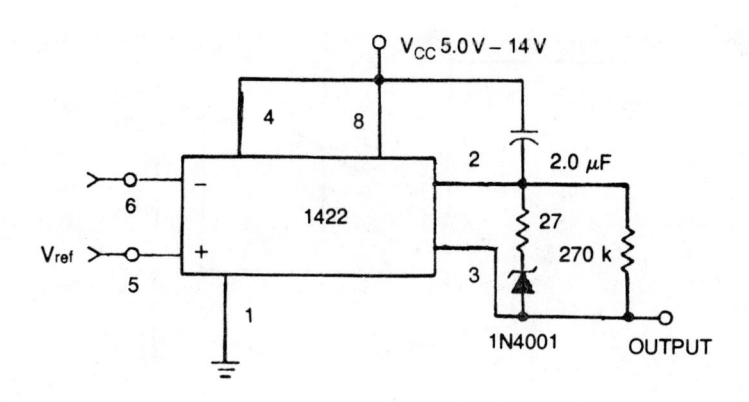

○ V_{CC} 5.0 V – 14 V

V_{ref}

2.0 µF
27
270 k
1N4001
OUTPUT

Comparator with time out

1436—HIGH VOLTAGE OPERATIONAL AMPLIFIER

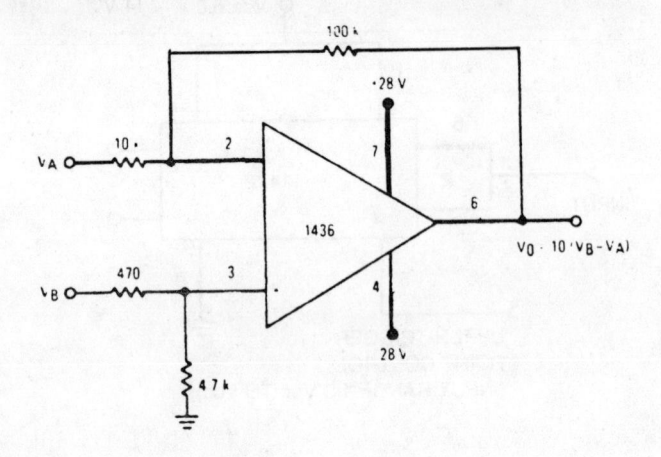

Differential amplifier with ±20 V common-mode input voltage range

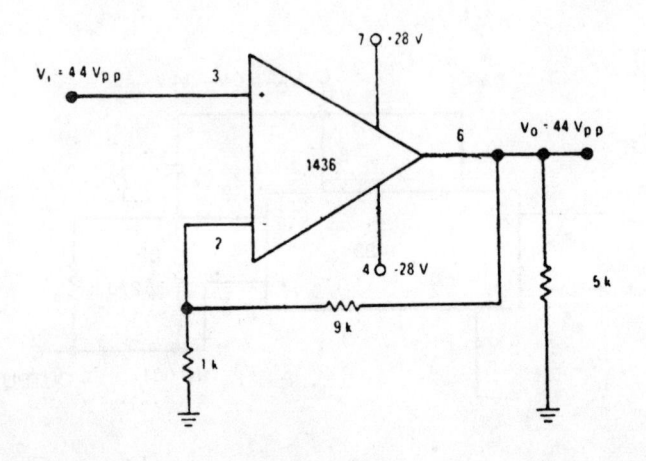

Typical noninverting X10 voltage amplifier

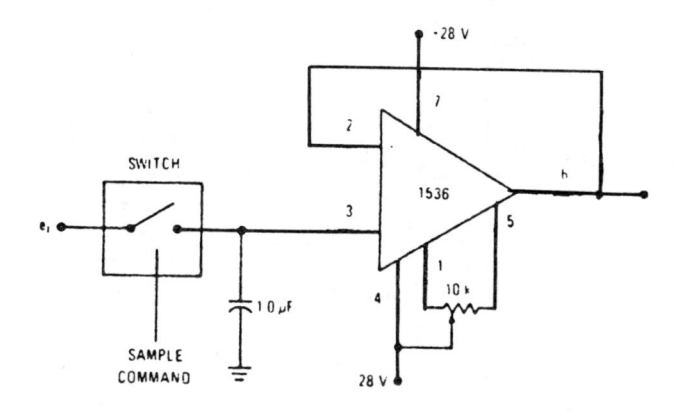

Low-drift sample and hold

1438—POWER BOOSTER OPERATIONAL AMPLIFIER

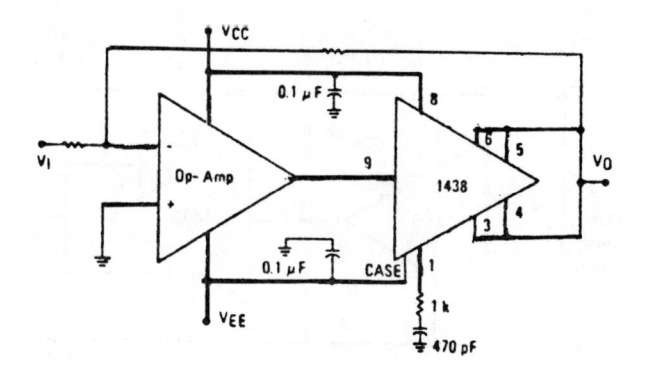

Operational amplifier boost circuit

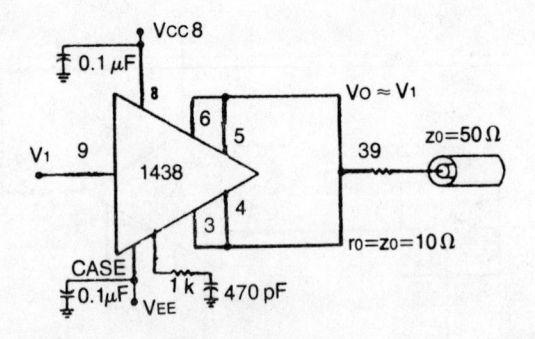

Digital or analog line driver

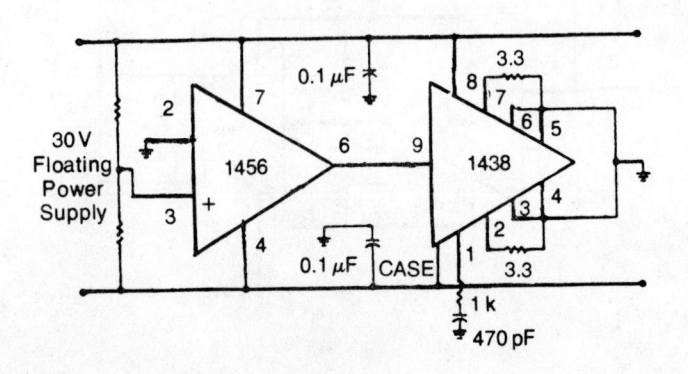

Power supply splitter

266

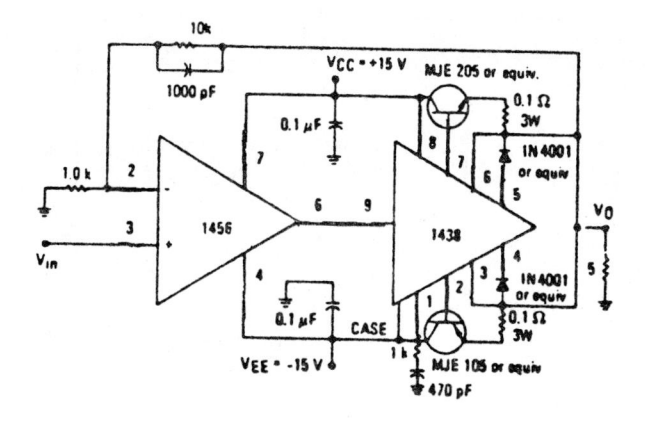

Servo/power amplifier

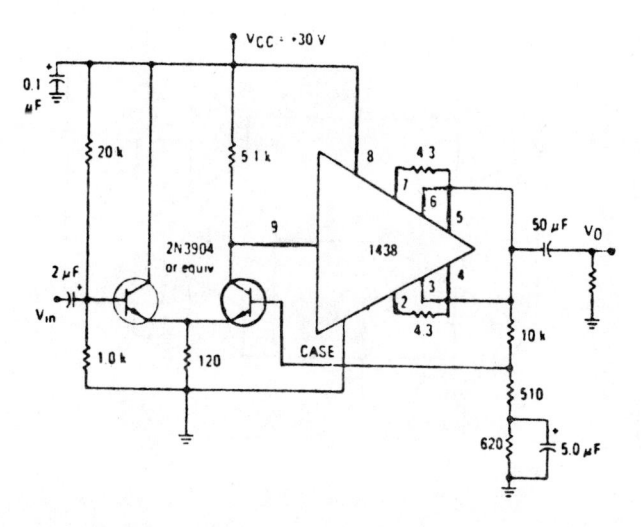

Noninverting ac power amplifier

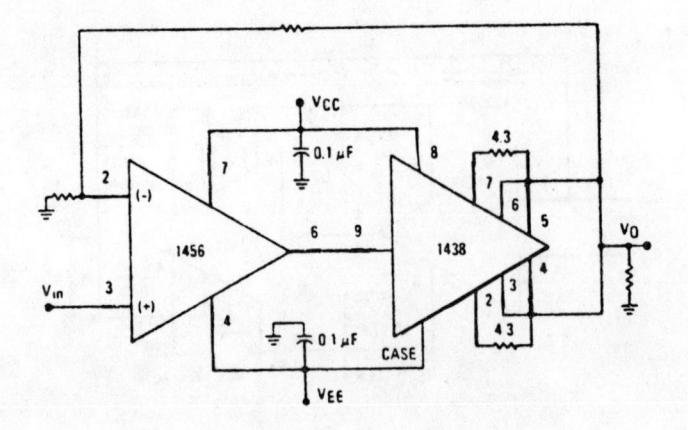

Noninverting power amplifier

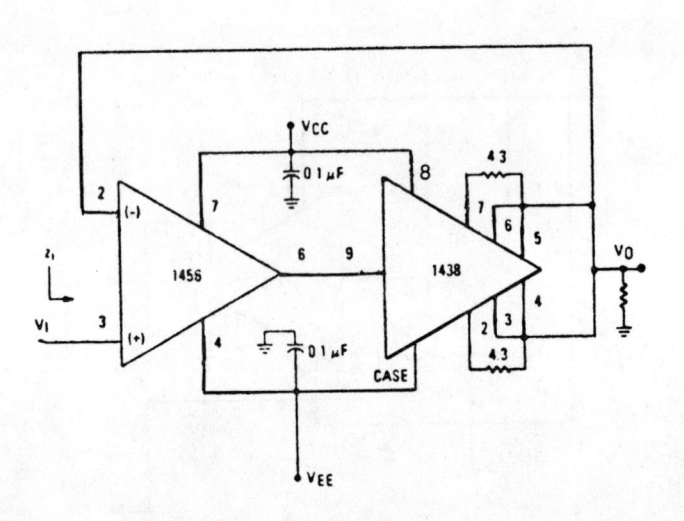

Noninverting voltage follower

268

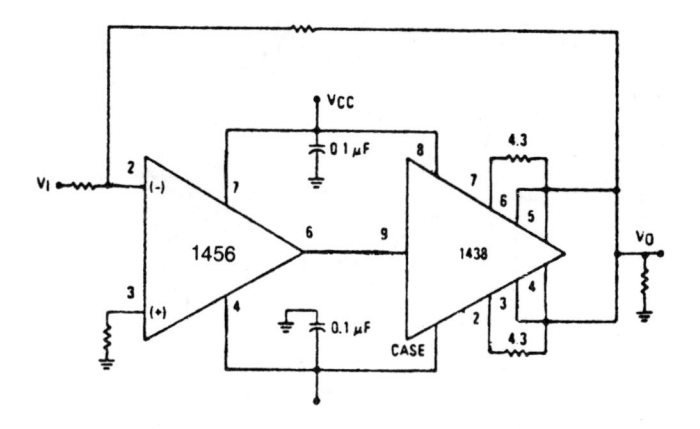

Inverting power amplifier

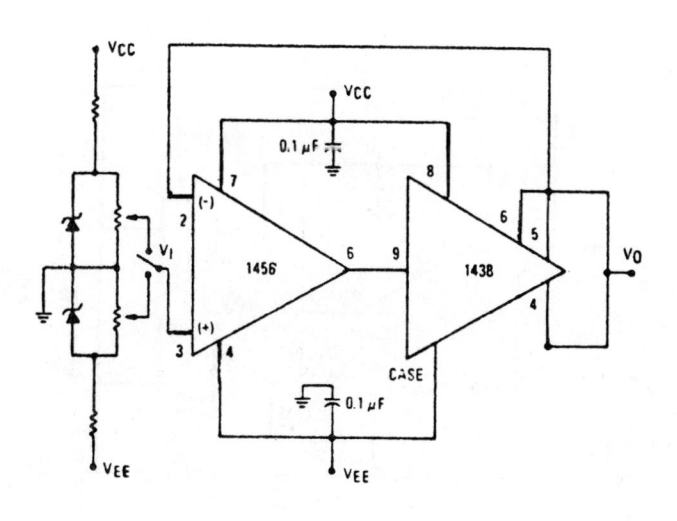

Programmable voltage source

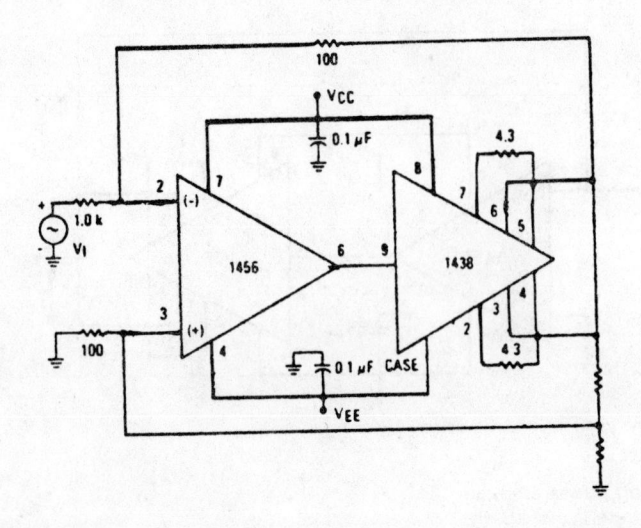

Transconductance amplifier

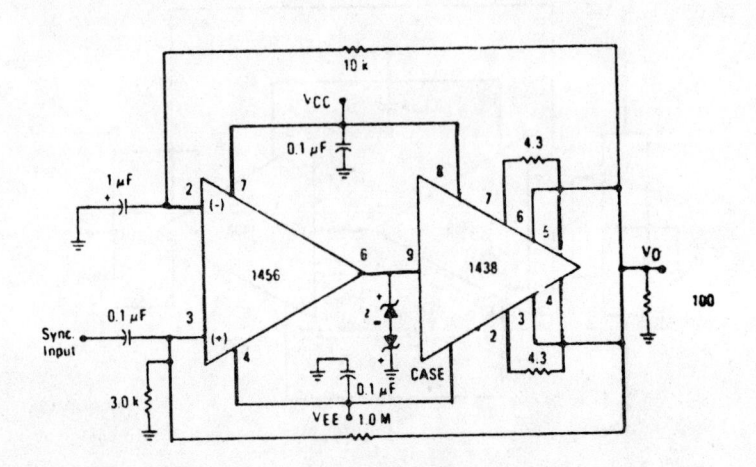

Astable multivibrator

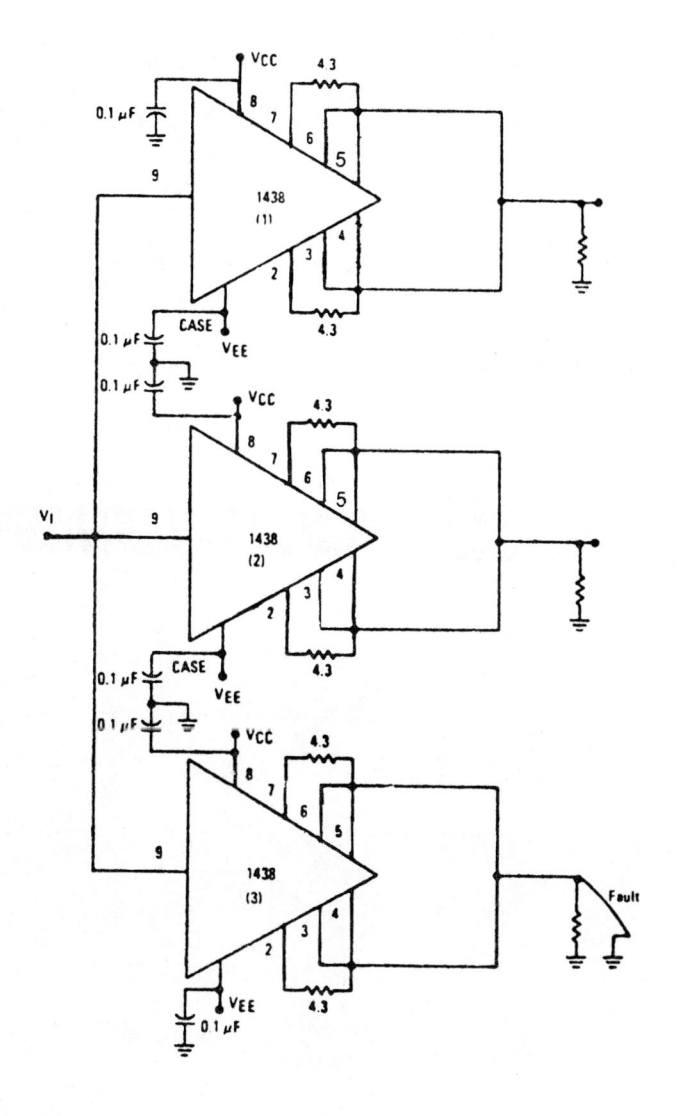

Signal distribution

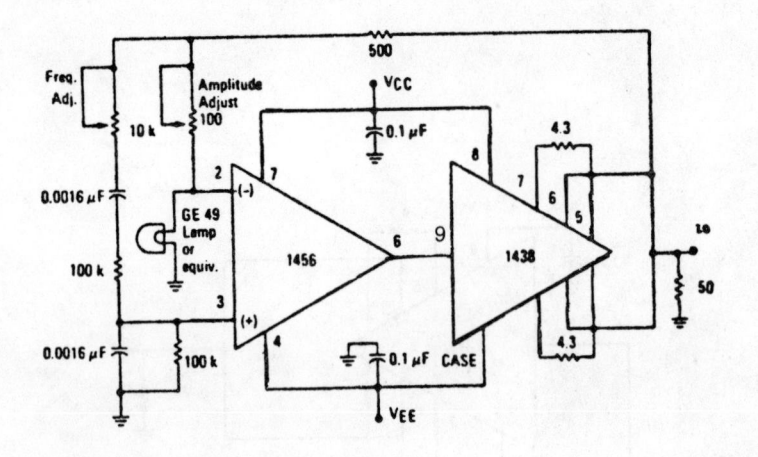

Wien bridge oscillator

1439—UNCOMPENSATED OPERATIONAL AMPLIFIER

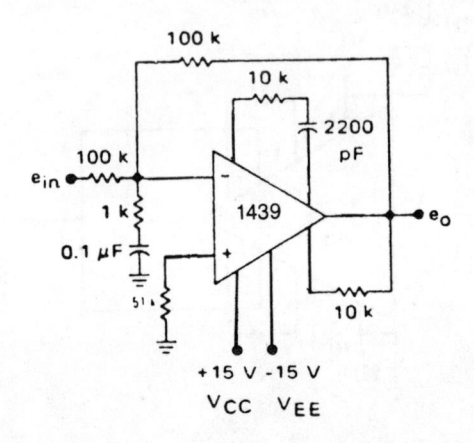

High slew rate inverter

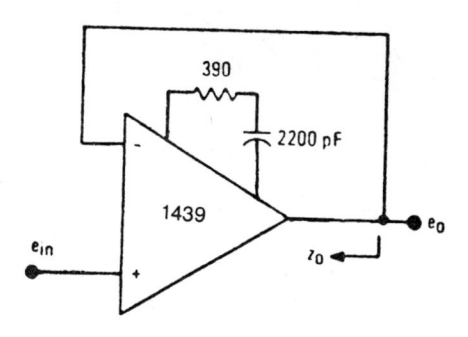

Voltage follower

+15 volt regulator

1458 — DUAL COMPENSATED OPERATIONAL AMPLIFIER

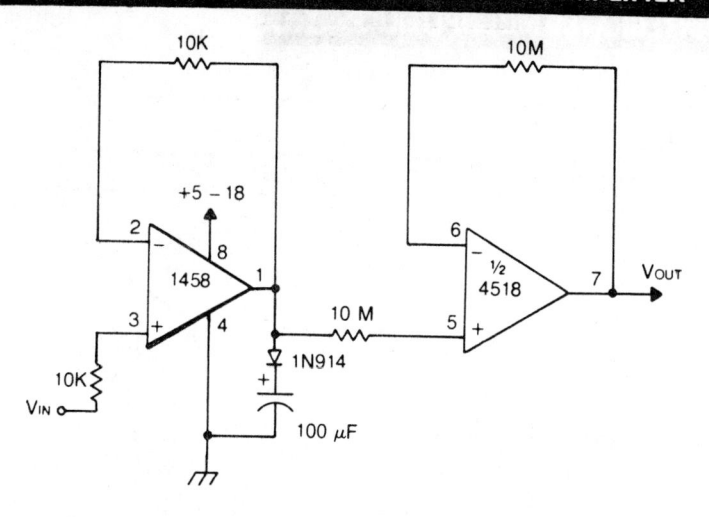

Peak detector

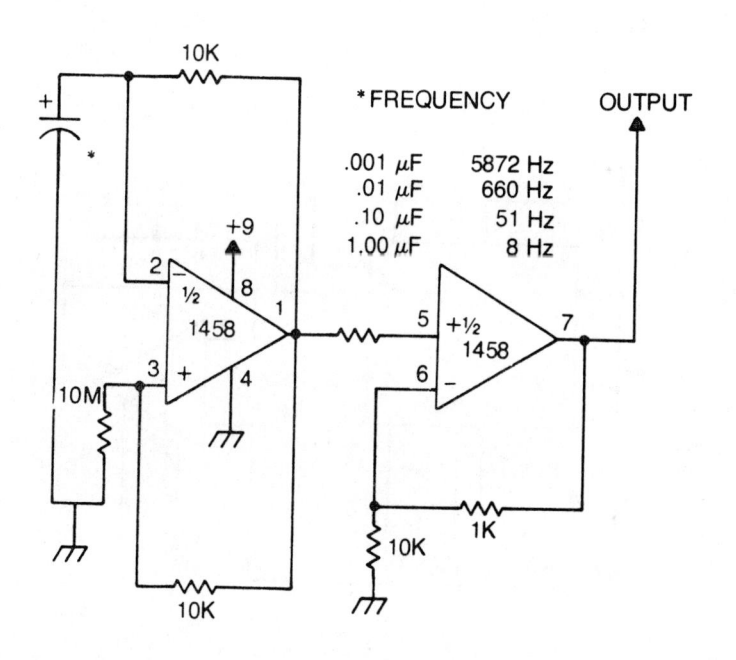

* FREQUENCY	
.001 µF	5872 Hz
.01 µF	660 Hz
.10 µF	51 Hz
1.00 µF	8 Hz

Pulse generator

275

1494 — LINEAR FOUR-QUADRANT INTEGRATED MULTIPLIER

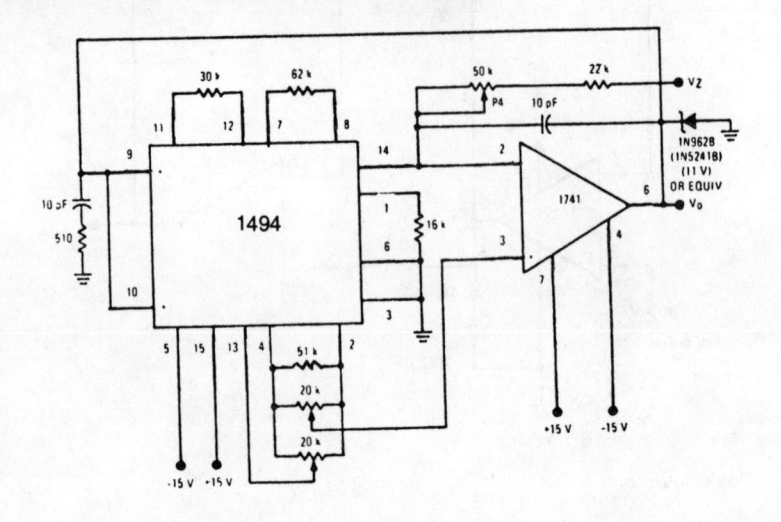

Square root circuit

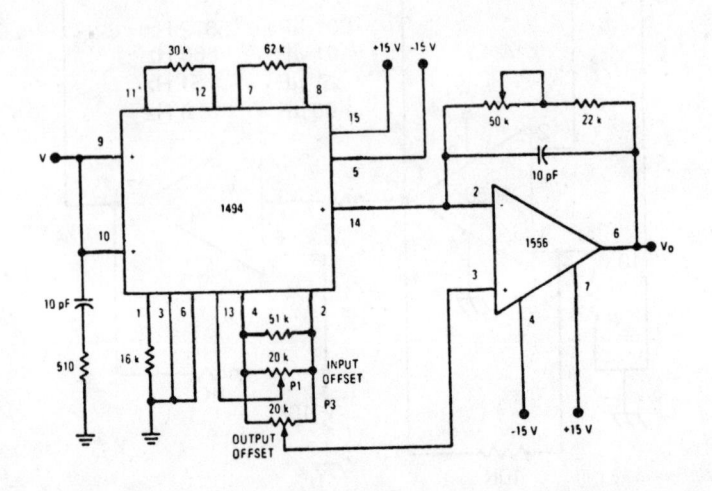

Squaring circuit

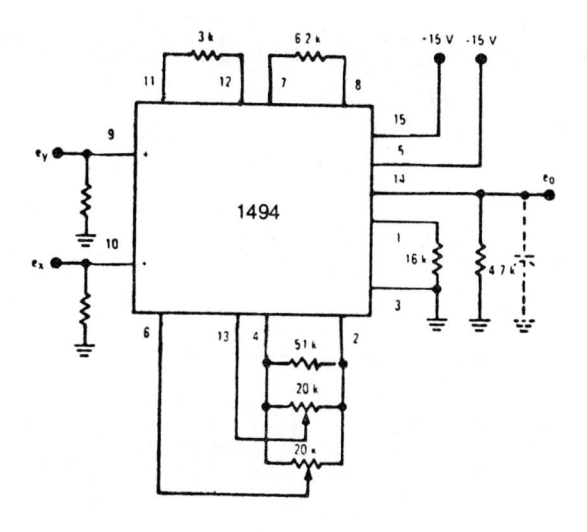

Wideband multiplier

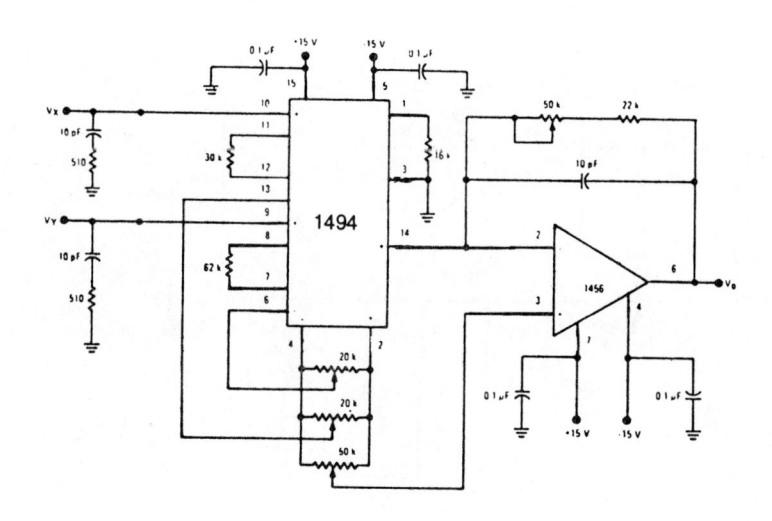

Typical multiplier

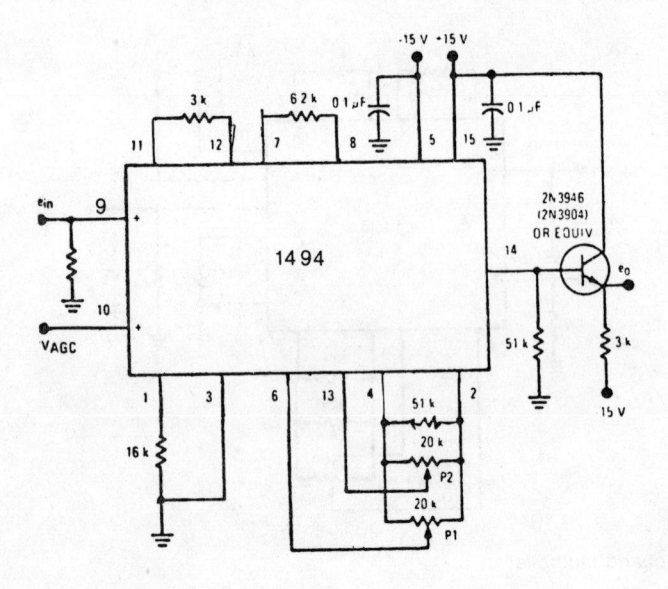

Wideband amplifier with linear AGC

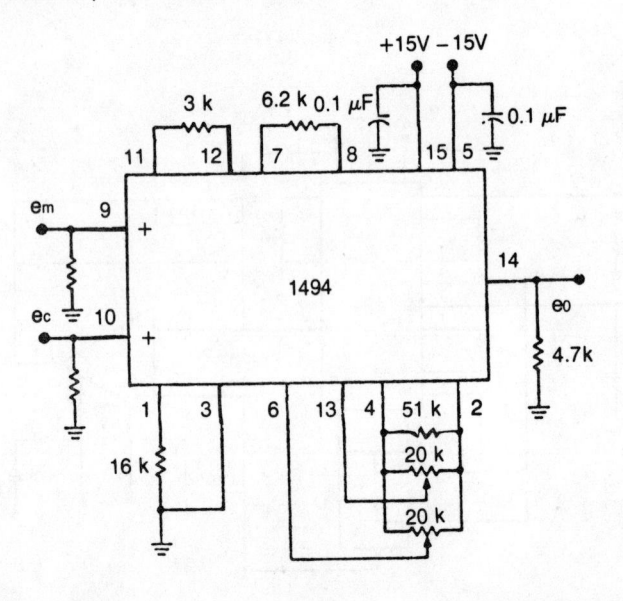

Balanced modulator

1495 — WIDEBAND LINEAR FOUR-QUADRANT INTEGRATED MULTIPLIER

Multiplier with op-amp level shift

Multiplier with improved linearity

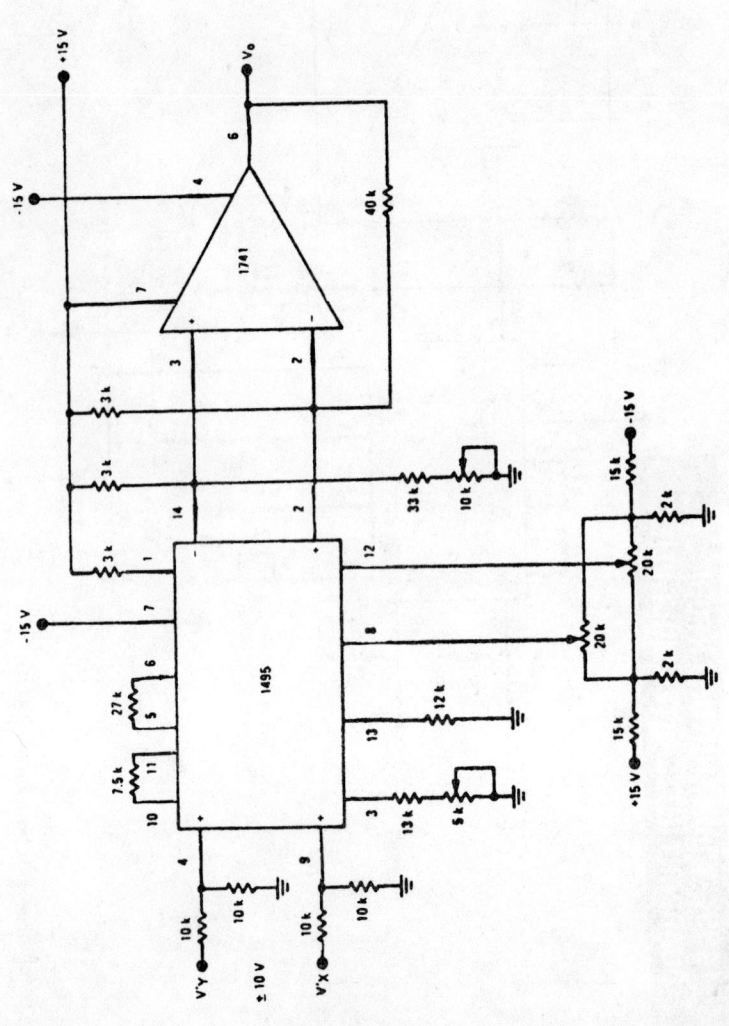

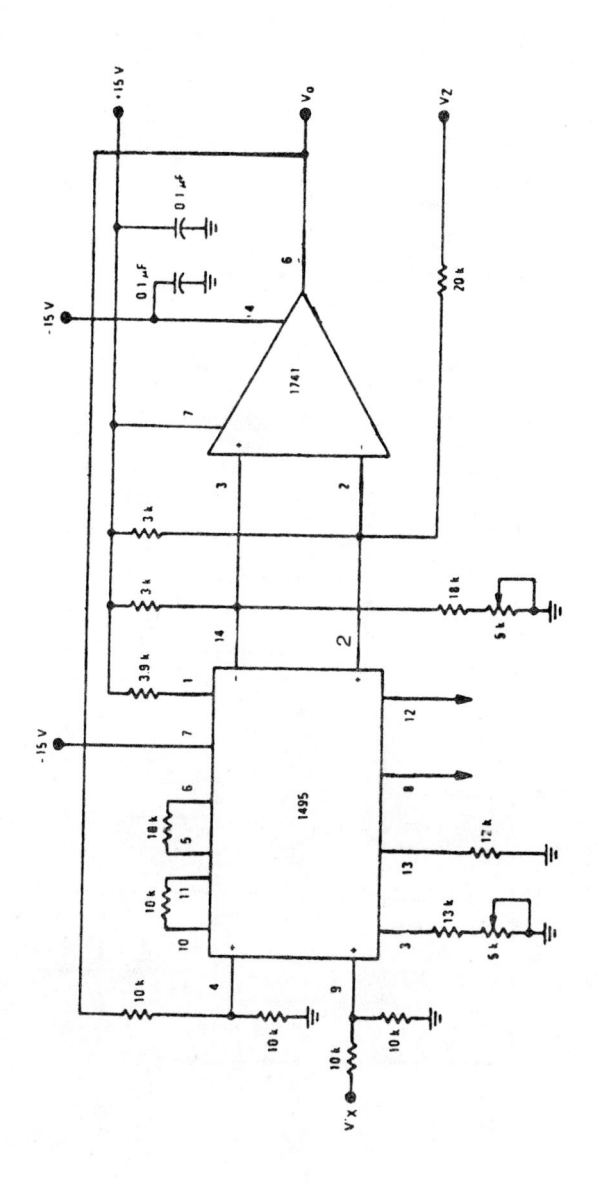

Divide circuit

Square root circuit

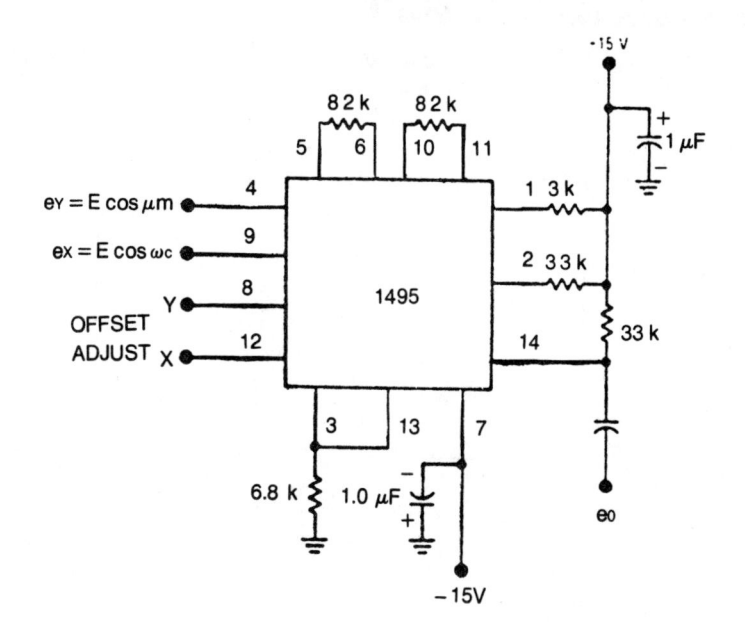

Balanced modulator

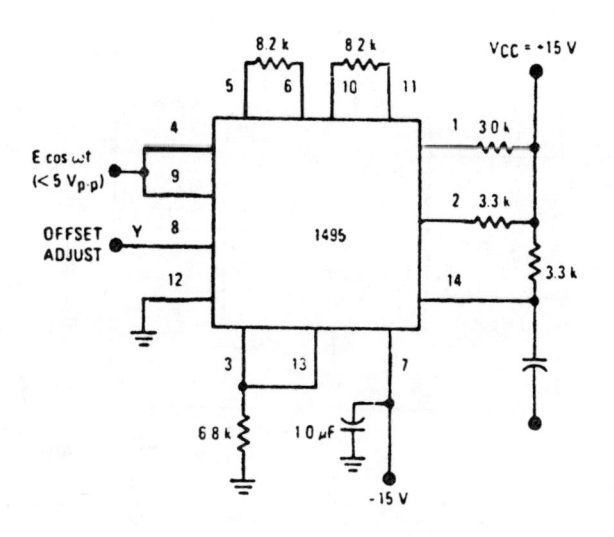

Frequency doubler

283

1555—INTEGRATED TIMER

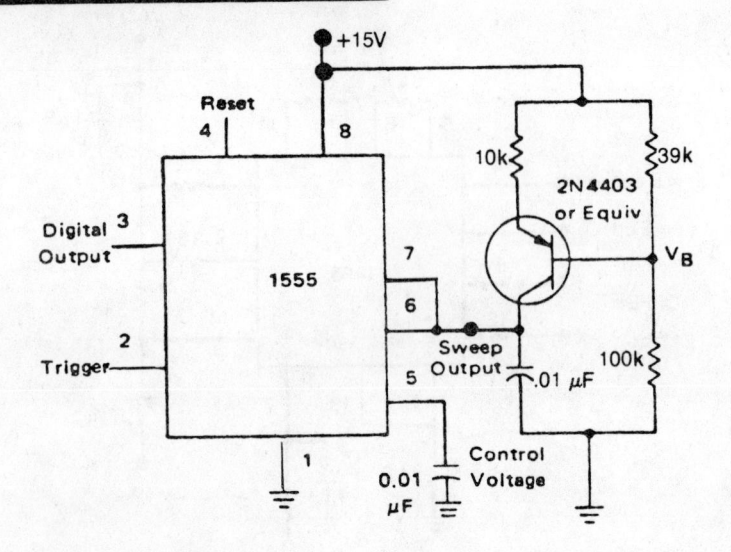

Linear voltage sweep circuit

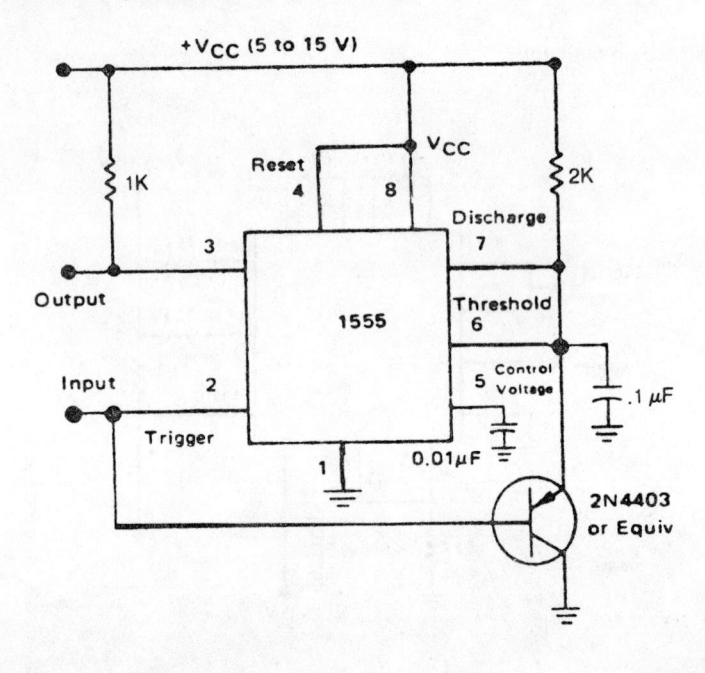

Missing pulse detector

284

Sequential timing circuit

285

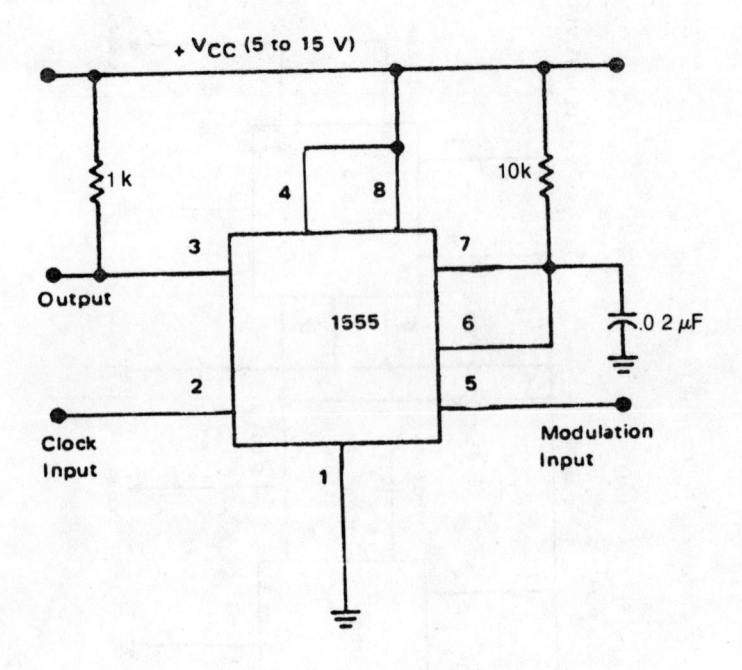

Pulse width modulator

1558—DUAL INTERNALLY COMPENSATED OPERATIONAL AMPLIFIER

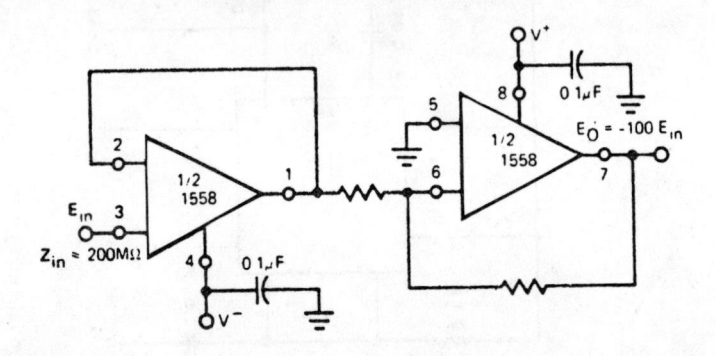

High-impedance, high-gain inverting amplifier

286

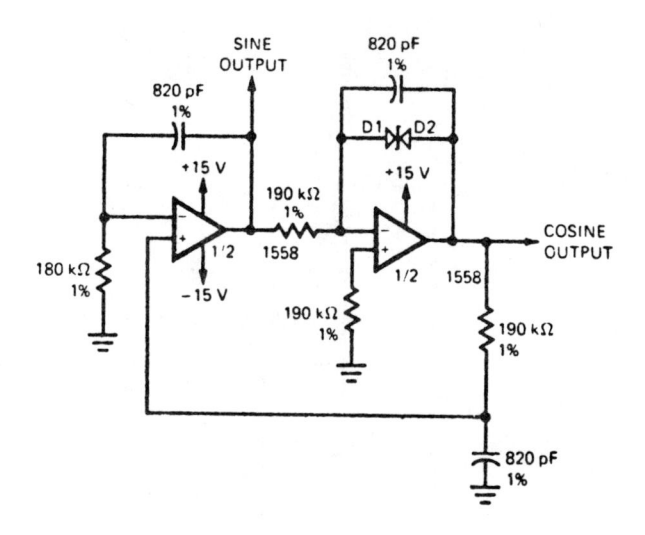

Quadrature oscillator

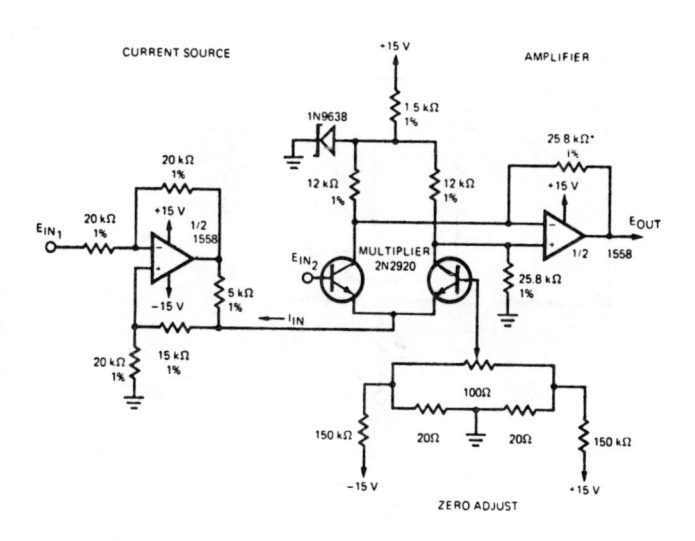

Analog multiplier

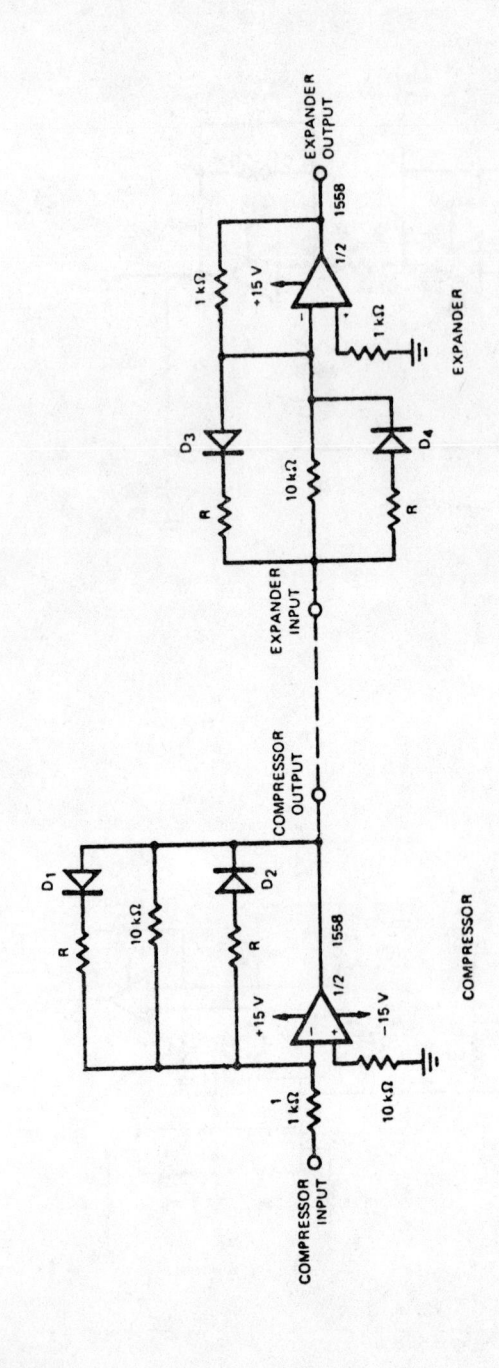

Compressor/expander amplifiers

1776 — MICROPOWER PROGRAMMABLE OPERATIONAL AMPLIFIER

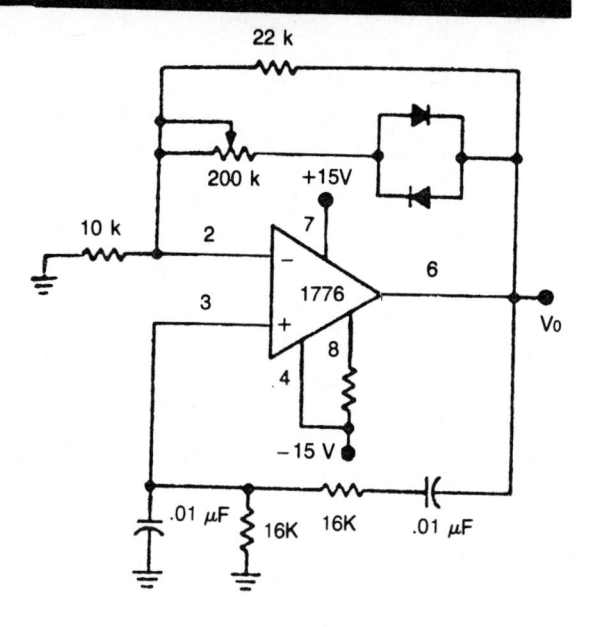

Wien bridge oscillator

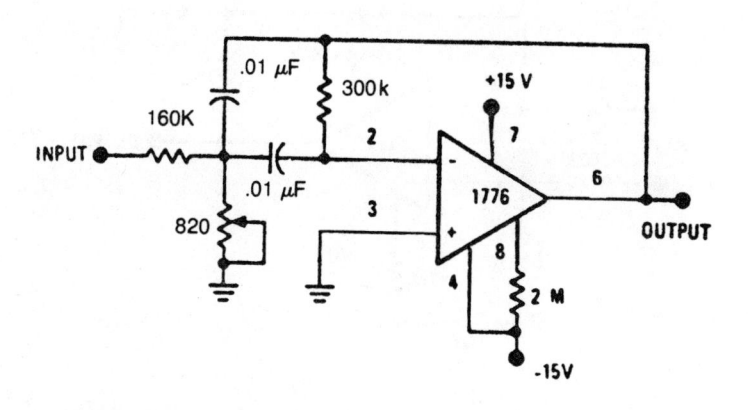

Multiple feedback bandpass filter (1.0kHz)

289

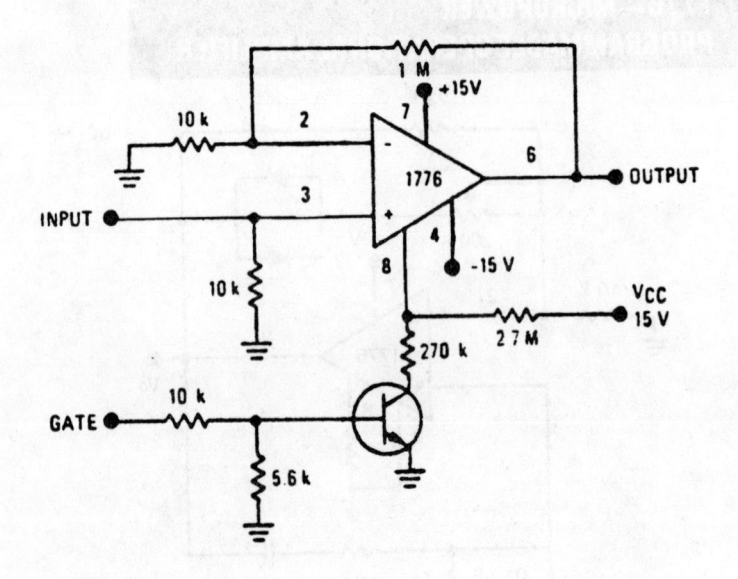

Gated amplifier

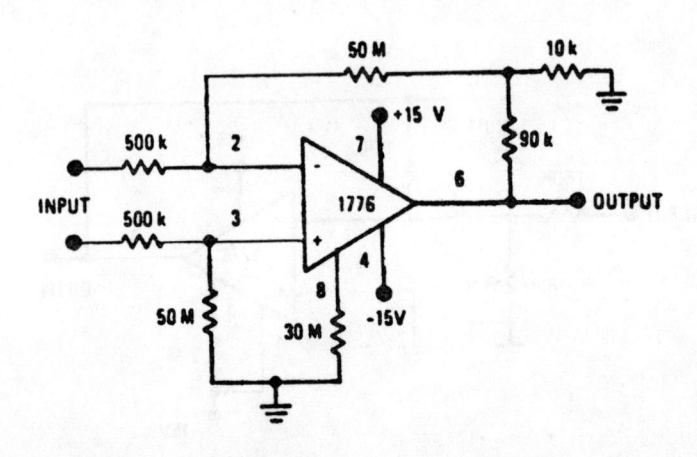

High input impedance amplifier

1877 — DUAL CHANNEL POWER AUDIO AMPLIFIER

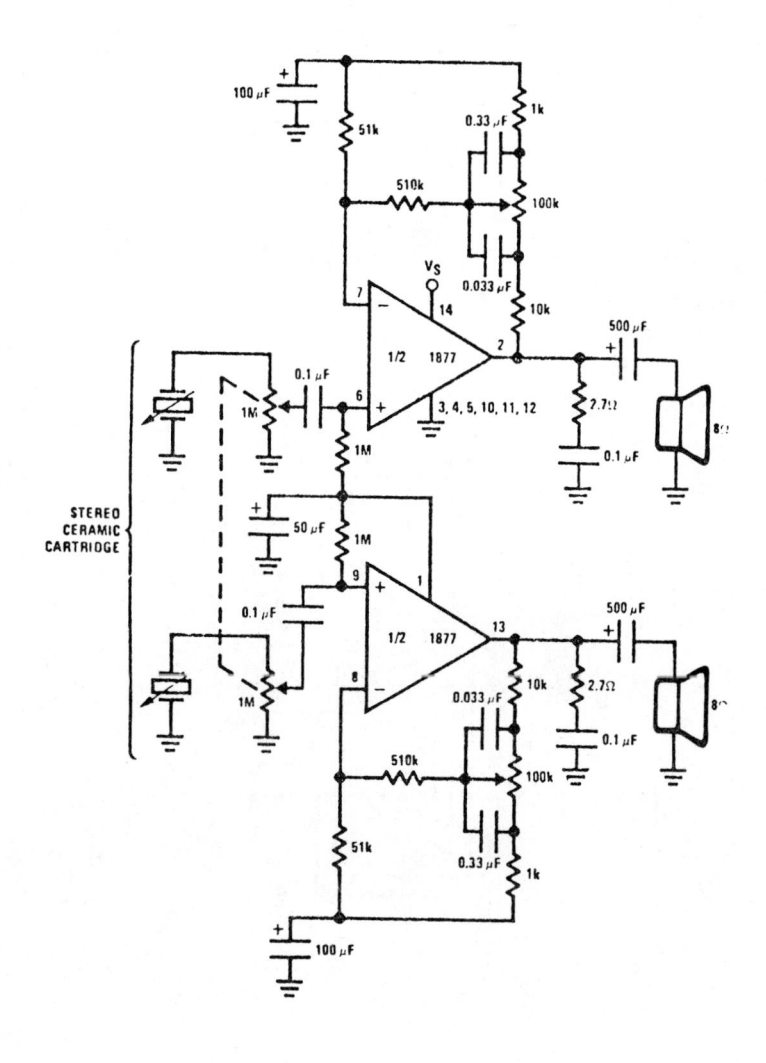

Stereo phonograph amplifier with bass tone control

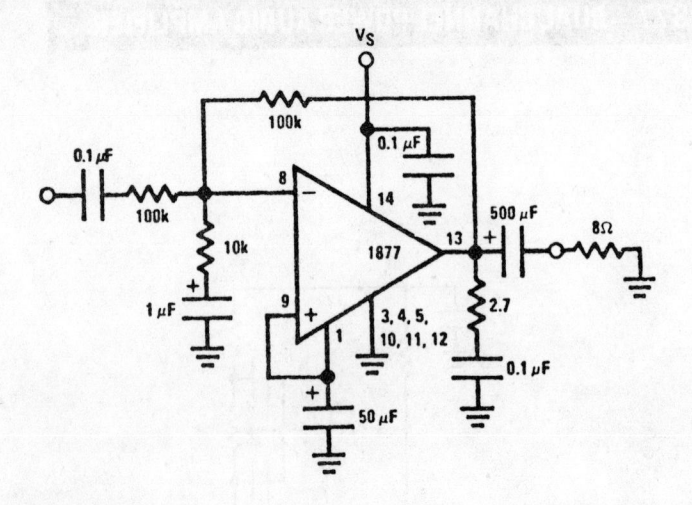

Inverting unity gain amplifier

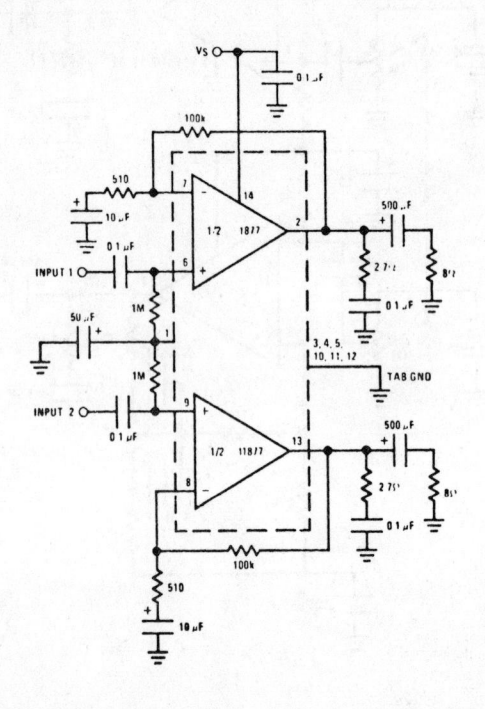

Stereo amplifier with $A_V = 200$

1900—QUAD OPERATIONAL AMPLIFIER

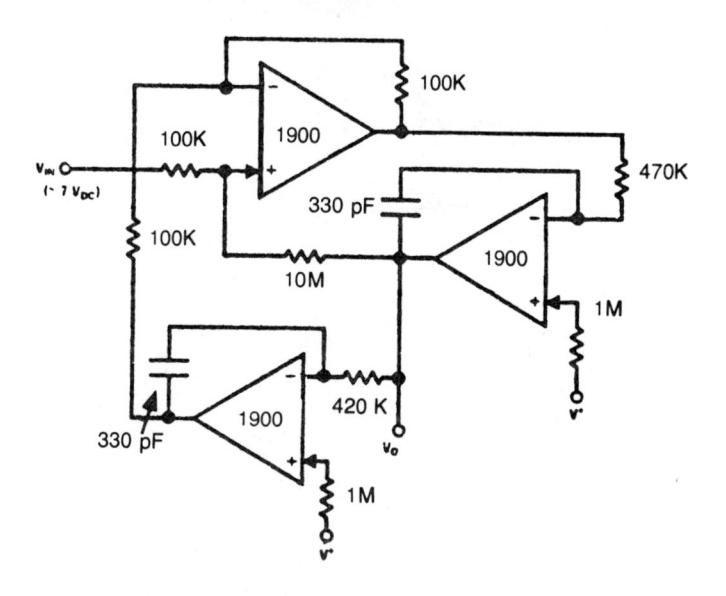

Bi-quad active filter

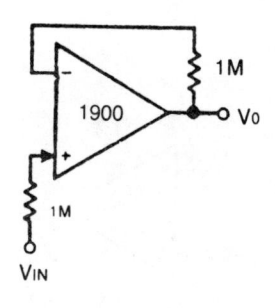

Buffer amplifier

293

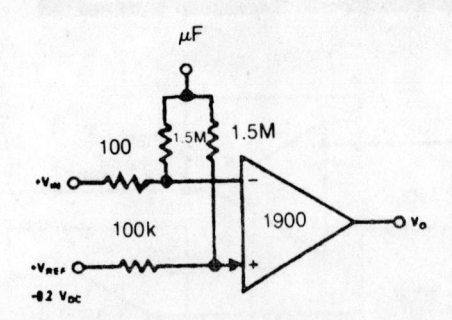

Low-voltage comparator

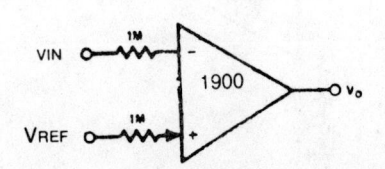

Comparator

294

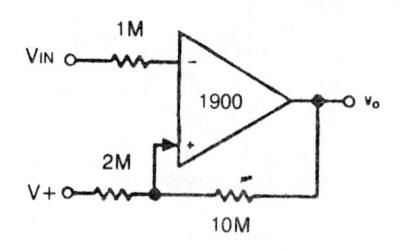

Schmitt-trigger

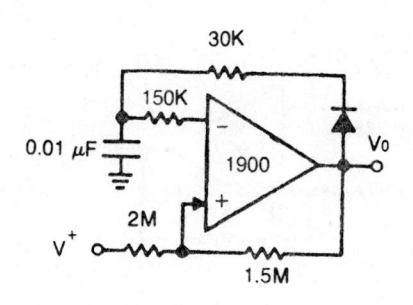

Pulse generator

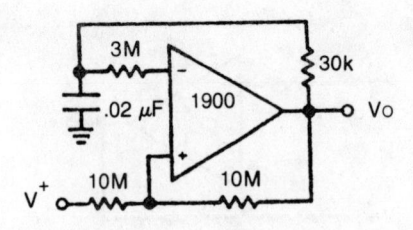

Square-wave oscillator

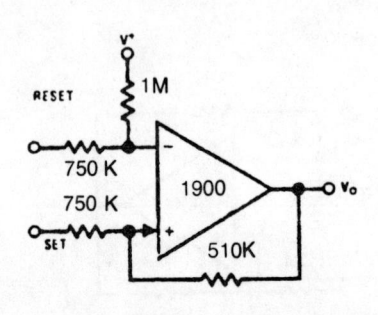

Bi-stable multivibrator

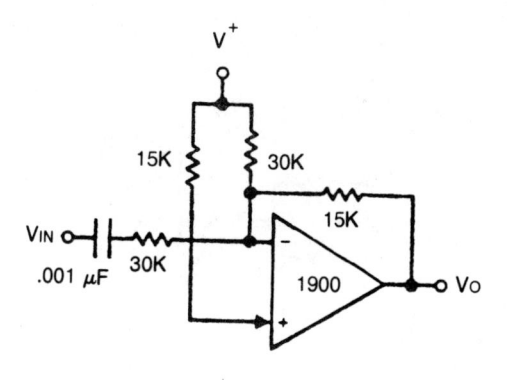

Differentiator

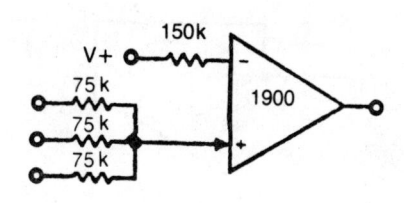

OR gate

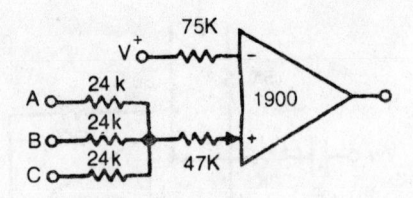

AND gate

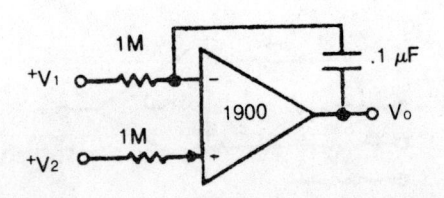

Difference integrator

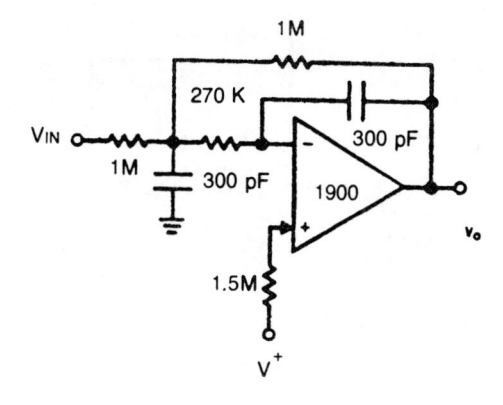

Low pass active filter

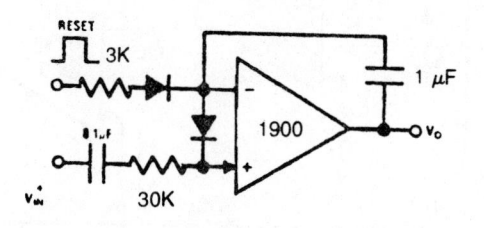

Staircase generator

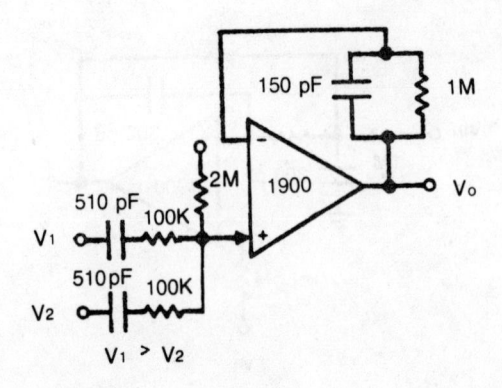

Low-frequency mixer

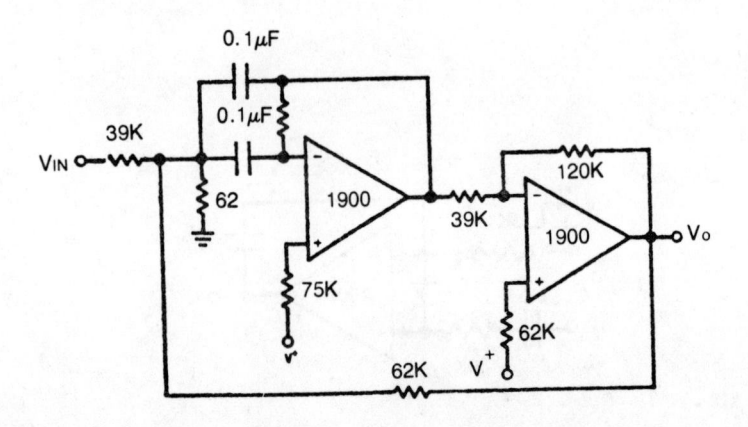

Bandpass active filter

300

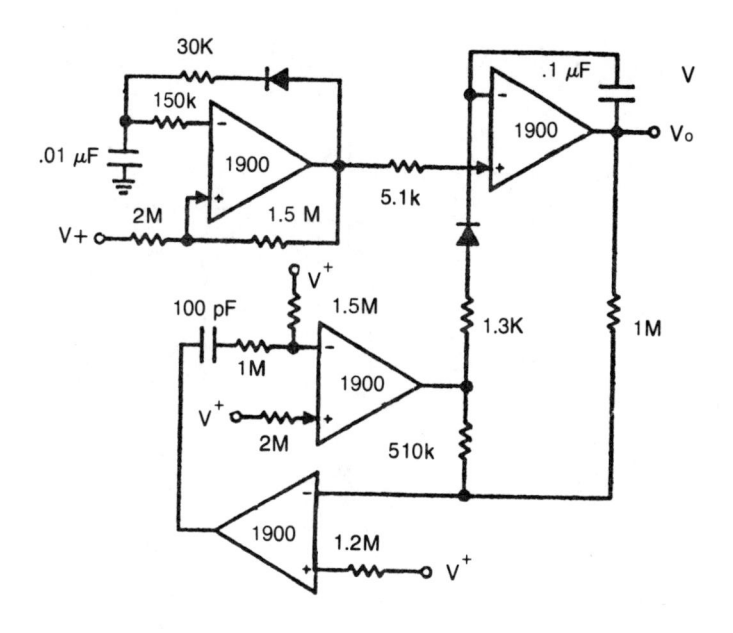

Free-running staircase generator/pulse counter

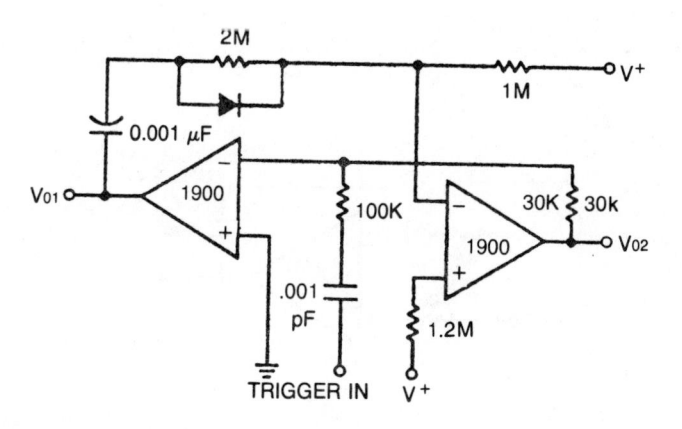

One-shot multivibrator

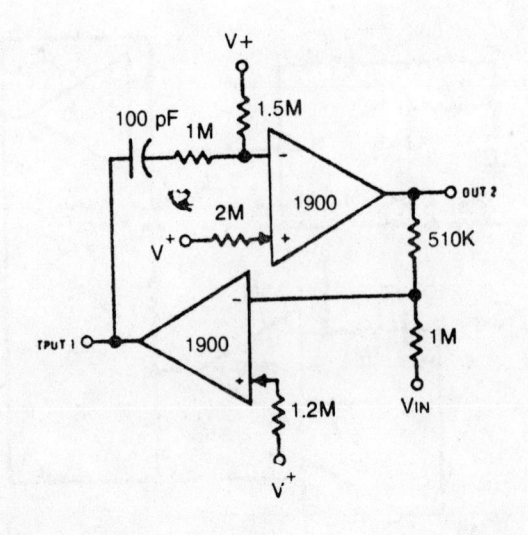

One-shot with dc input comparator

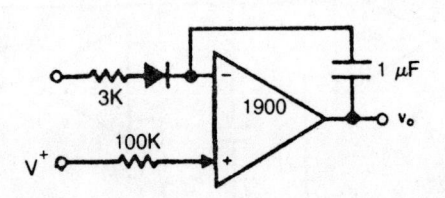

Sawtooth generator

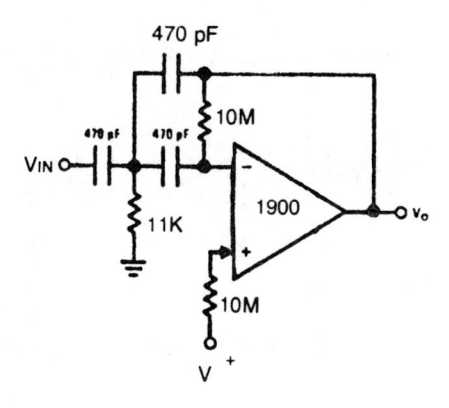

High pass active filter

3301—QUAD SINGLE SUPPLY OPERATIONAL AMPLIFIER

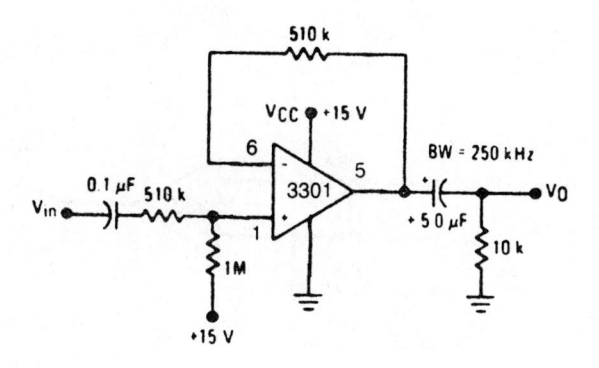

Noninverting amplifier

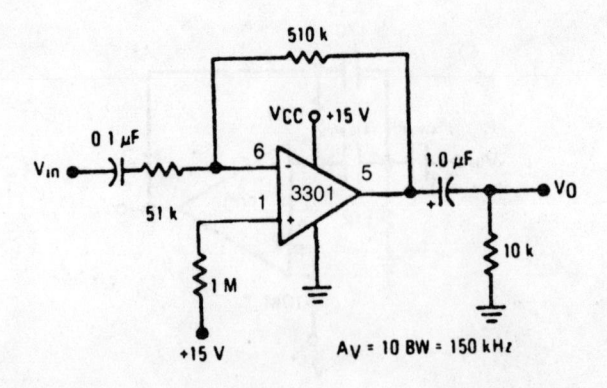

Inverting amplifier

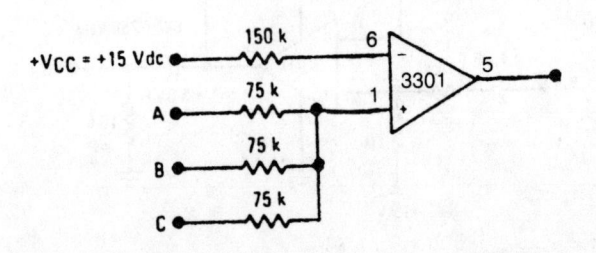

$A_V = 10$ BW = 150 kHz

Logic OR gate

304

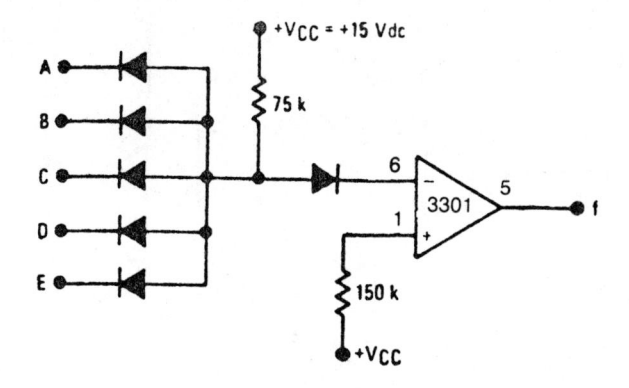

Logic NAND gate (Large Fan-In)

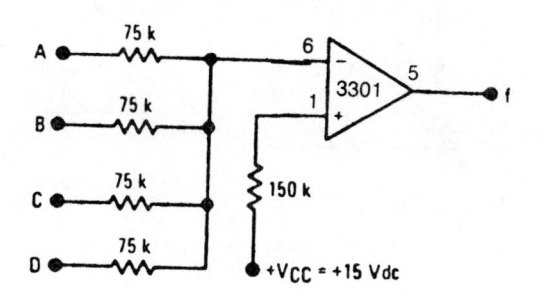

Logic NOR gate

305

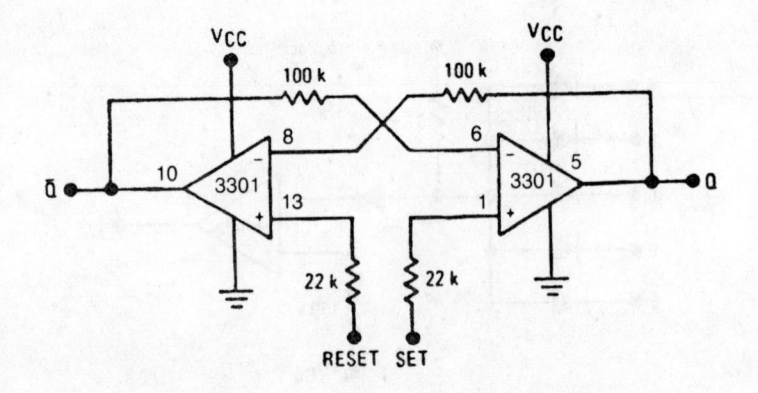

R-S flip-flop

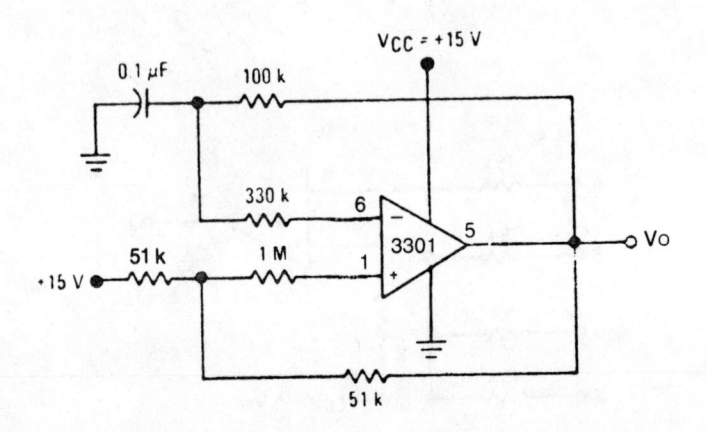

Astable multivibrator

306

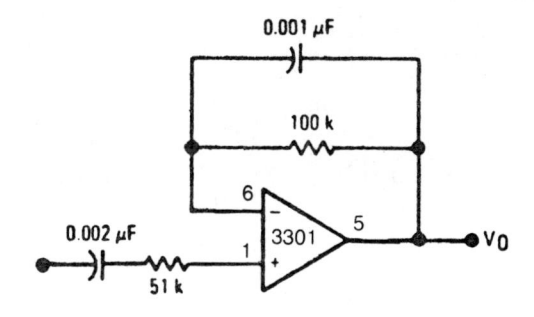

Positive-edge differentiator

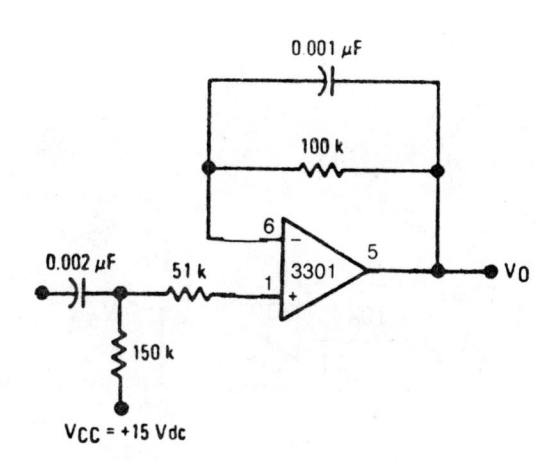

Negative-edge differentiator

3302—QUAD COMPARATOR

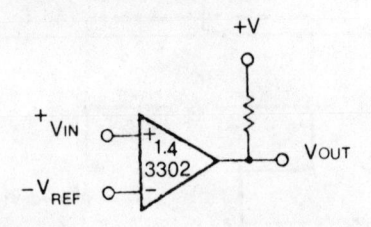

Basic comparator

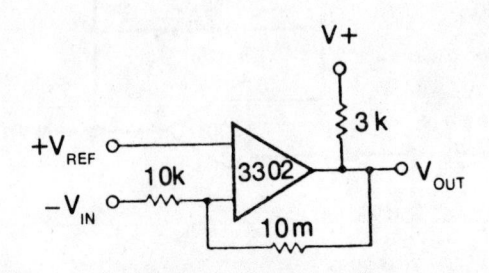

Noninverting comparator with hysteresis

Zero crossing detector

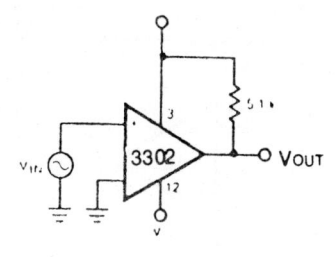

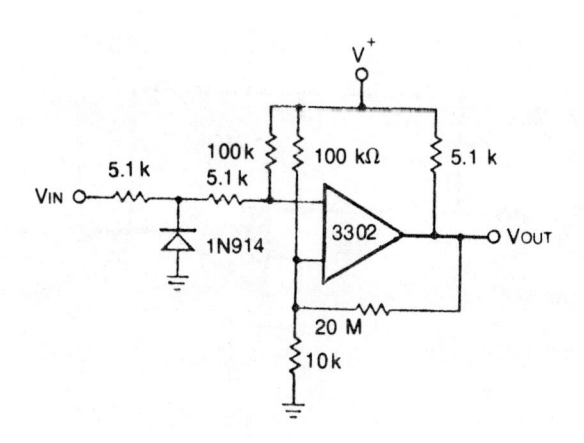

Zero crossing detector (single power supply)

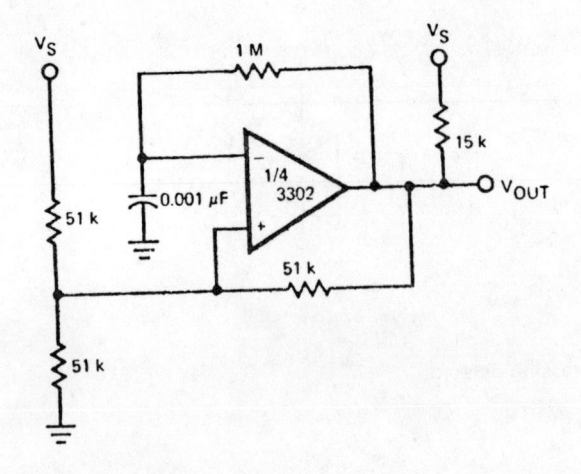

Free running square wave oscillator

3303 — QUAD OPERATIONAL AMPLIFIER

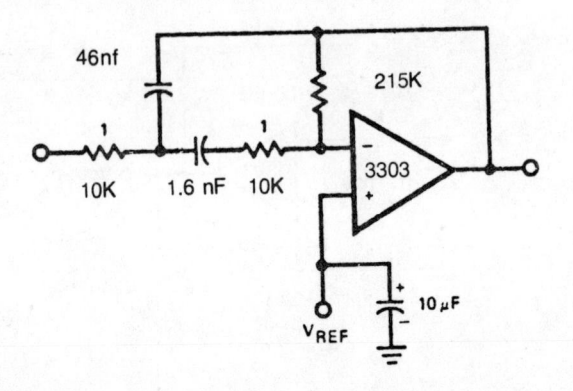

Multiple feedback bandpass filter

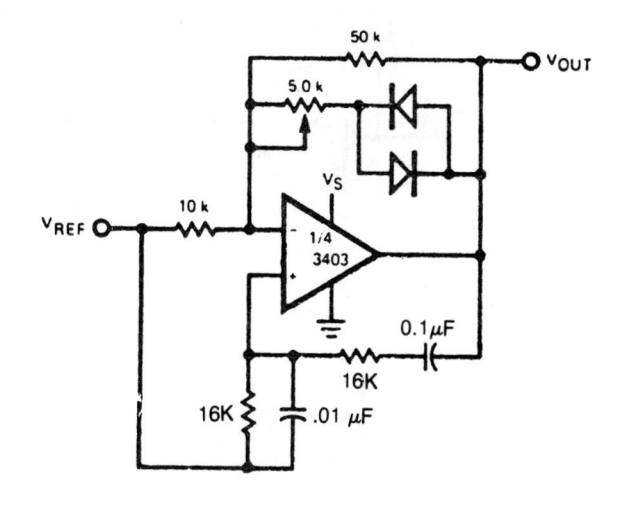

Wien bridge oscillator

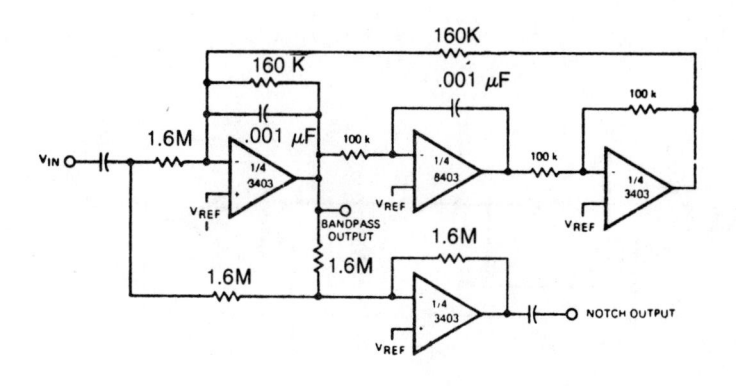

Bi-quad filter

311

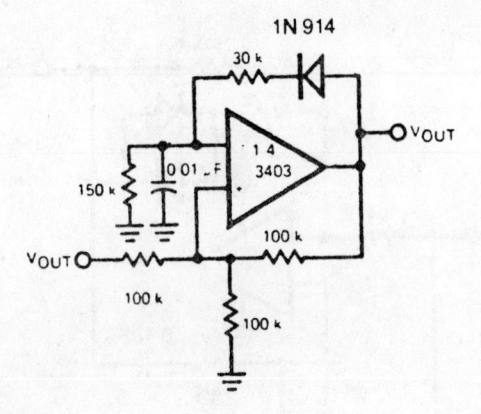

Pulse generator

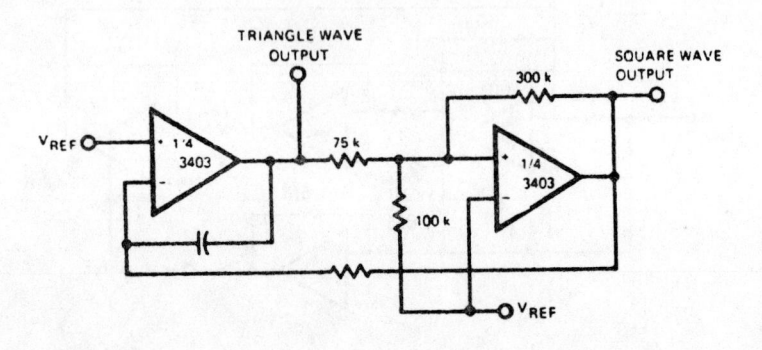

Function generator

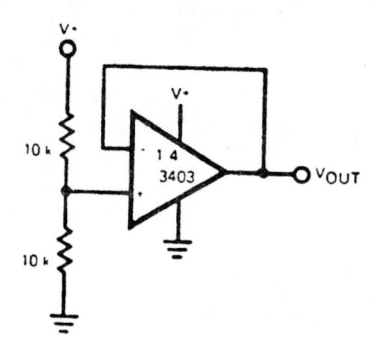

Voltage reference

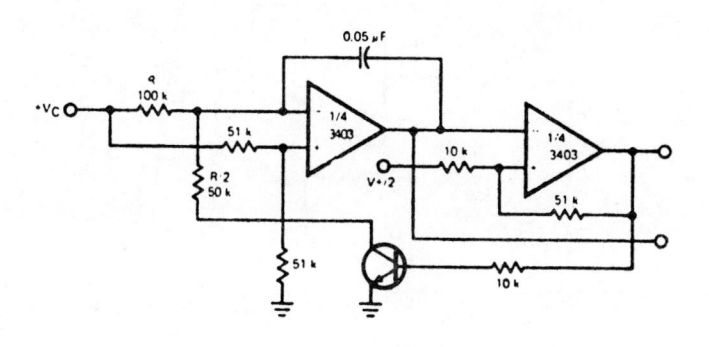

Voltage controlled oscillator

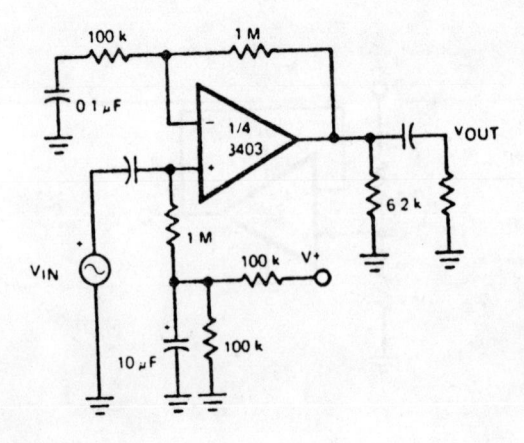

AC coupled noninverting amplifier

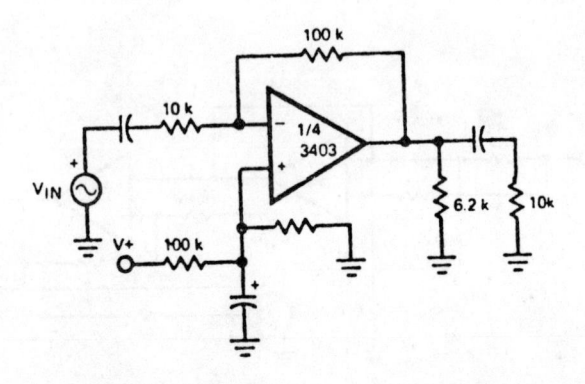

AC coupled inverting amplifier

3380—EMITTER COUPLED INTEGRATED ASTABLE MULTIVIBRATOR

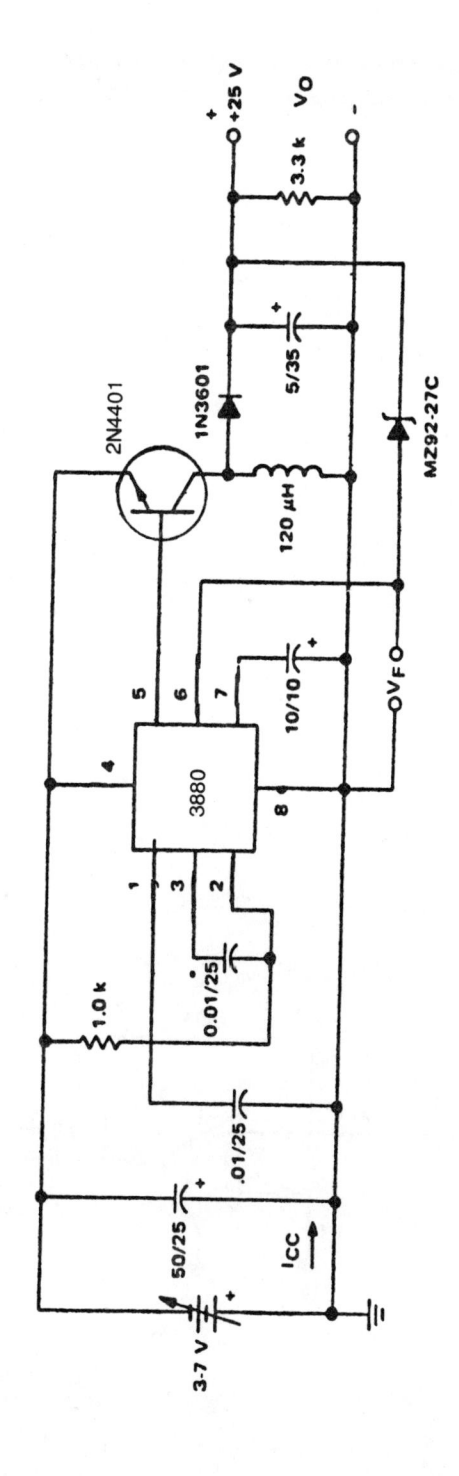

Typical application in 3 - 25 V dc-dc converter configuration

3401 — QUAD OPERATIONAL AMPLIFIER

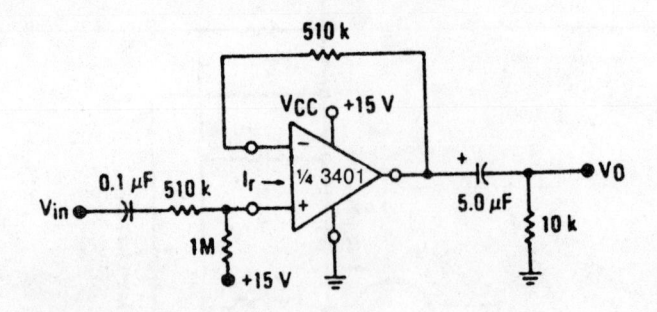

Noninverting amplifier

Inverting amplifier

316

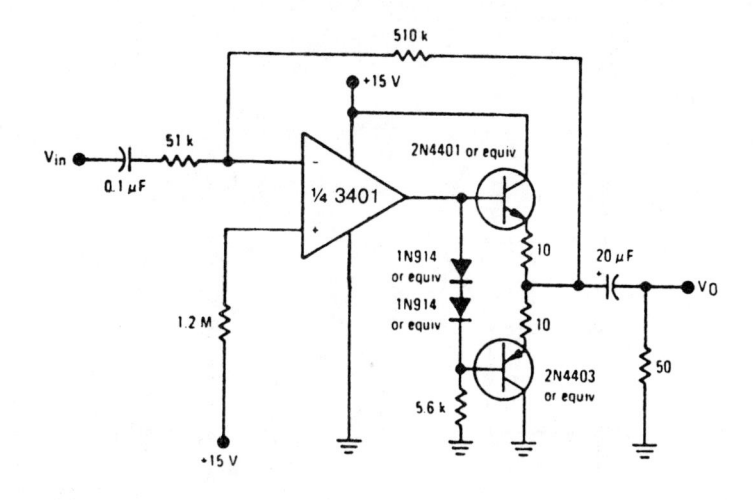

Amplifier and driver for a 50-ohm line

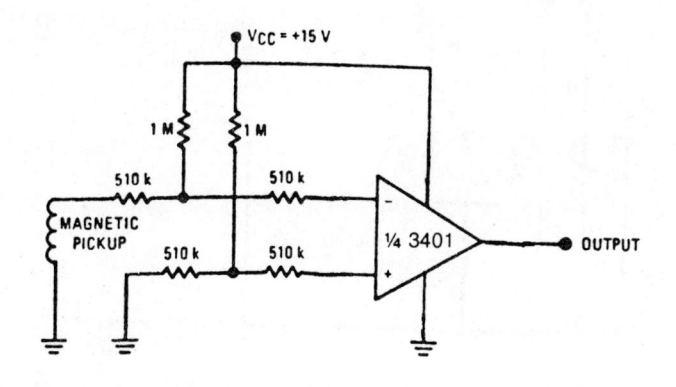

Zero crossing detector

Bandpass and notch filter

3403 — LOW DISTORTION QUAD OPERATIONAL AMPLIFIER

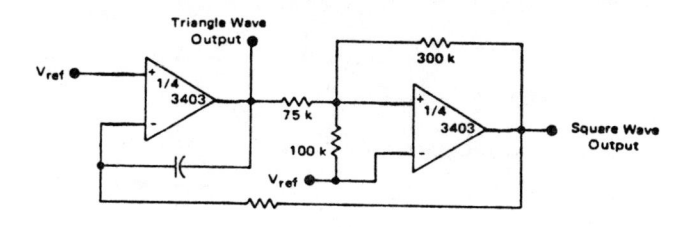

Function generator

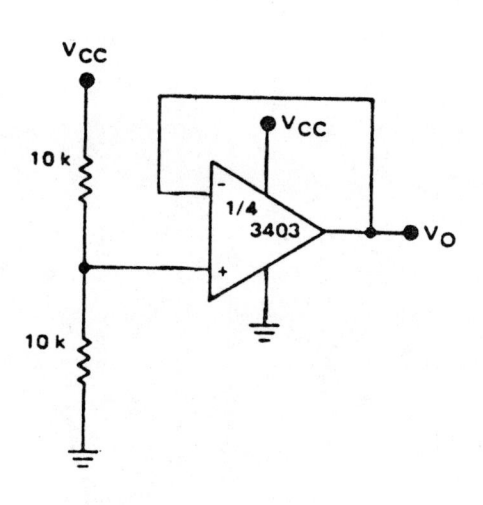

Voltage reference

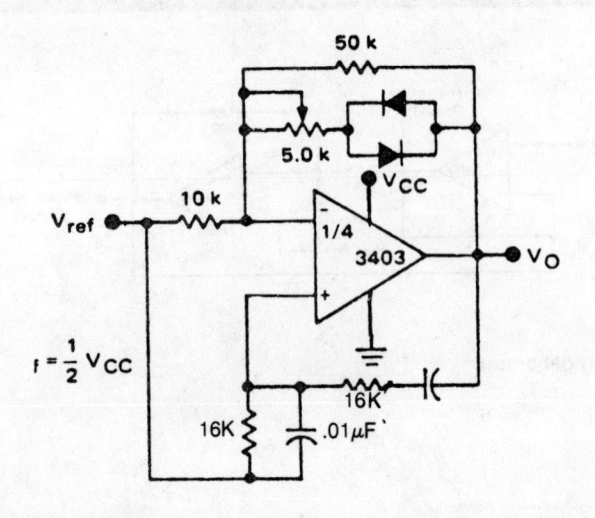

Wien bridge oscillator

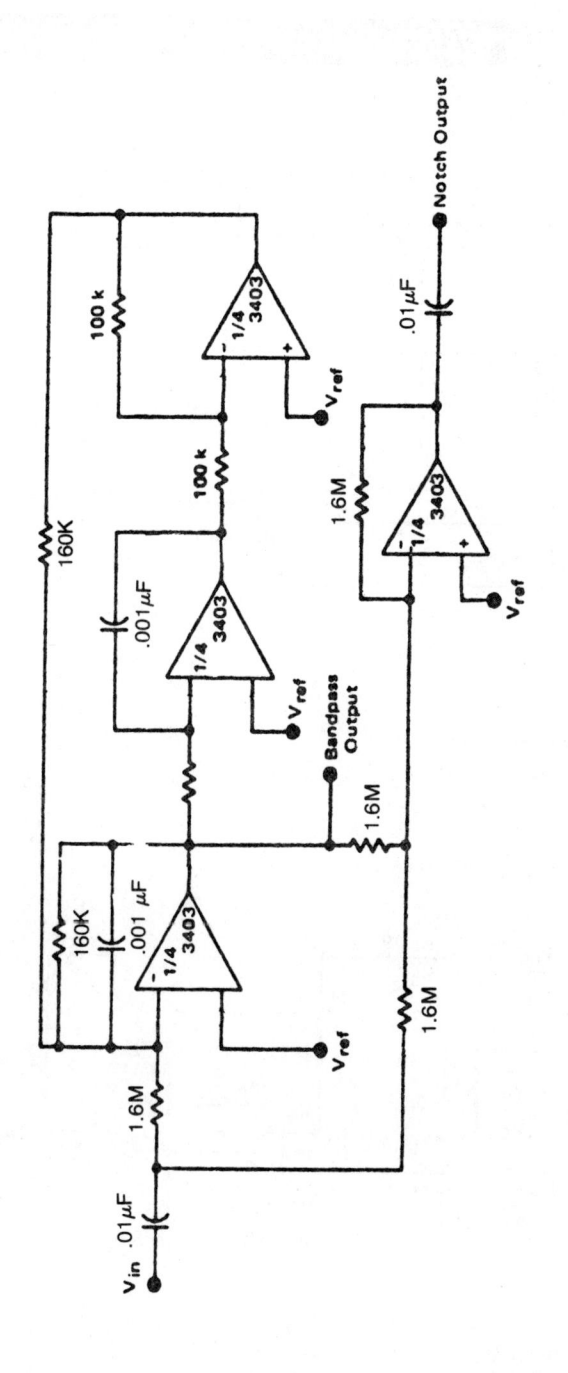

Bi-quad filter

3405 — DUAL OPERATIONAL AMPLIFIER/COMPARATOR

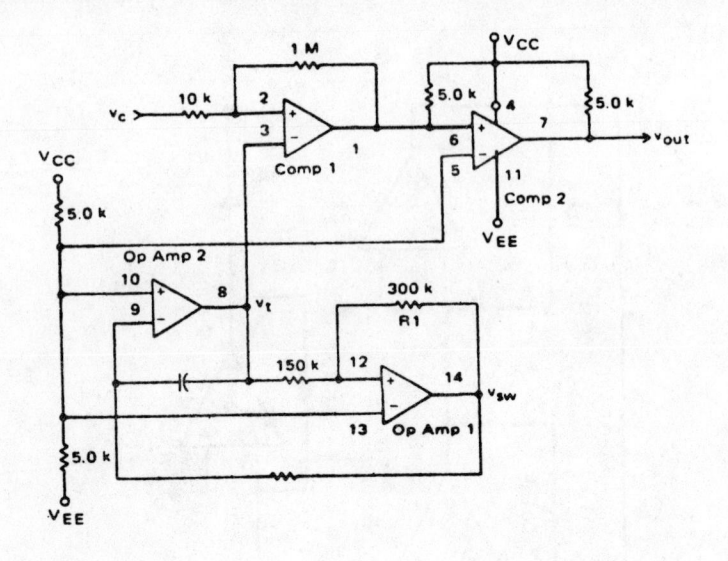

Pulse width modulator

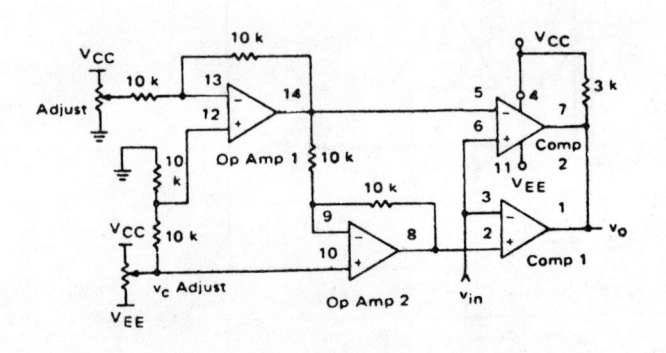

Window comparator

322

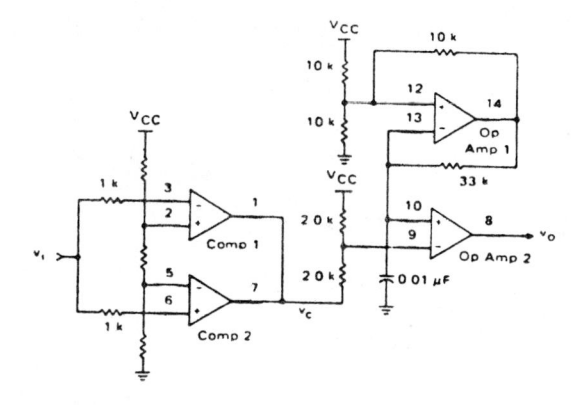

High/low limit alarm

3458—COMPENSATED DUAL OPERATIONAL AMPLIFIER

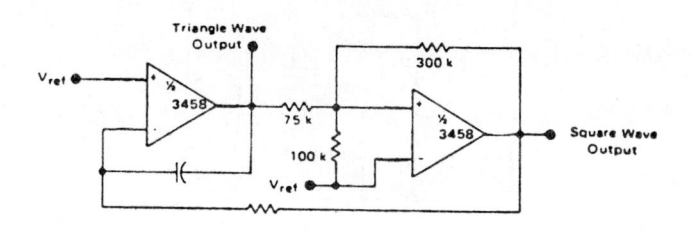

Function generator

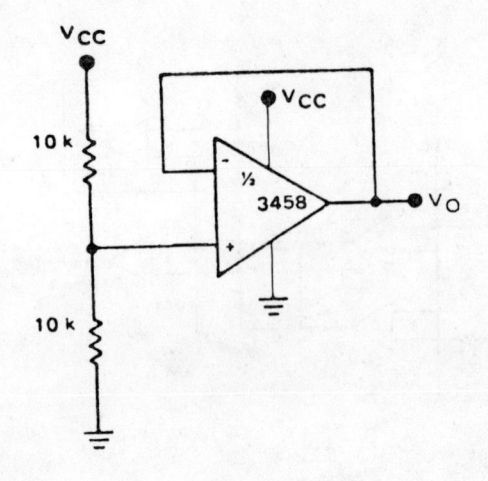

Voltage reference

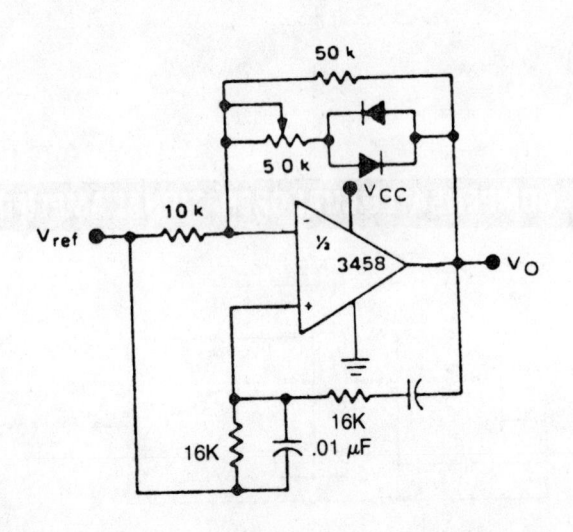

Wien bridge oscillator

Bi-quad filter

3556 — DUAL INTEGRATED TIMER

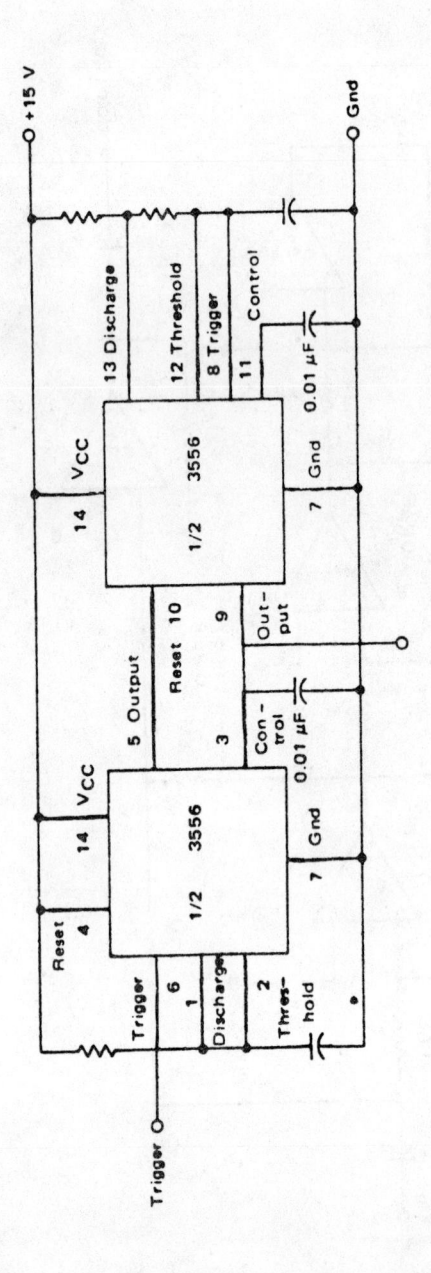

Tone burst generator

Dual astable multivibrator

Sequential timing

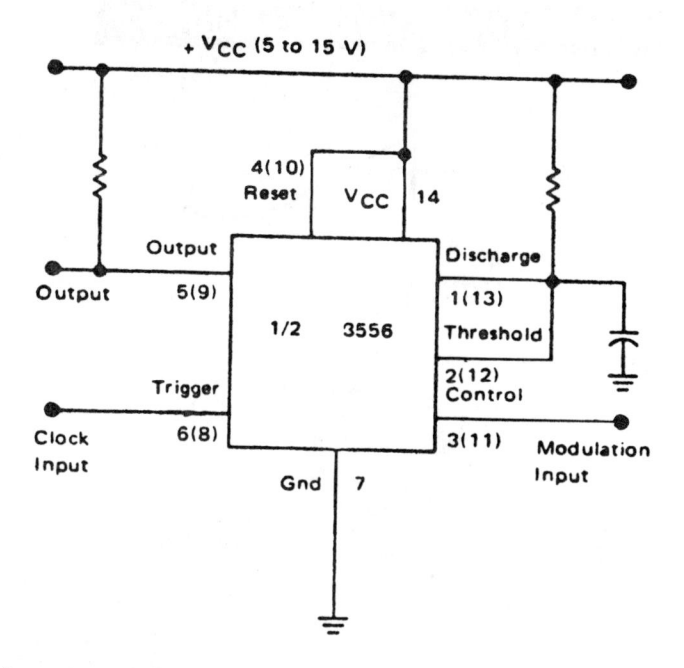

Pulse width modulator

3900—INTEGRATED QUAD AMPLIFIER

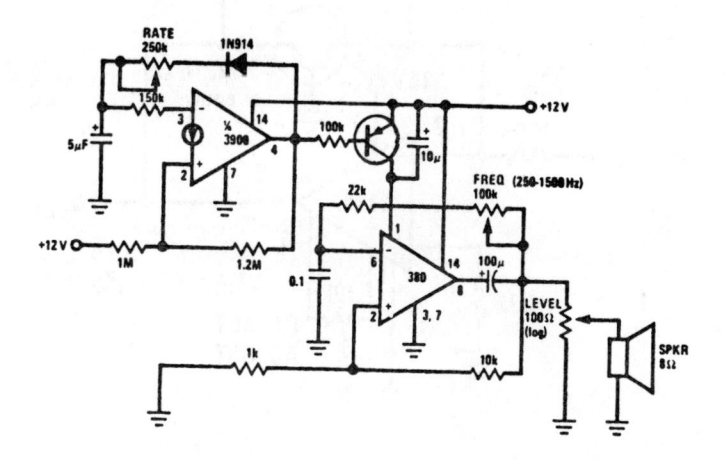

Siren with programmable frequency and rate adjustment

4131 — HIGH GAIN OPERATIONAL AMPLIFIER

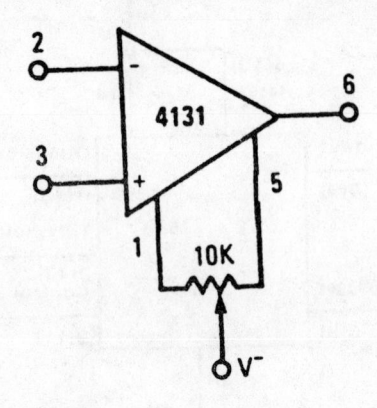

Voltage offset null circuit

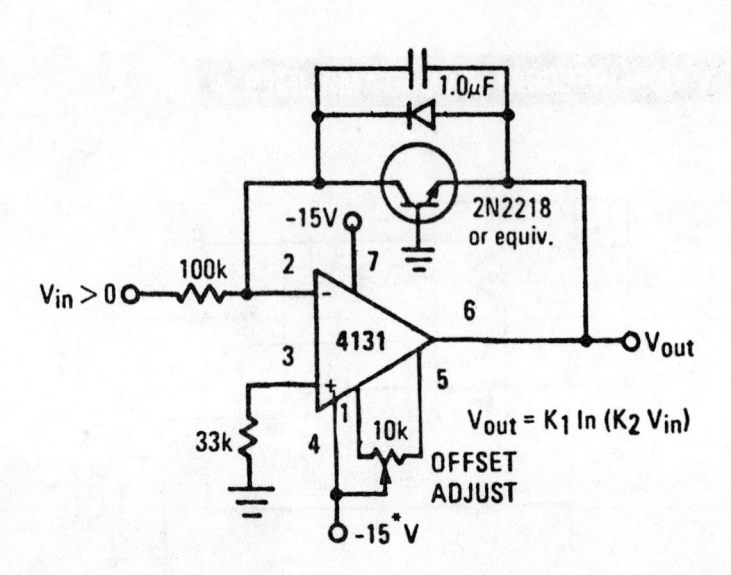

$V_{out} = K_1 \ln (K_2 V_{in})$

Logarithmic amplifier

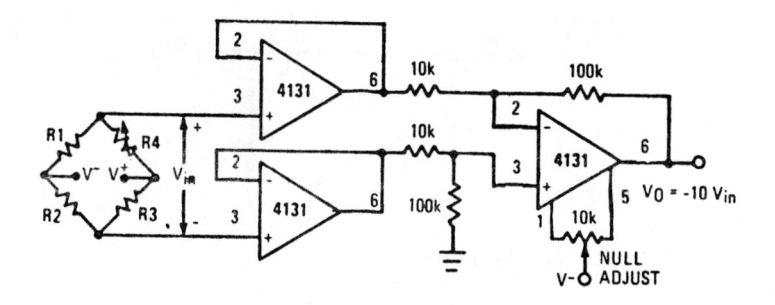

High impedance bridge amplifier

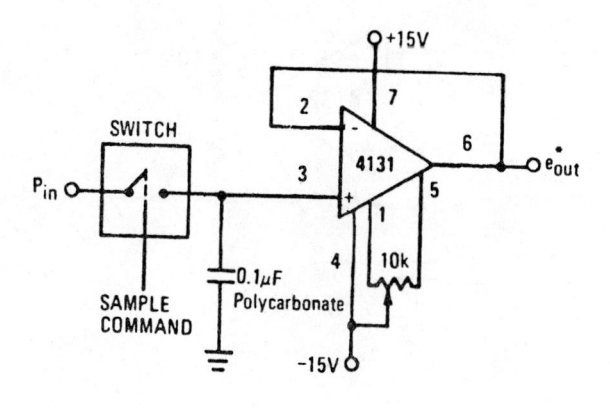

Low drift sample and hold

4132 — MICROPOWER OPERATIONAL AMPLIFIER

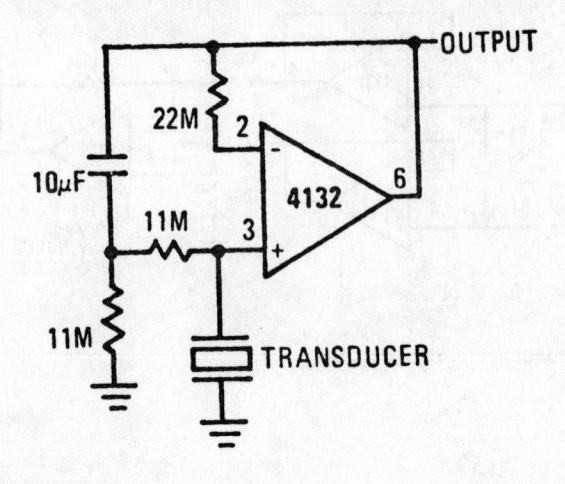

Amplifier for Piezoelectric transducers

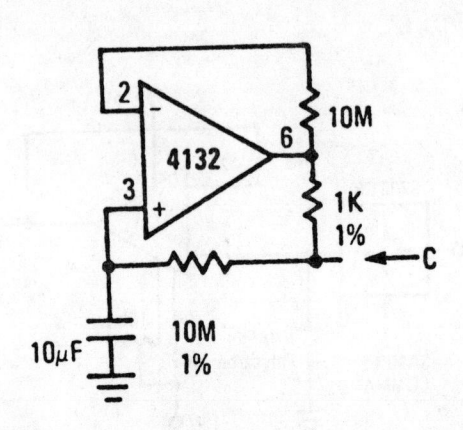

Capacitance multiplier

332

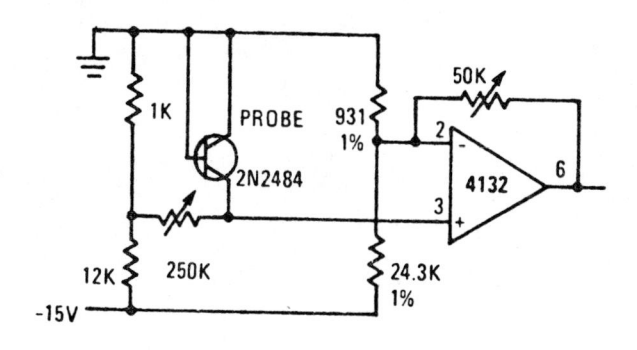

Temperature probe

4136—QUAD 741 OPERATIONAL AMPLIFIER

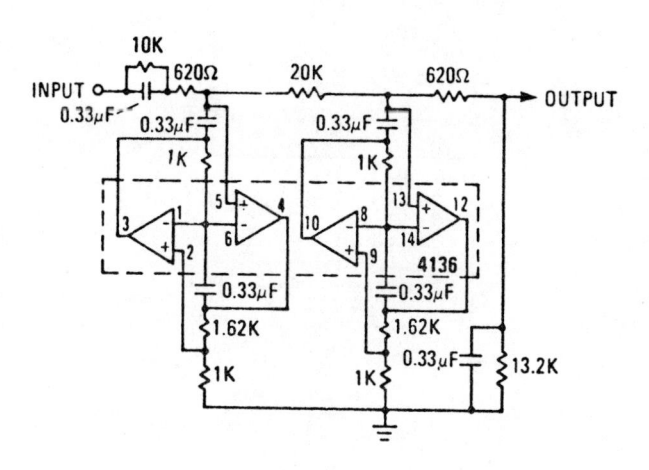

400 Hz lowpass butterworth active filter

Stereo tone control

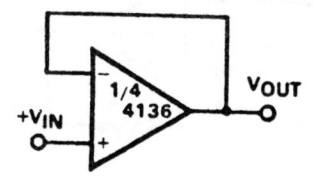

Voltage follower

Lamp driver

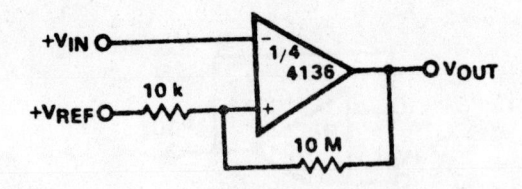

Comparator with hysteresis

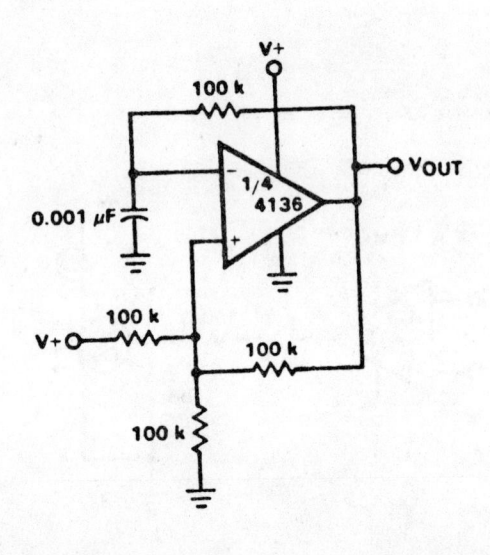

Squarewave oscillator

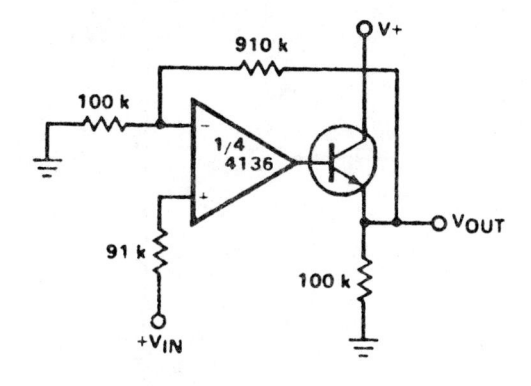

Power amplifier

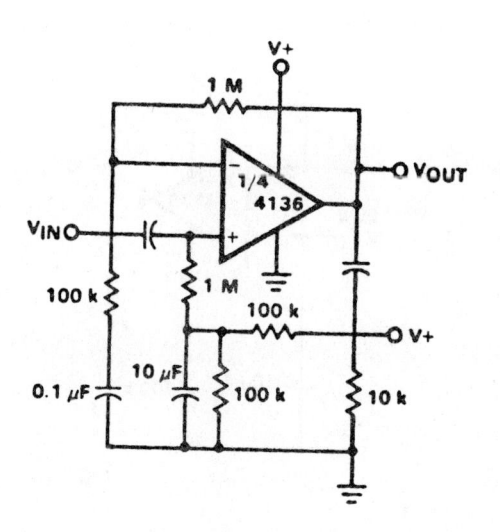

AC coupled noninverting amplifier

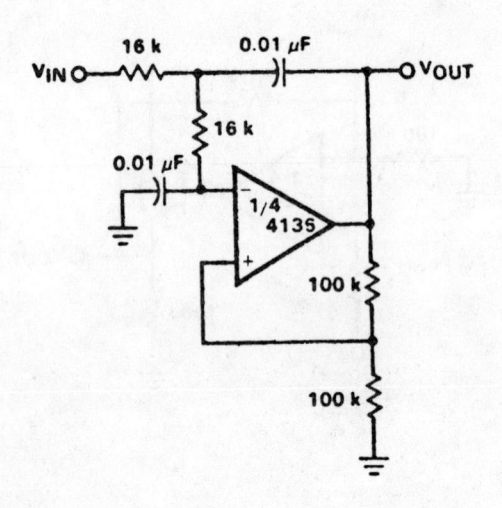

DC coupled 1 kHz low-pass active filter

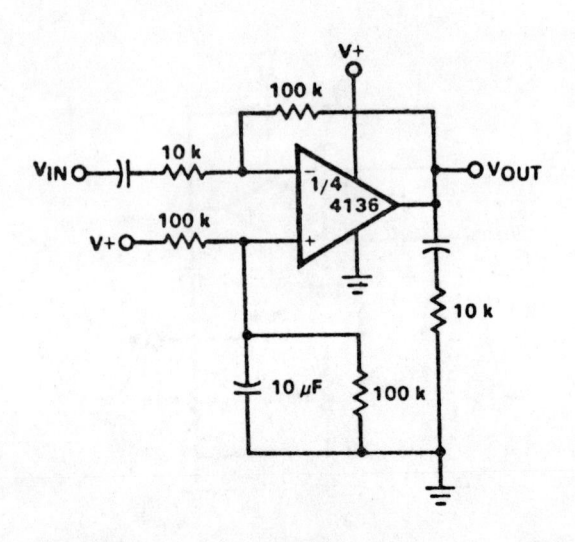

AC coupled inverting amplifier

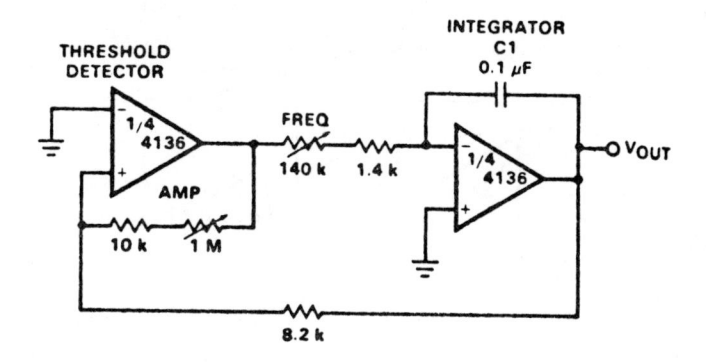

Triangular-wave generator

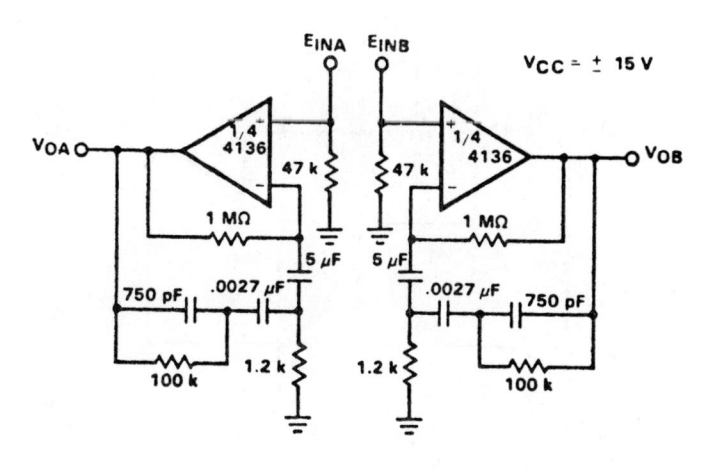

RIAA preamplifier

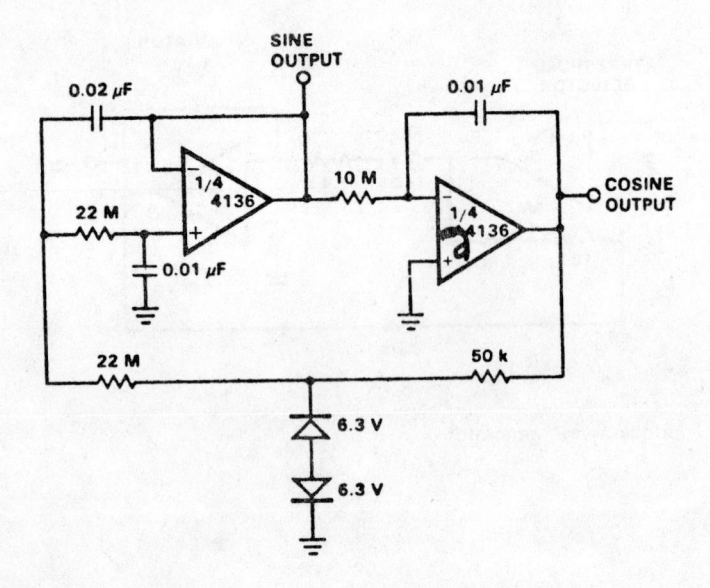

Low frequency sine wave generator with quadrature output

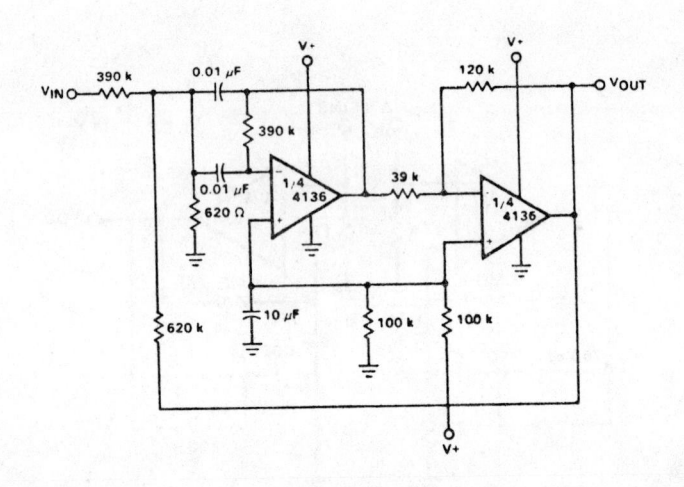

1 kHz bandpass active filter

340

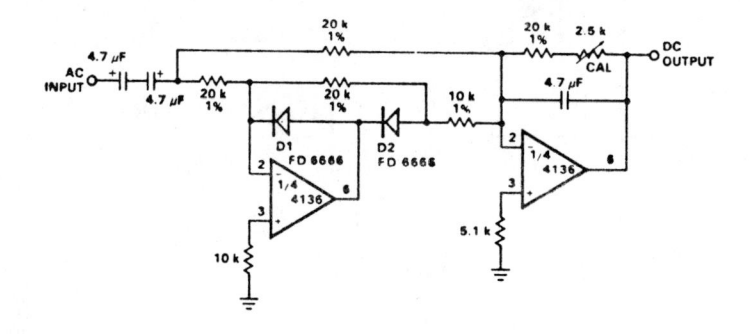

Full-wave rectifier and averaging filter

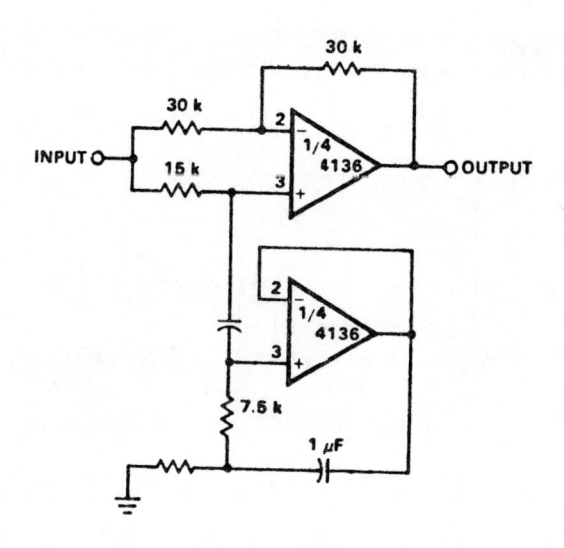

Notch filter

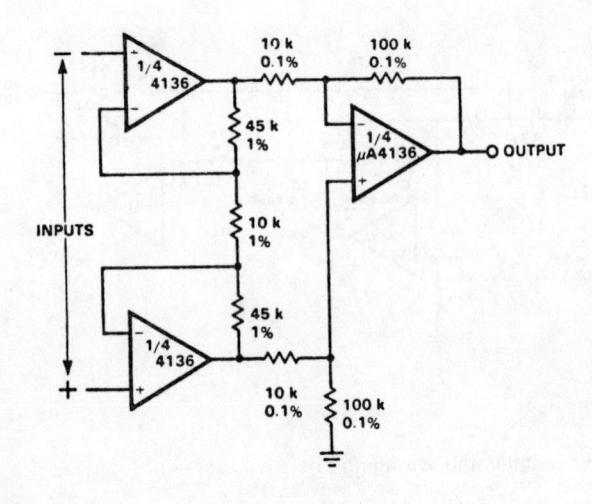

Differential input instrumentation amplifier with high common mode rejection

4250—PROGRAMMABLE OPERATIONAL AMPLIFIER

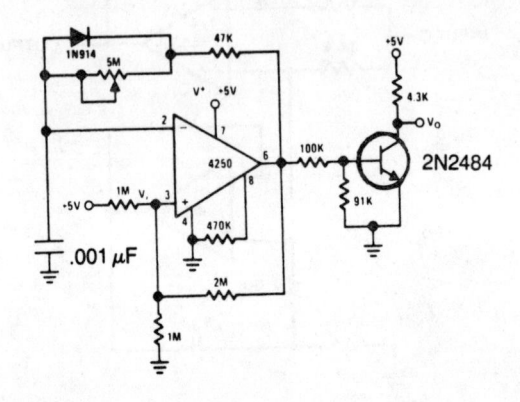

Pulse generator

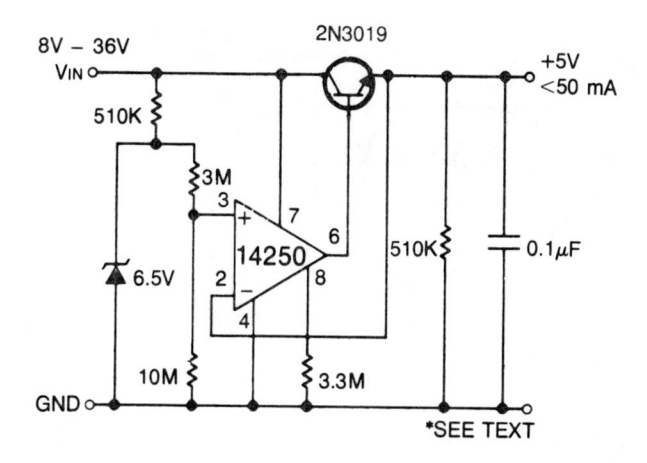

.350 µW quiescent drain 5 volt regulator

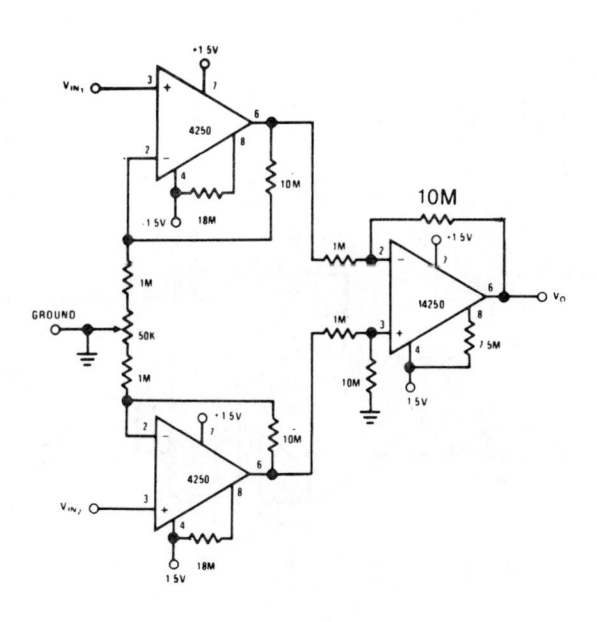

Instrumentation amplifier

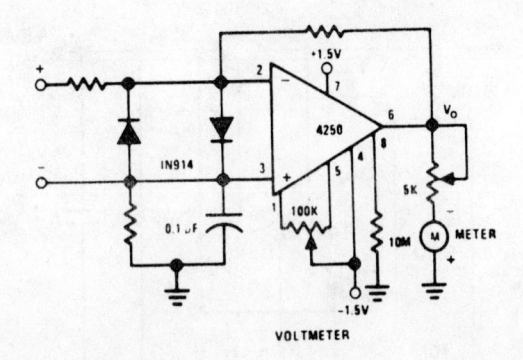

Voltmeter

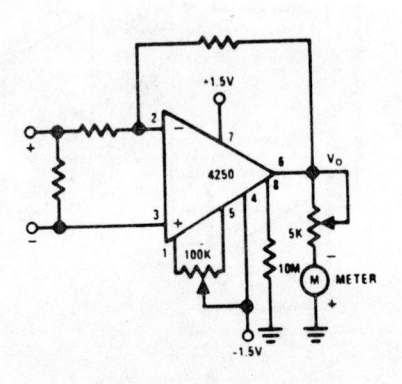

Ammeter

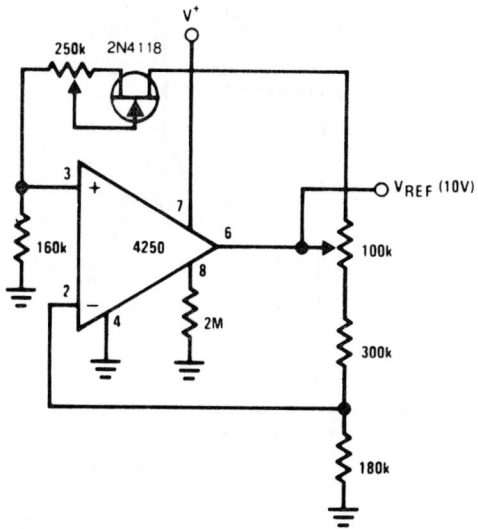

Basic voltage reference

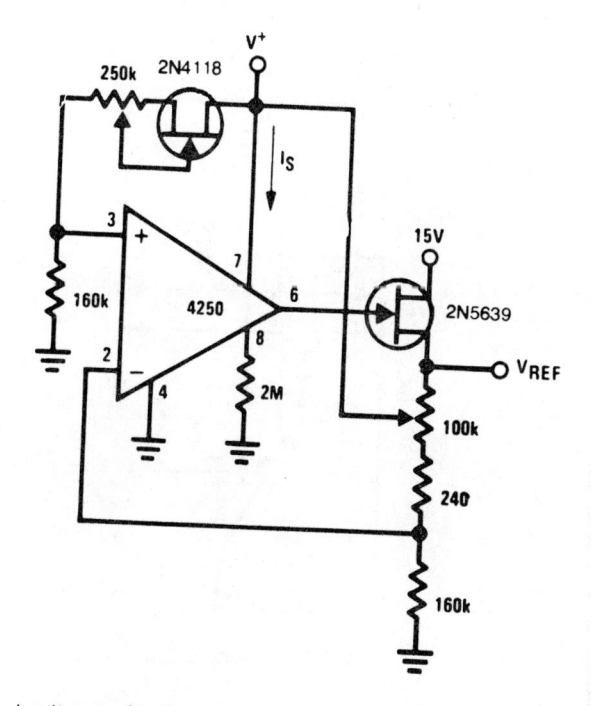

Improved voltage reference

345

4739 — DUAL LOW-NOISE OPERATIONAL AMPLIFIER

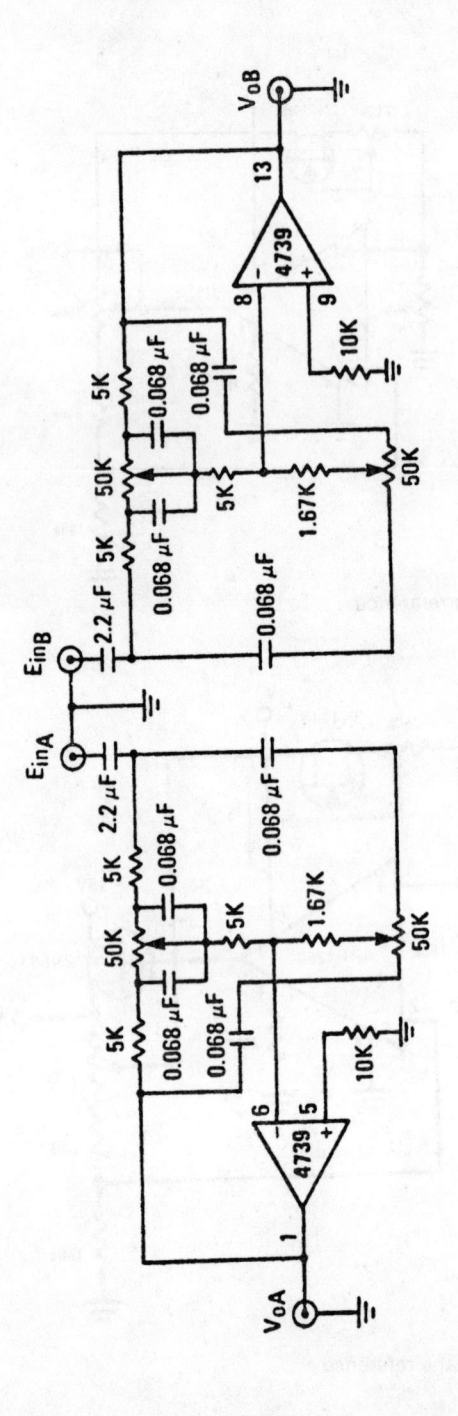

Stereo tone control

7514—HIGH GAIN OPERATIONAL AMPLIFIER

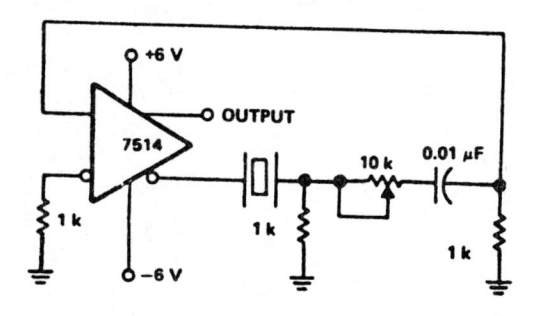

Crystal oscillator with variable feedback

8015—INTEGRATED SCHMITT TRIGGER

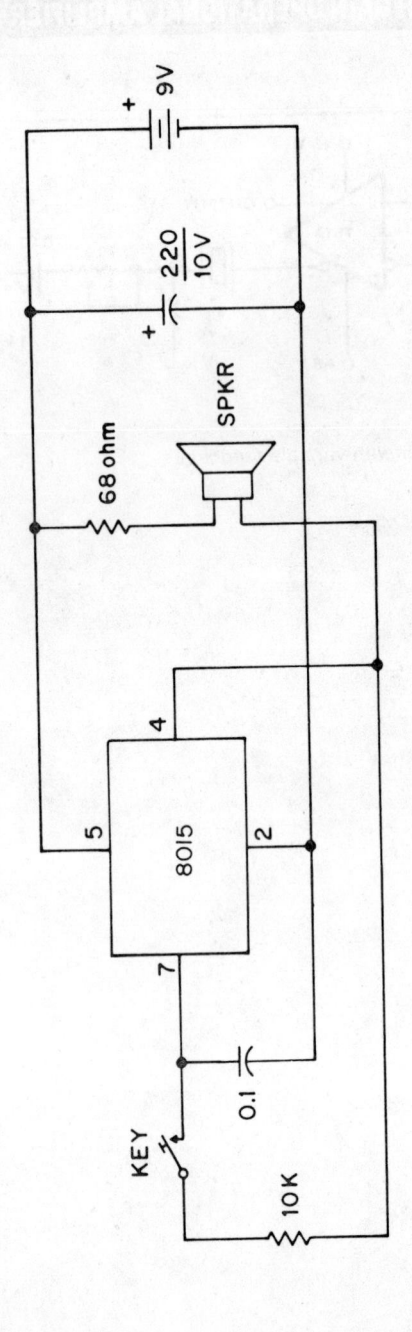

Morse code practice oscillator

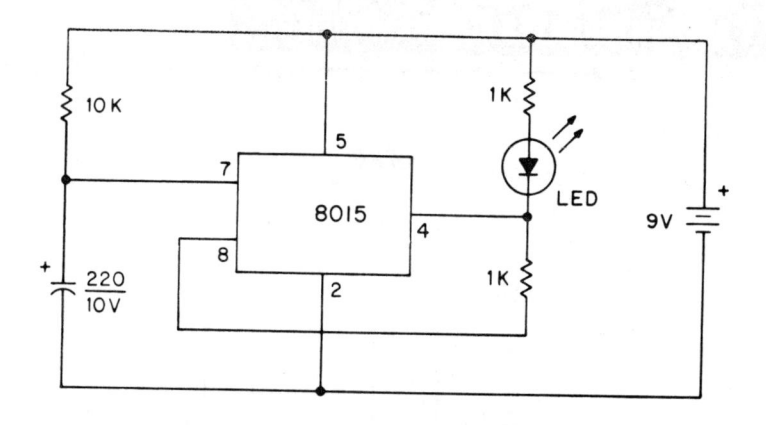

LED flasher

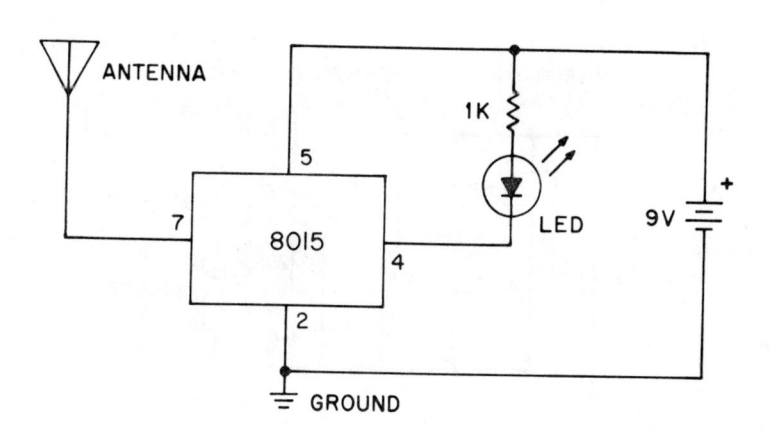

RF energy detector

8341—HIGH INPUT IMPEDANCE INTEGRATED AUDIO AMPLIFIER

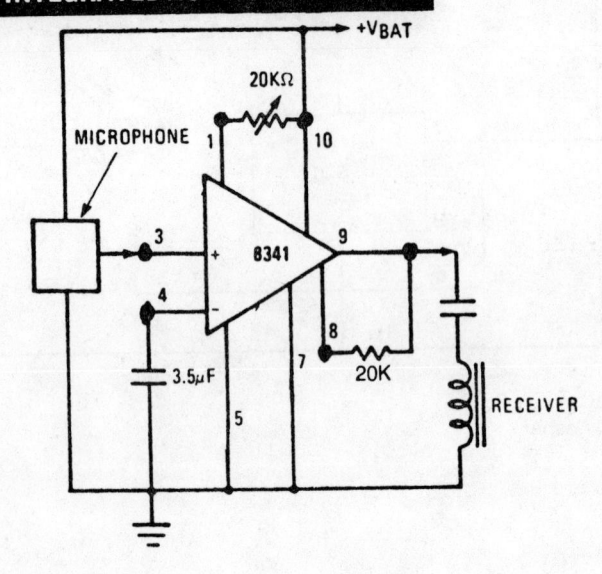

Basic amplifier

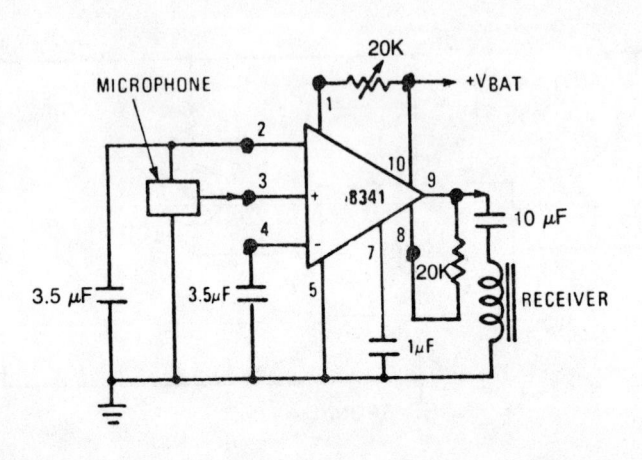

Amplifier with AGC

52107 — LOW INPUT OFFSET OPERATIONAL AMPLIFIER

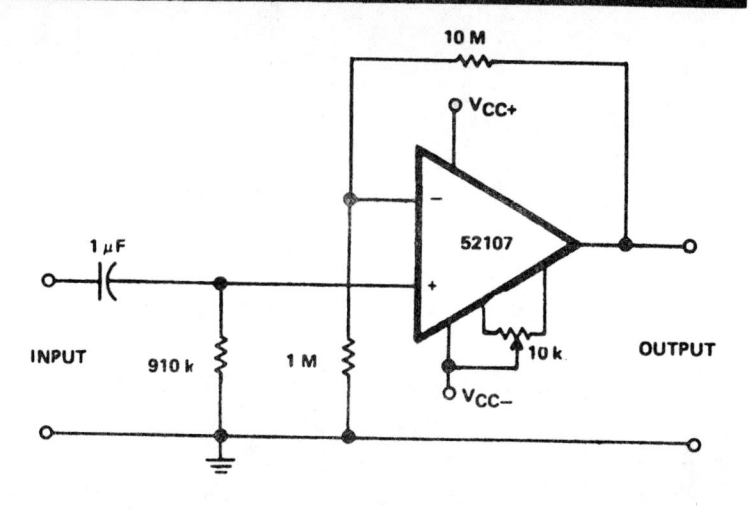

Noninverting amplifier

52702 — INTERNALLY COMPENSATED OPERATIONAL AMPLIFIER

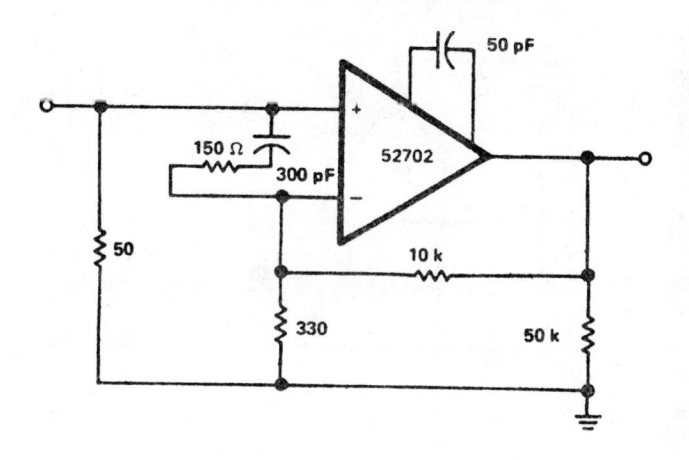

30-dB, 10-MHz amplifier

72301—INTERNALLY COMPENSATED OPERATIONAL AMPLIFIER

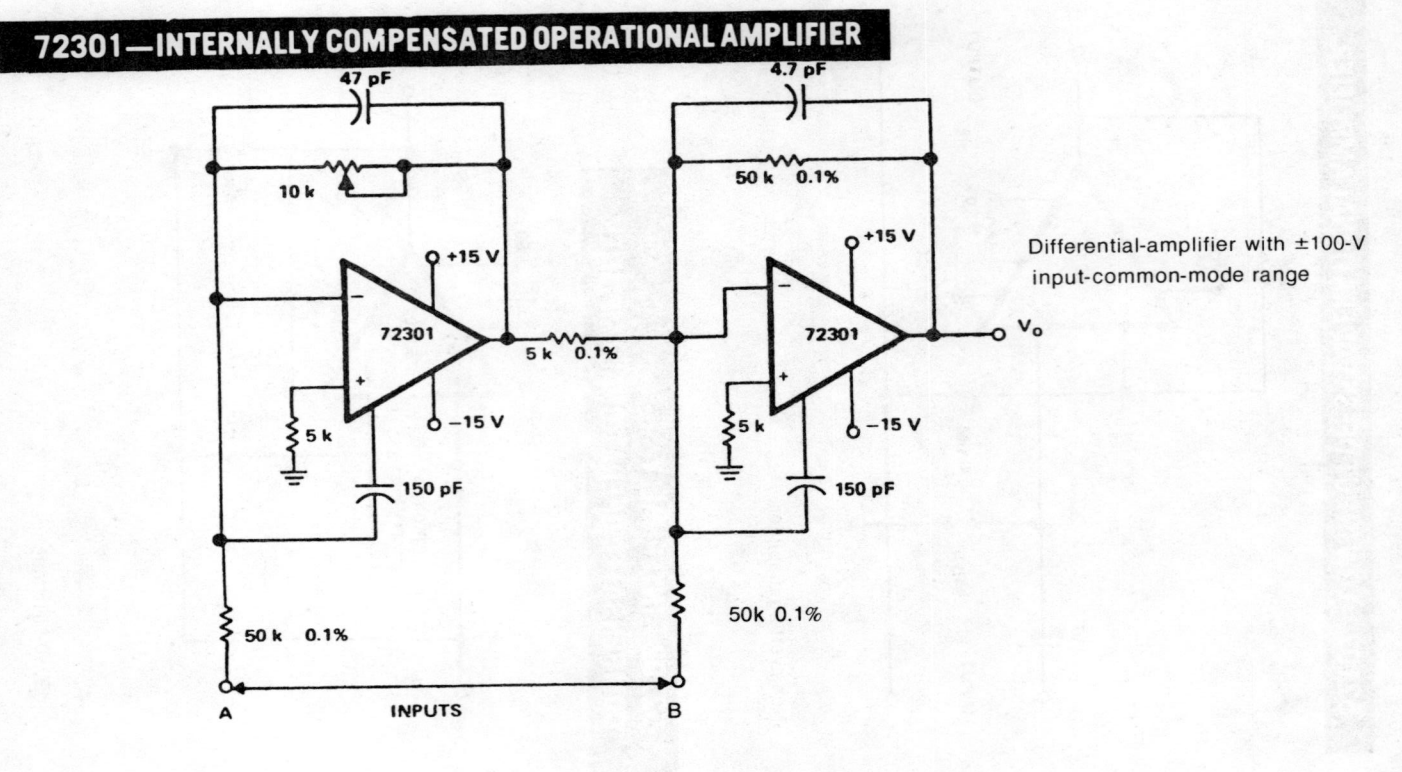

Differential-amplifier with ±100-V input-common-mode range

72709 — LOW INPUT NOISE OPERATIONAL AMPLIFIER

60-dB preamplifier for a magnetic tape output

Band-reject active filter

AC millivoltmeter

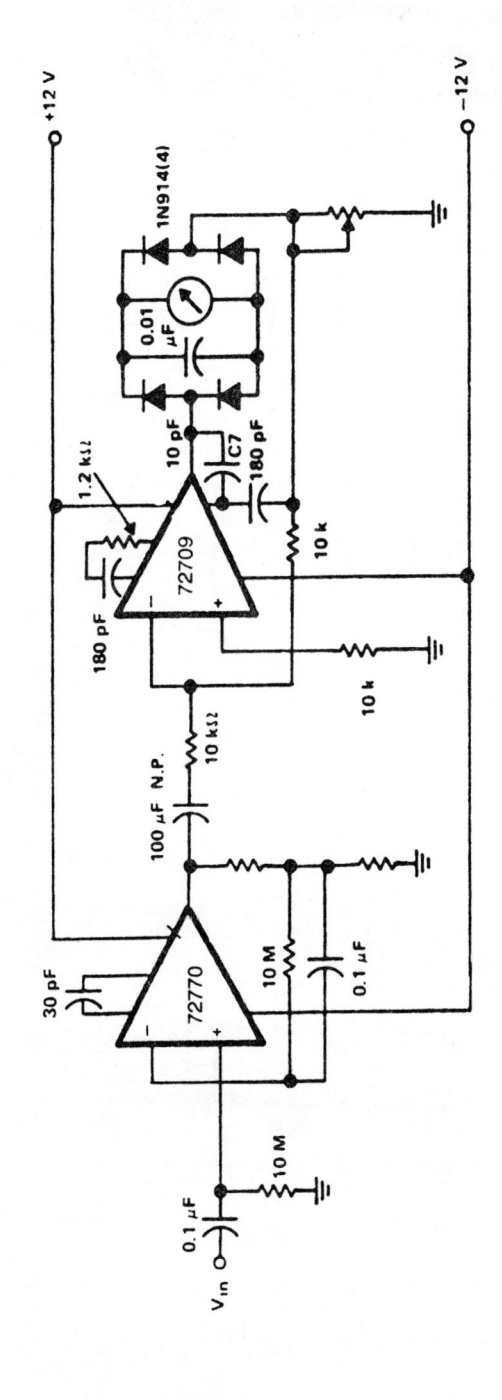

72733 — WIDEBAND HIGH INPUT IMPEDANCE OPERATIONAL AMPLIFIER

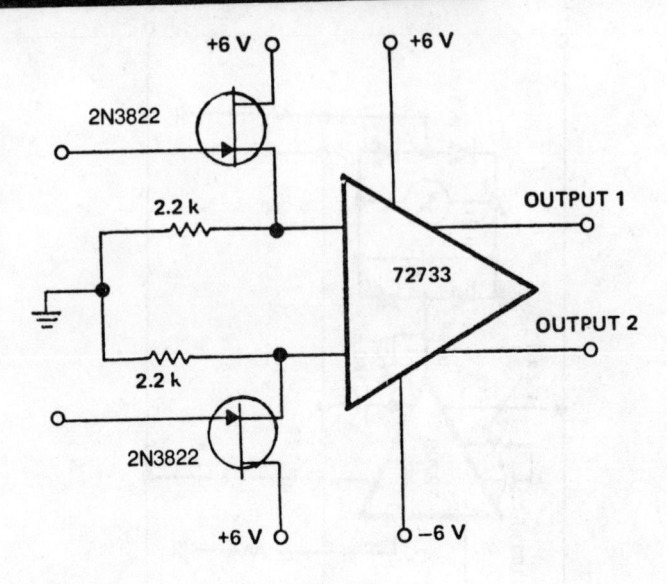

High-input-impedance wideband amplifier

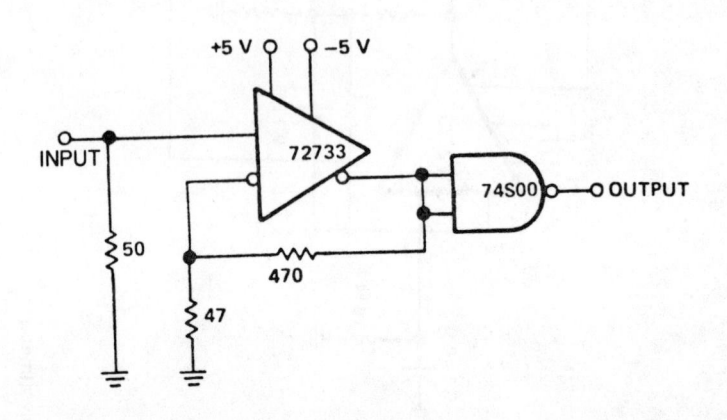

Fast Schmitt trigger using a video amplifier

72741 — INTERNALLY COMPENSATED OPERATIONAL AMPLIFIER

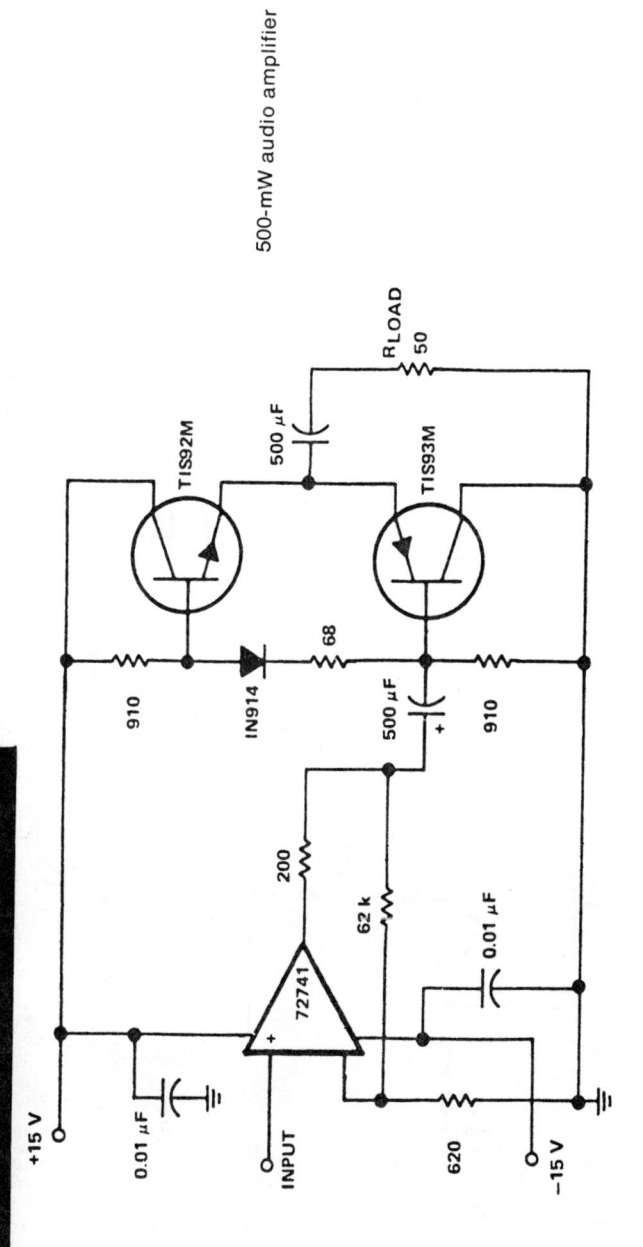

500-mW audio amplifier

Audio amplifier with AGC

Multivibrator with voltage-controlled frequency

72770 — SUPER BETA OPERATIONAL AMPLIFIER

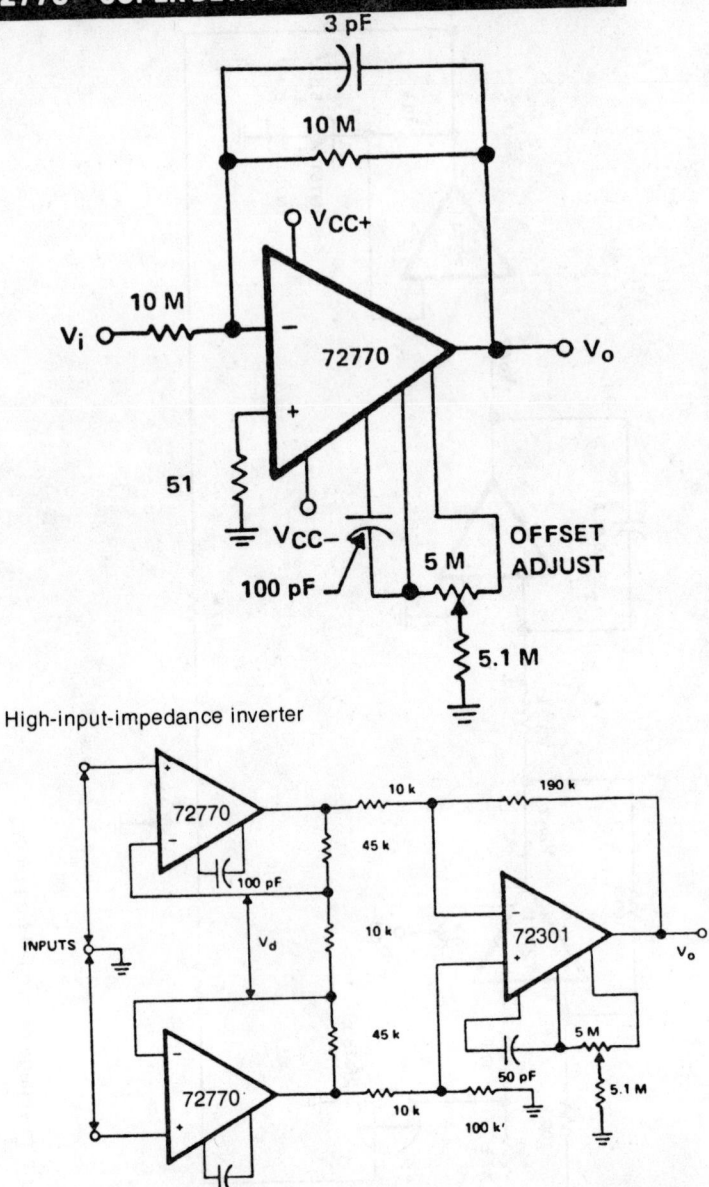

High-input-impedance inverter

High-input-impedance differential amplifier with improved common-mode rejection

72810—5 MHZ OPERATIONAL AMPLIFIER

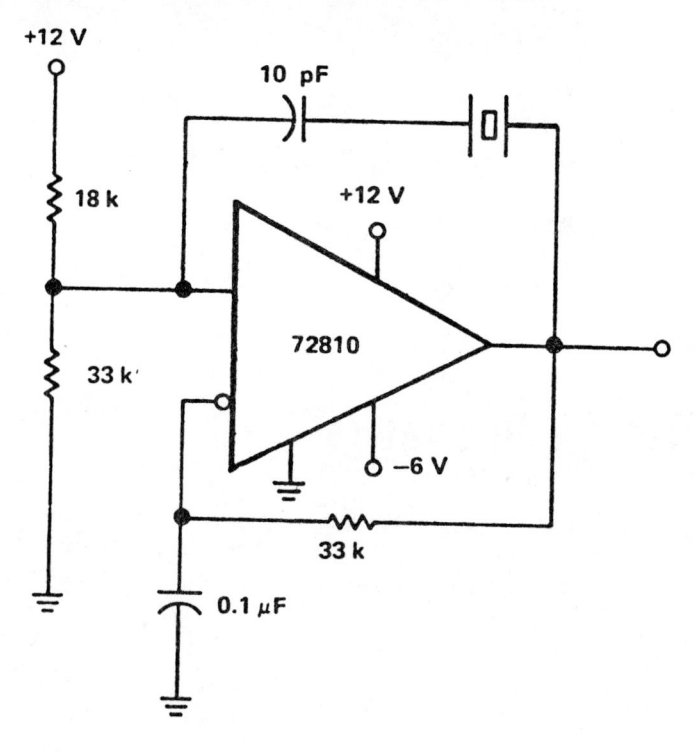

Crystal-controlled logic generator

Part II

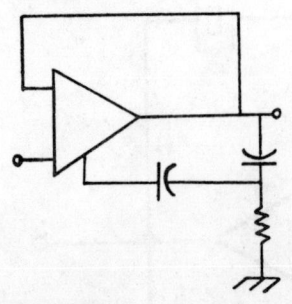

Voltage Regulators

It wasn't too long ago when voltage regulator circuits involved several transistors (or tubes!), a few diodes, resistors, capacitors and a lot of space and cost. Now a single chip replaces all that.

100—INTEGRATED VOLTAGE REGULATOR

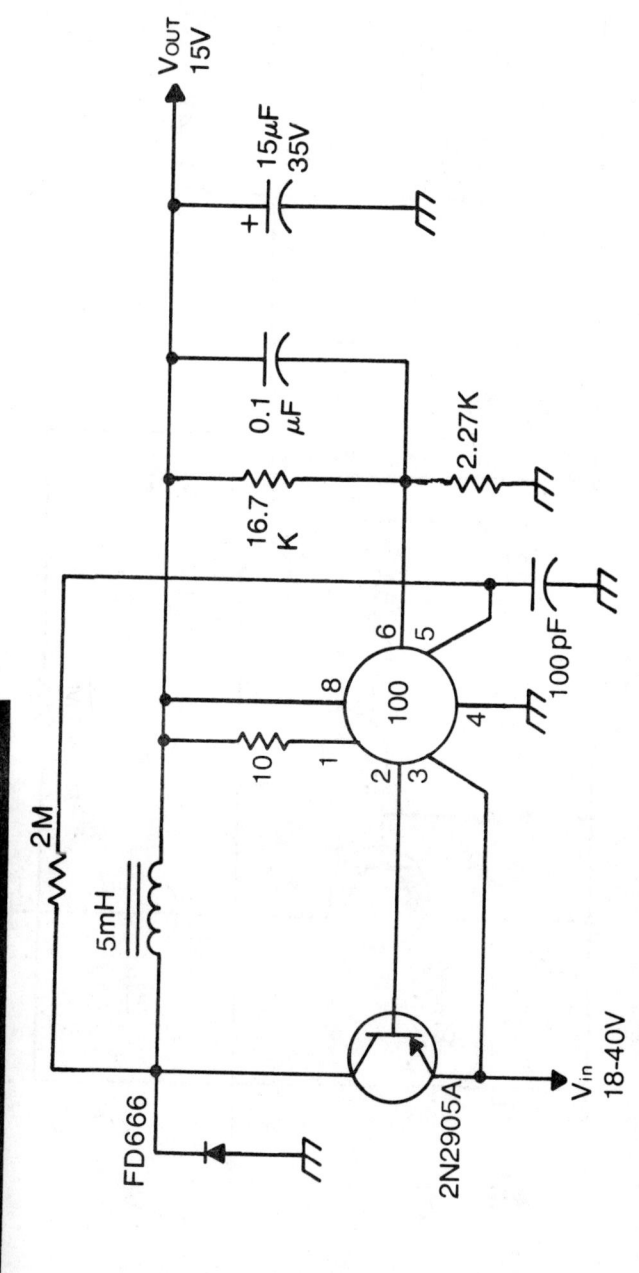

Switching regulator delivering up to 500 mA

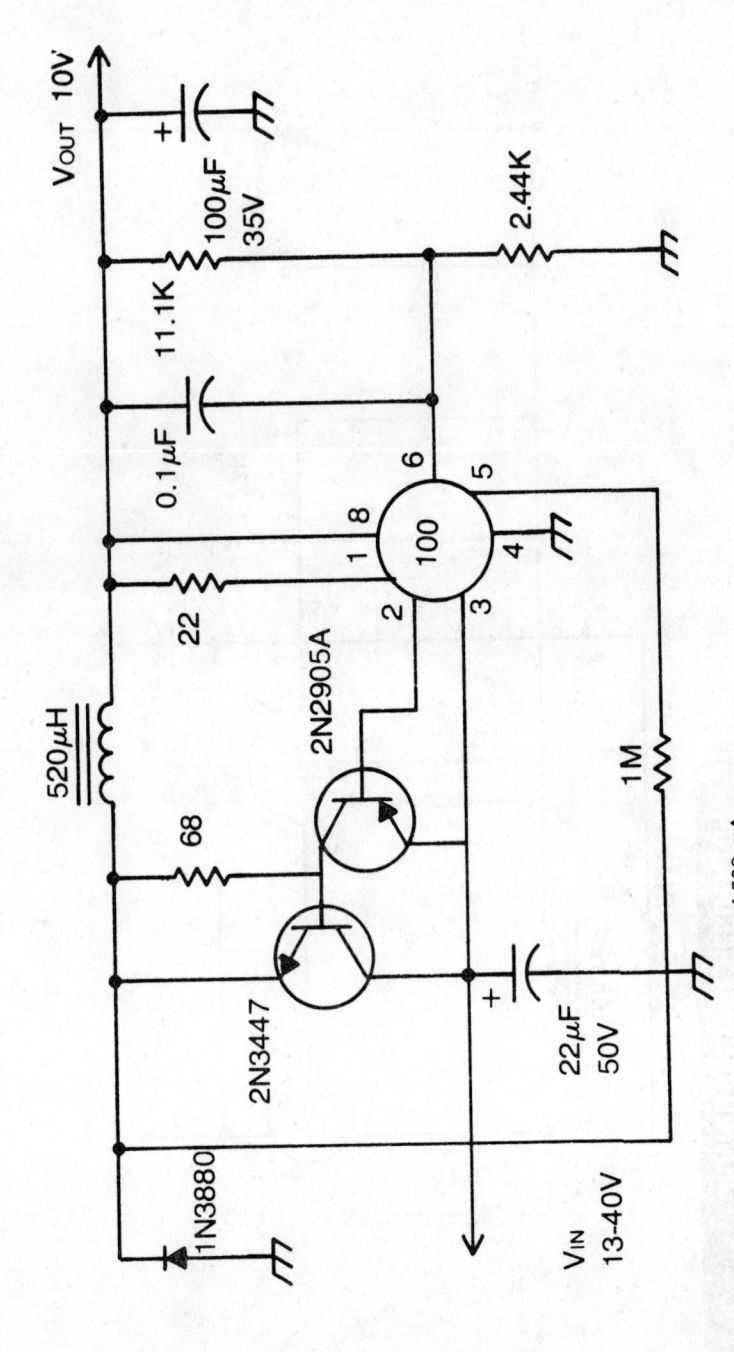

Switching regulator delivering in excess of 500 mA

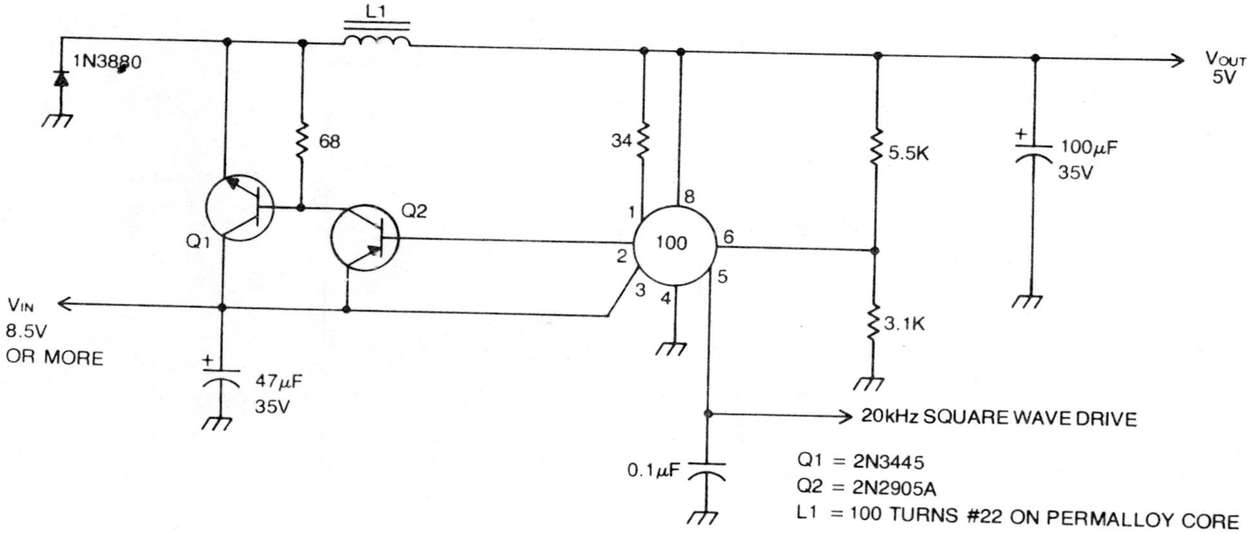

Synchronized switching regulator using 20 kHz square wave signal for drive

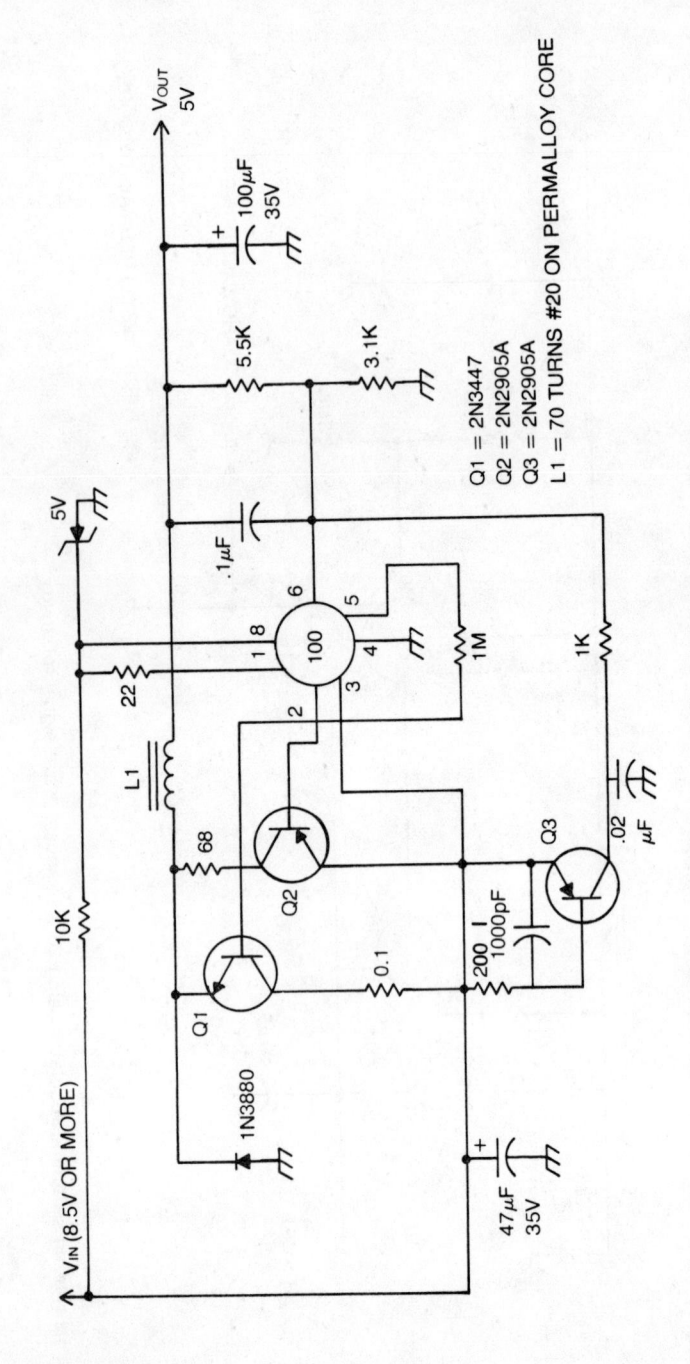

Switching regulator with overload/short-circuited output protection

104—NEGATIVE VOLTAGE REGULATOR

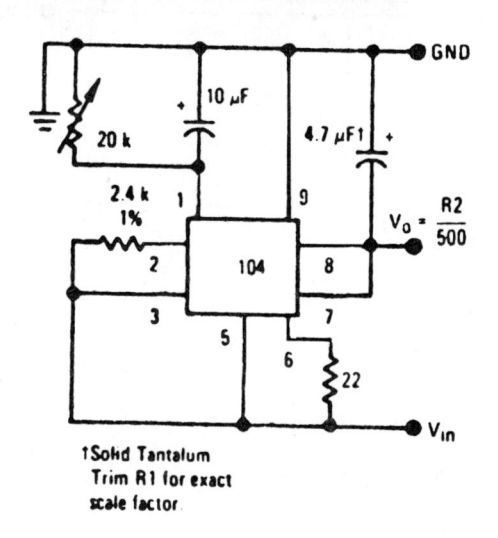

$V_o = \dfrac{R2}{500}$

†Solid Tantalum
Trim R1 for exact scale factor.

Basic regulator circuit

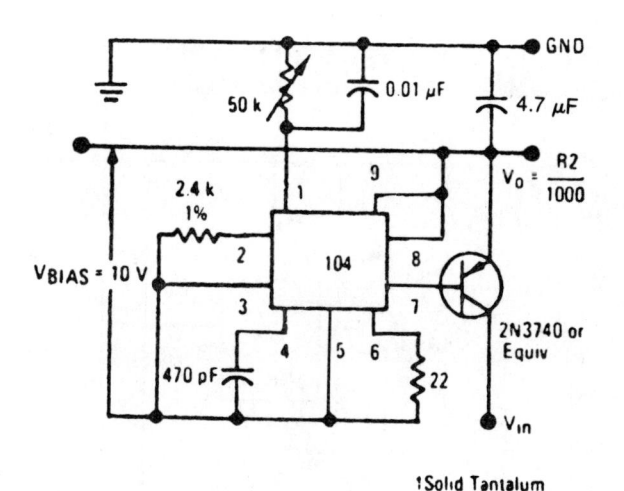

$V_o = \dfrac{R2}{1000}$

$V_{BIAS} = 10\ V$

†Solid Tantalum

Separate bias supply operation

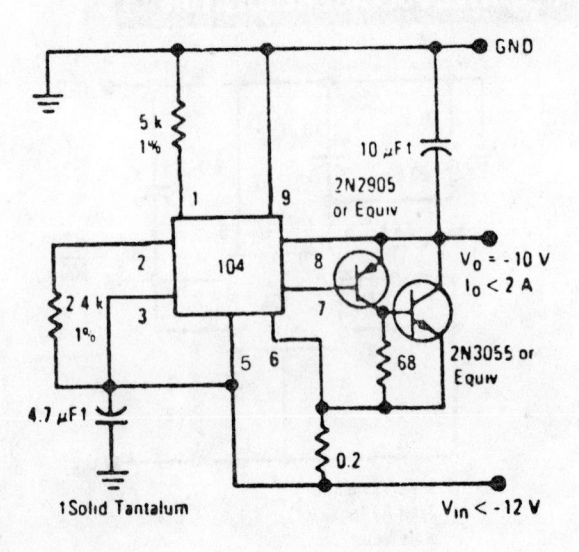

High current regulator

105—POSITIVE VOLTAGE REGULATOR

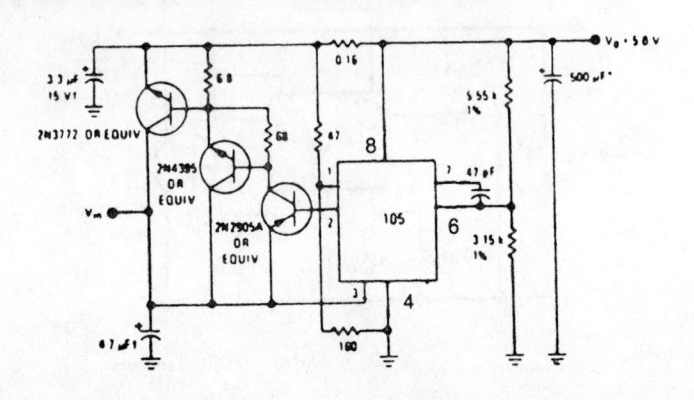

10 amp regulator with foldback current limiting

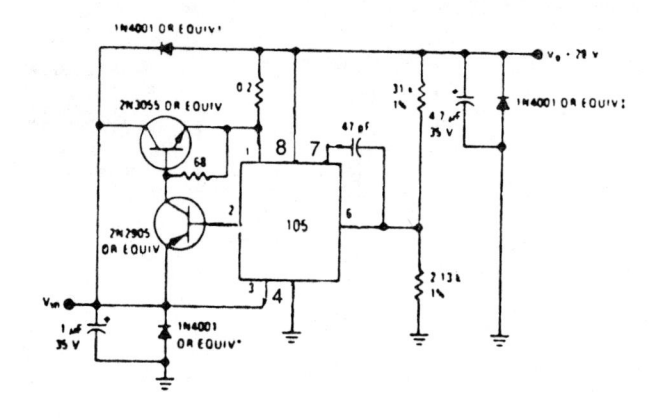

1.0 amp regulator with protective diodes

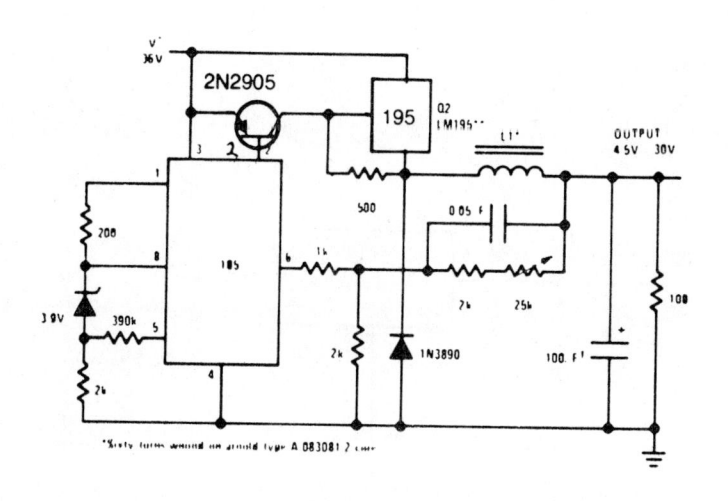

6 amp variable output switching regulator

369

109—5 VOLT INTEGRATED VOLTAGE REGULATOR

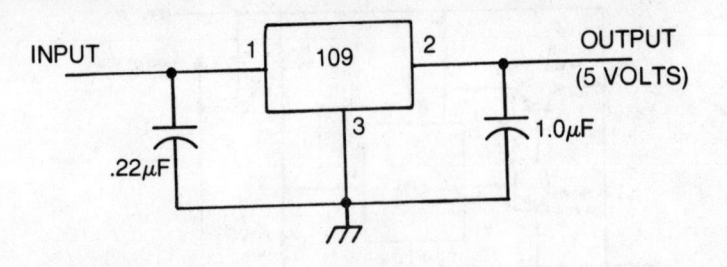

Fixed 5 volt regulator

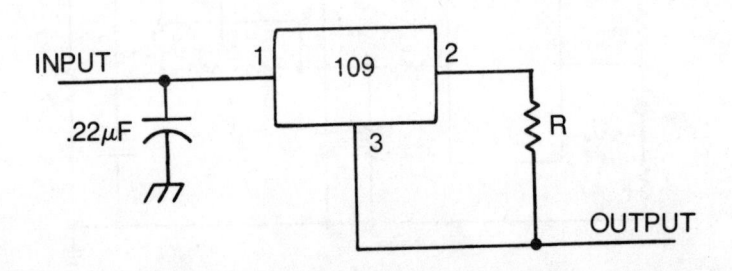

Current regulator (output current varies with value of resistor R)

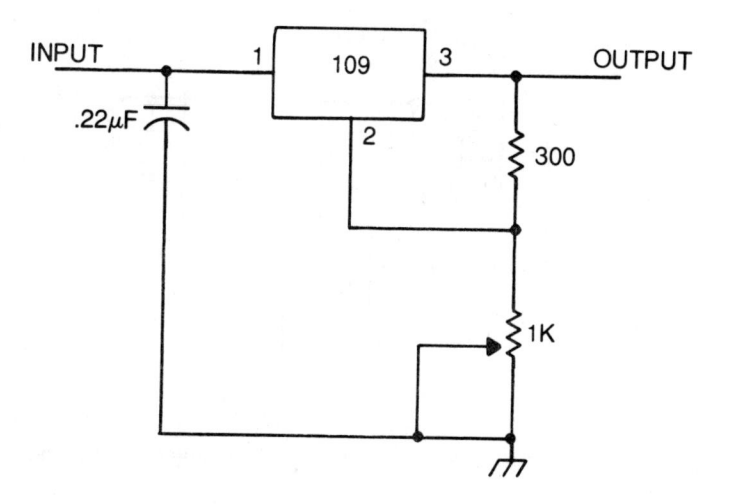

Regulator with adjustable output

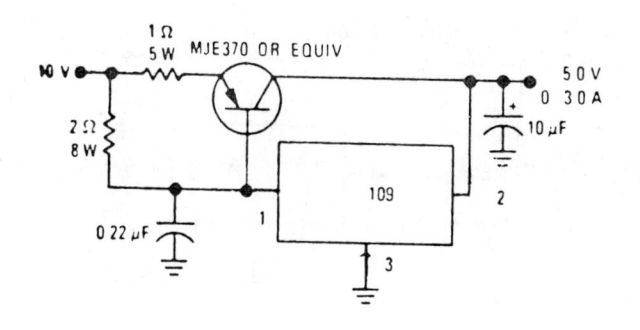

5.0-volt, 3.0-ampere regulator (with plastic boost transistor)

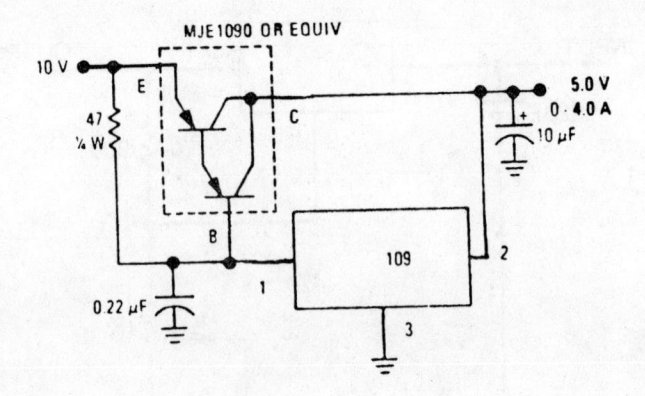

5.0 volt, 4.0-ampere transistor (with plastic Darlington boost transistor)

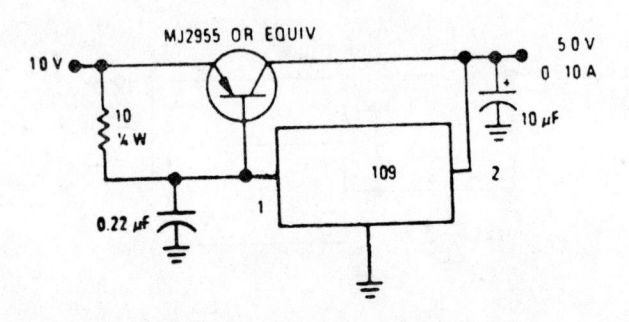

5.0-volt, 10-ampere regulator

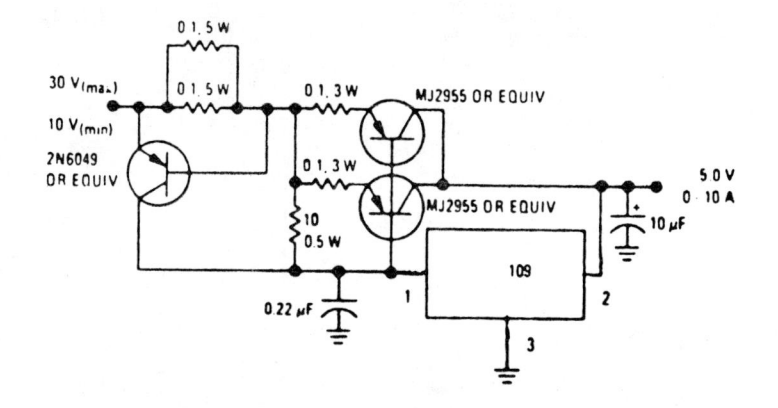

5.0-volt, 10-ampere regulator (with short-circuit current limiting for safe-area protection of pass transistors)

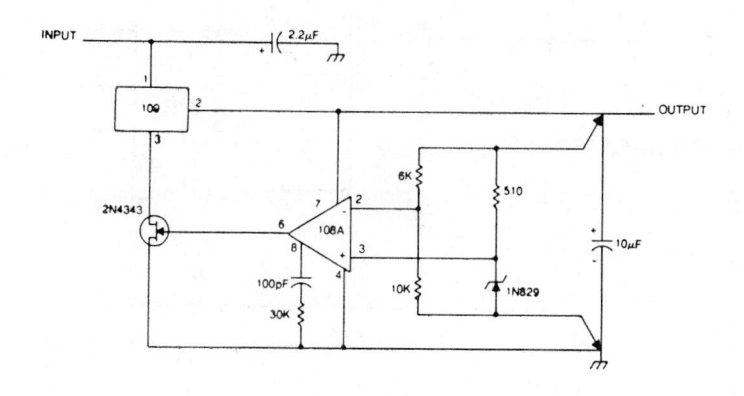

High stability (0.01%) regulator

373

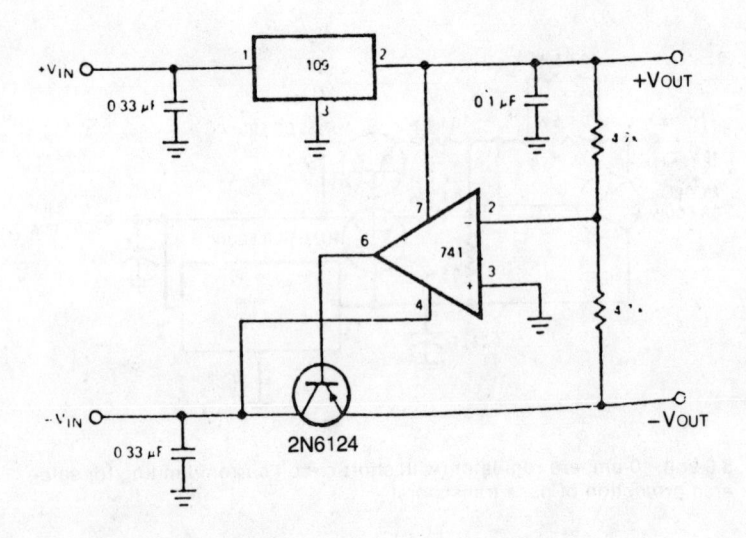

Tracking voltage regulator

117—INTEGRATED ADJUSTABLE VOLTAGE REGULATOR

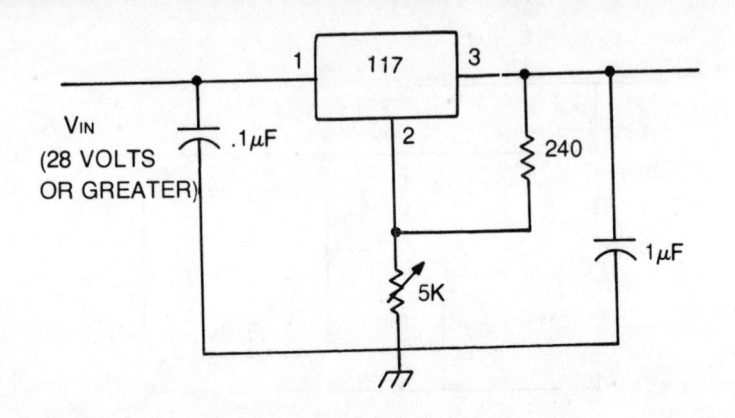

1.2-25 volt adjustable output regulator

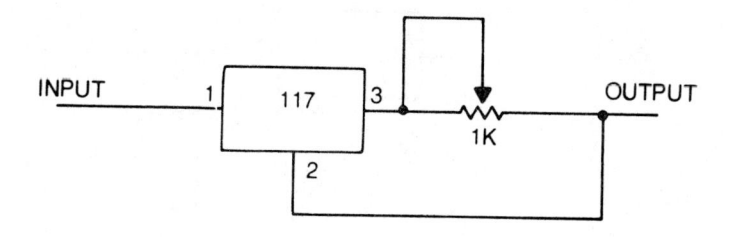

Precision current limiter

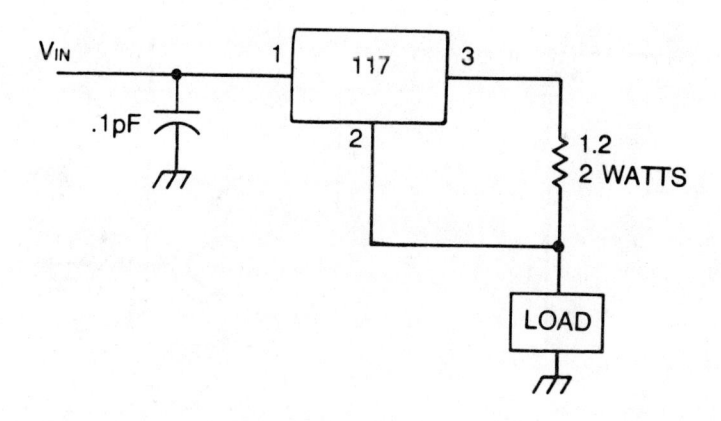

One amp current regulator

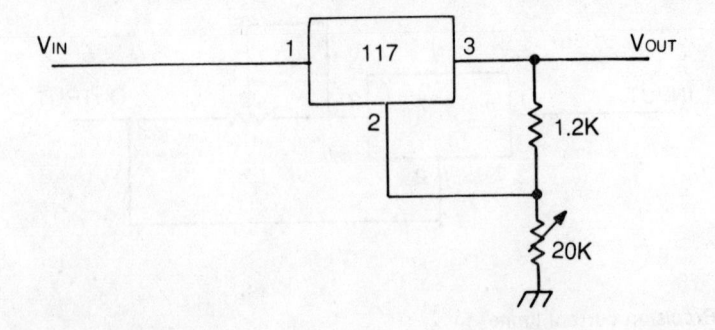

1.2-20 volt adjustable regulator with 4 mA minimum load current

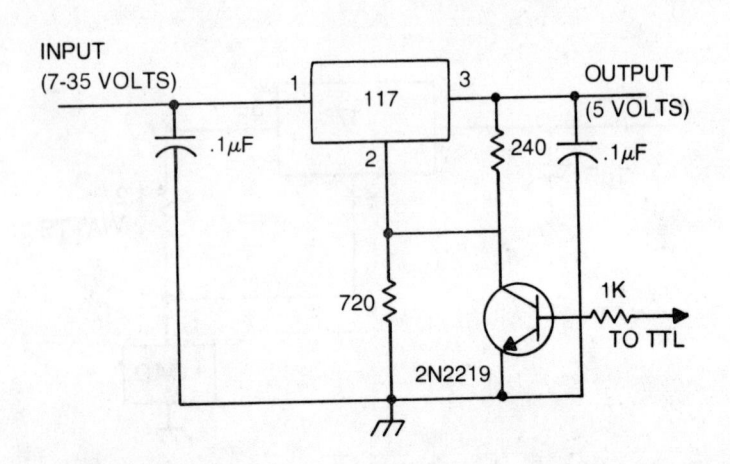

Five volt logic regulator with electronic shutdown

376

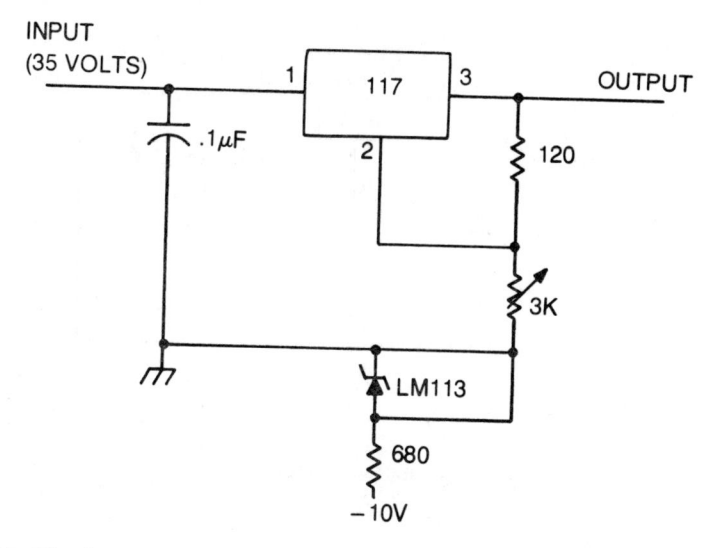

0 to 30 volt regulator

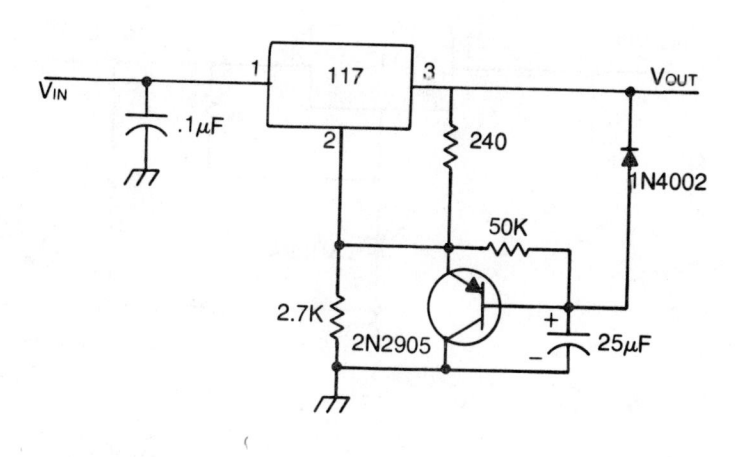

Slow turn on 15 volt regulator

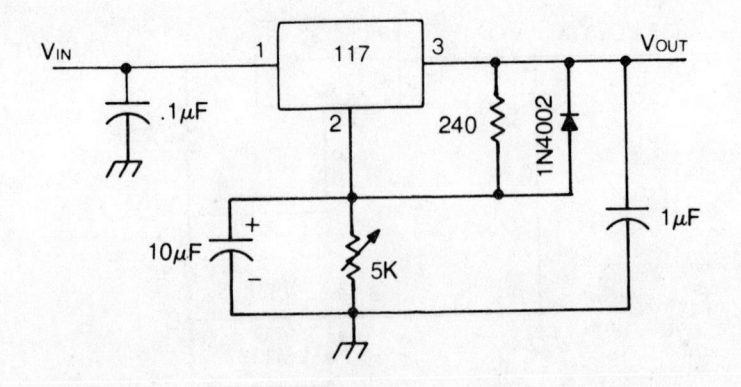

Adjustable regulator with improved ripple suppression

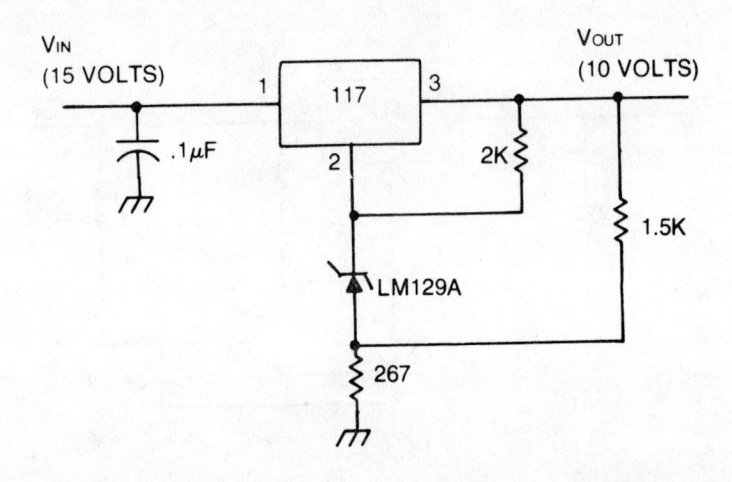

Highly stable 10 volt regulator

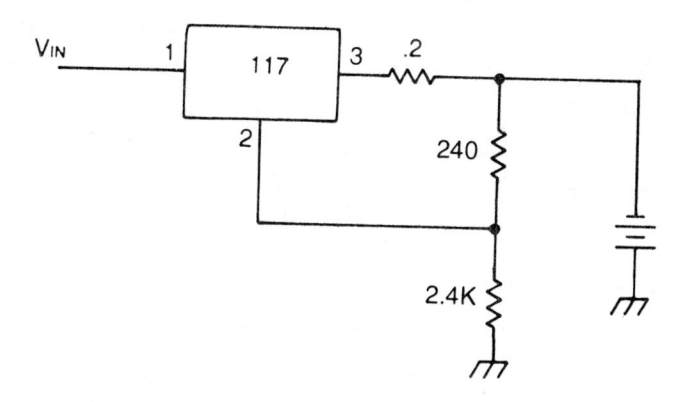

12 volt battery charger

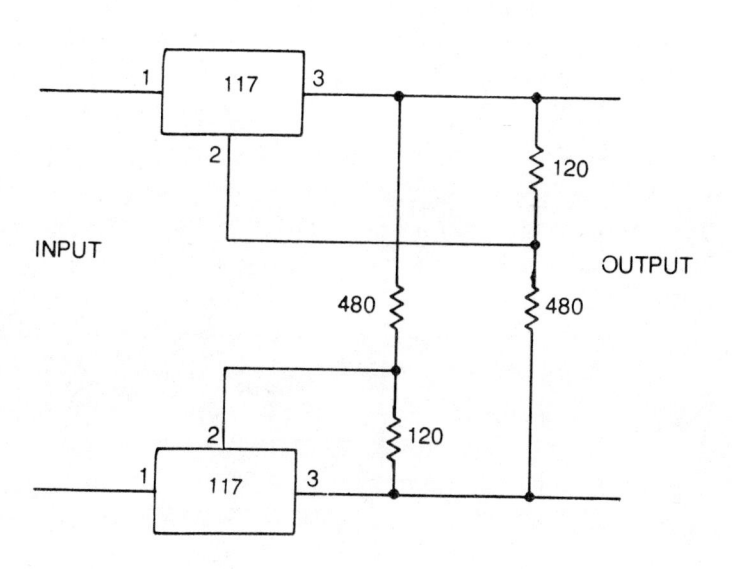

AC voltage regulator

120 — INTEGRATED NEGATIVE VOLTAGE REGULATOR

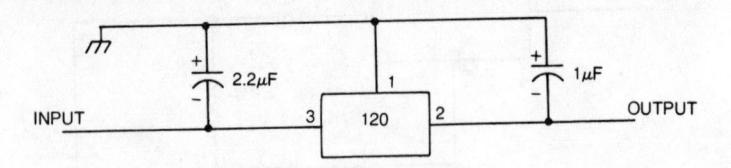

Fixed voltage regulator

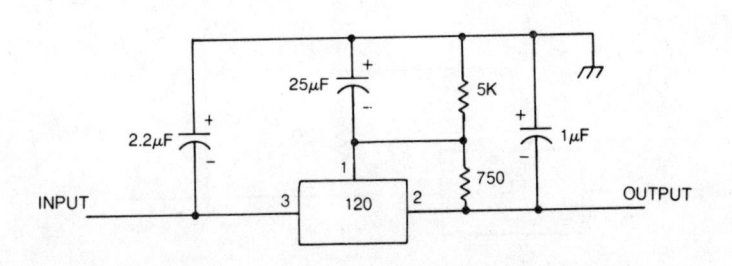

Variable output regulator

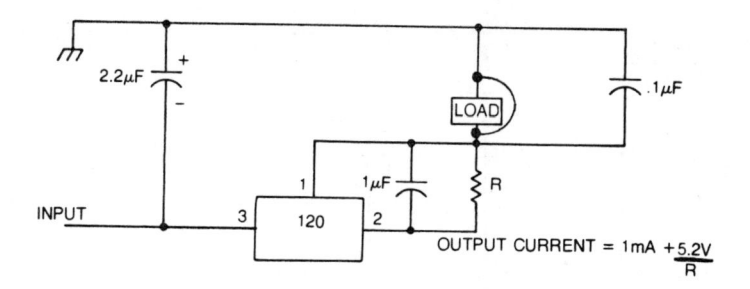

Precision current source

$$\text{OUTPUT CURRENT} = 1\text{mA} + \frac{5.2\text{V}}{R}$$

123 — 5 VOLT 3 AMP POSITIVE INTEGRATED REGULATOR

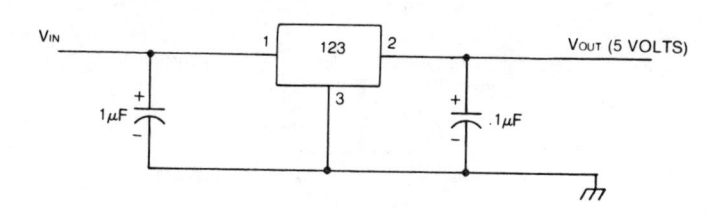

Basic 3 amp regulator

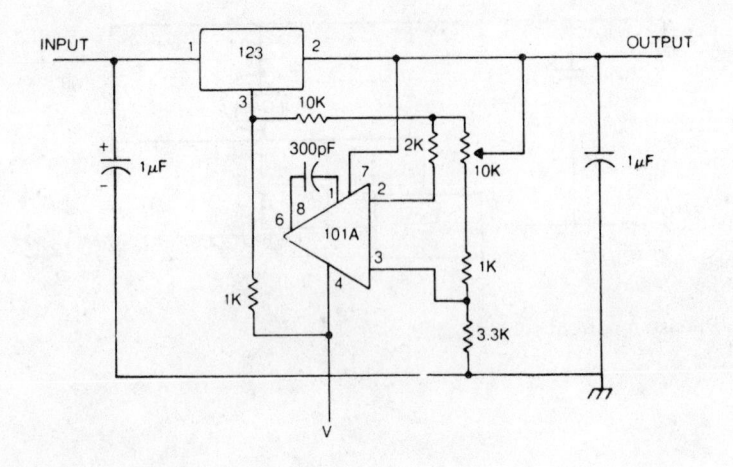

0-10 volt adjustable regulator

145—NEGATIVE THREE AMP INTEGRATED REGULATOR

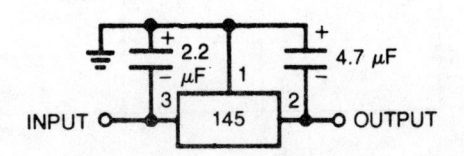

Fixed regulator

382

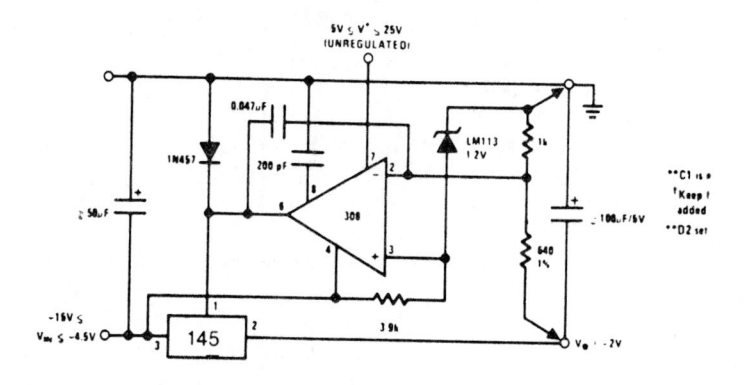

−2V ECL termination regulator

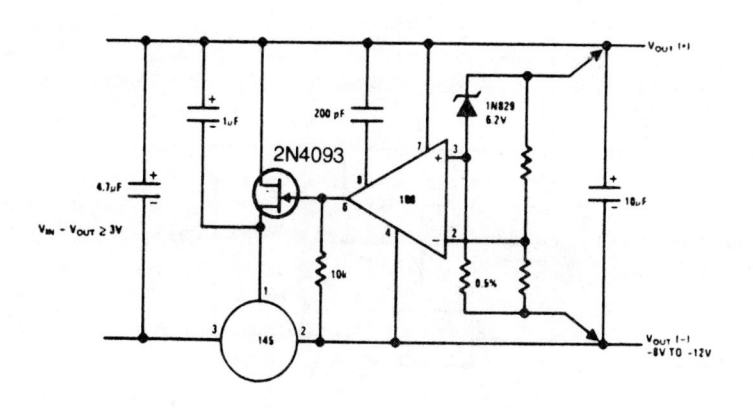

High stability regulator

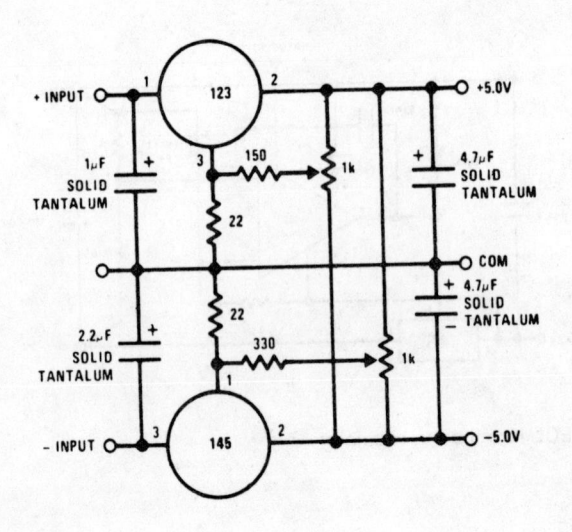

Dual 3 amp trimmed supply

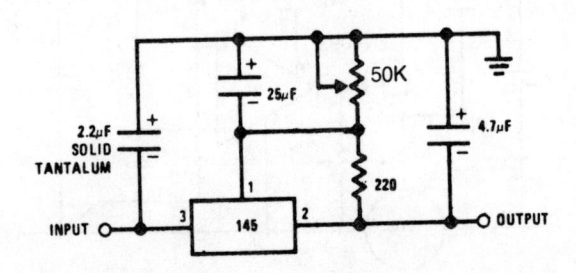

Variable output (−5.0V to −15V)

340—SERIES VOLTAGE REGULATOR

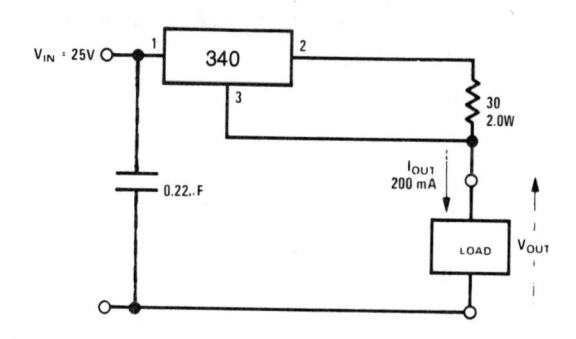

Current source

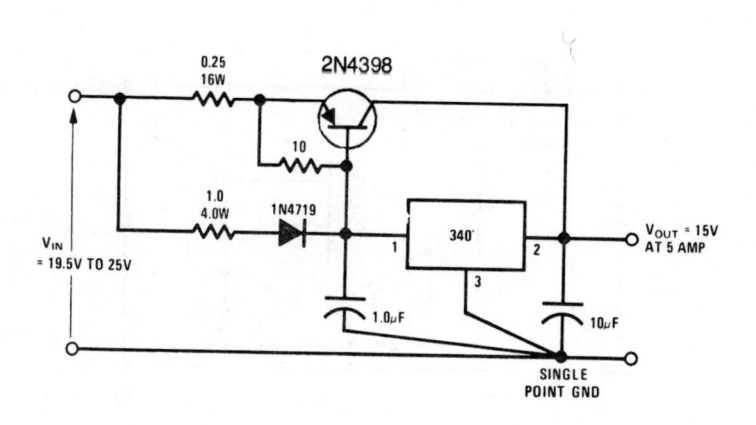

15V 5.0A regulator with short circuit current limit

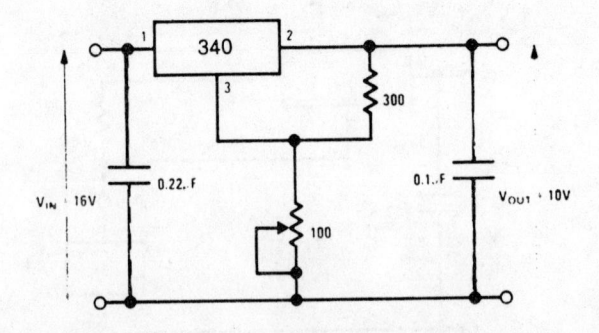

10V regulator

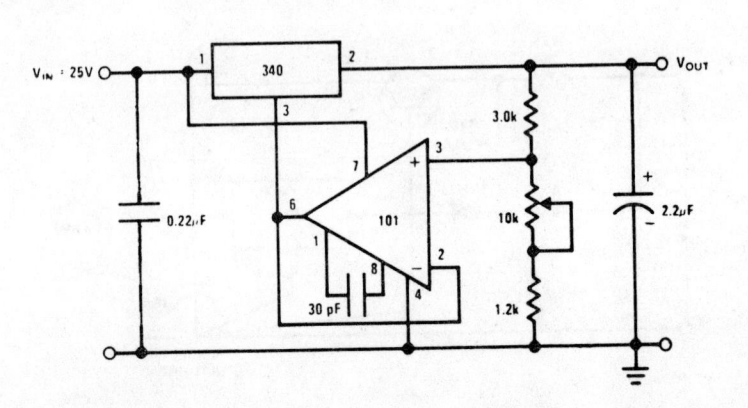

Variable output regulator

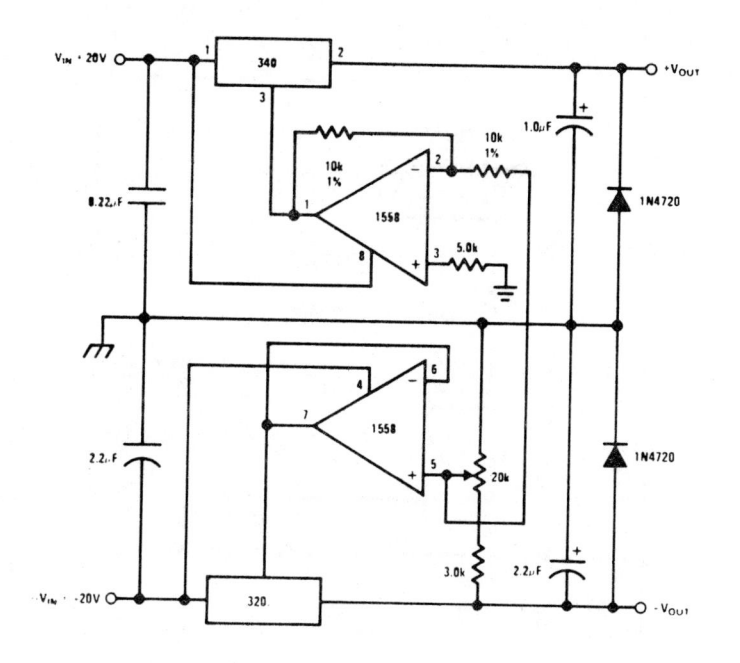

Tracking dual supply ±5.0V - ±18V

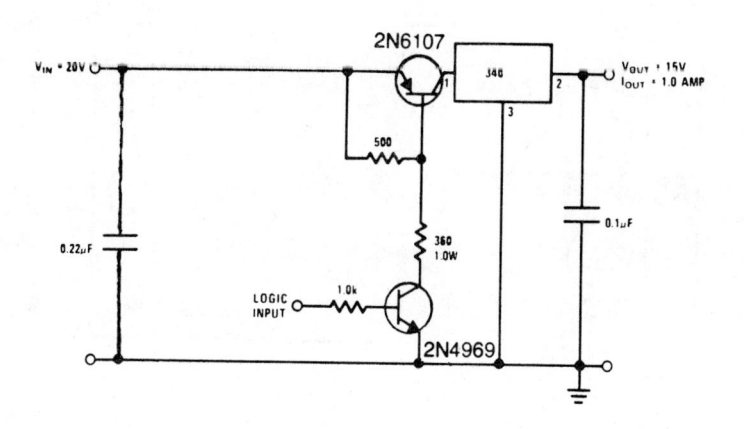

Electronic shutdown circuit

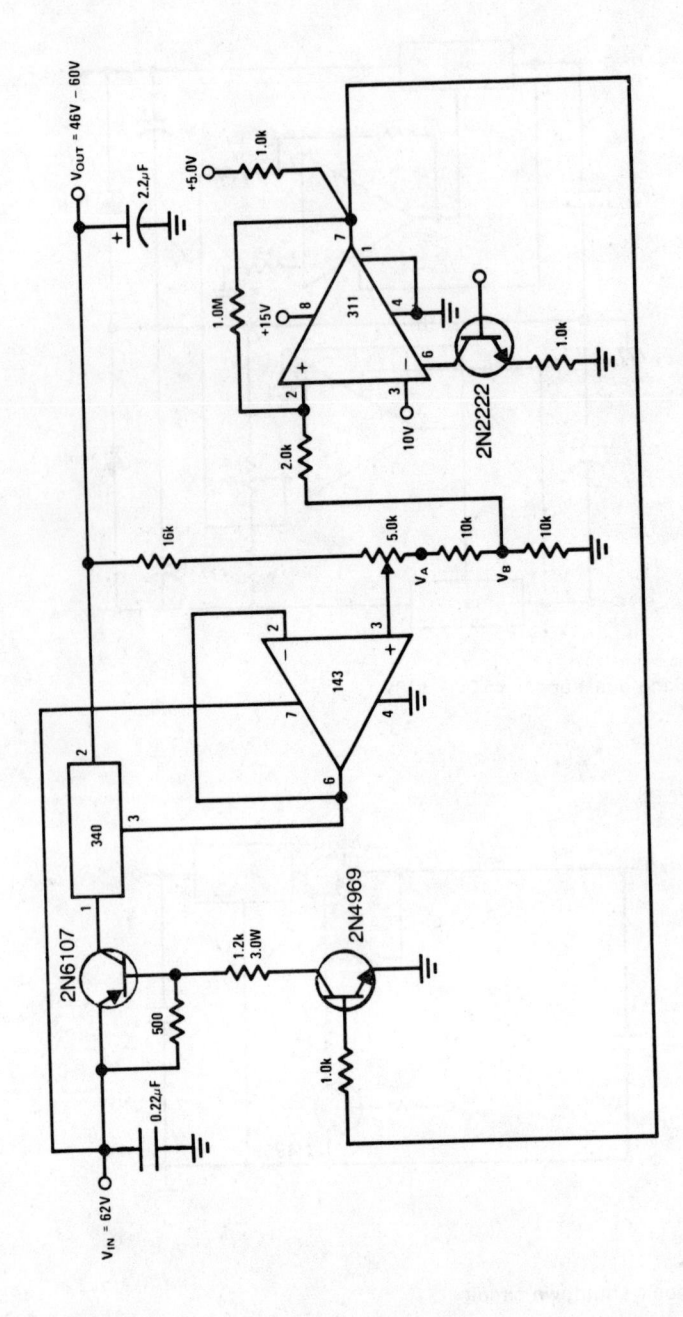

Variable high voltage regulator with shortcircuit and overvoltage protection

342—INTEGRATED VOLTAGE REGULATOR

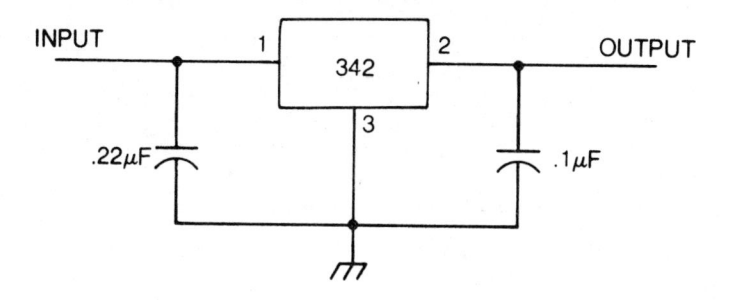

Fixed output regulator

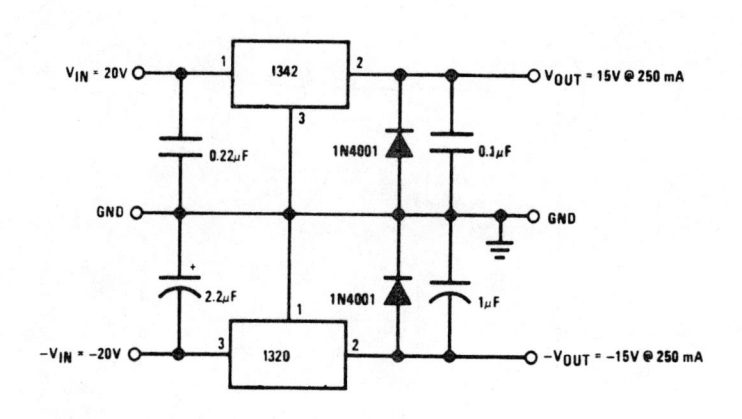

±15V, 250 mA dual power supply

High output regulator

0070—PRECISION BCD BUFFERED VOLTAGE REFERENCE

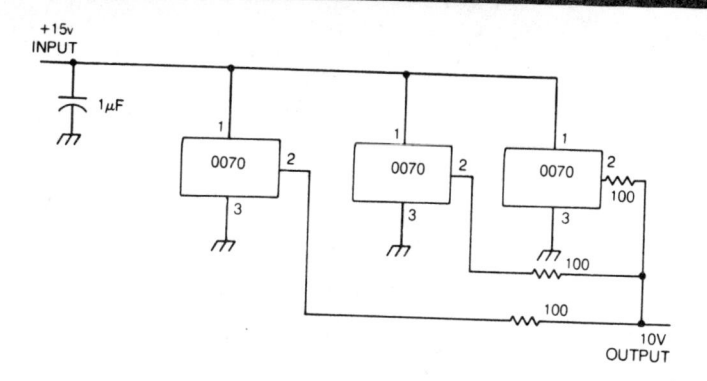

Statistical voltage standard

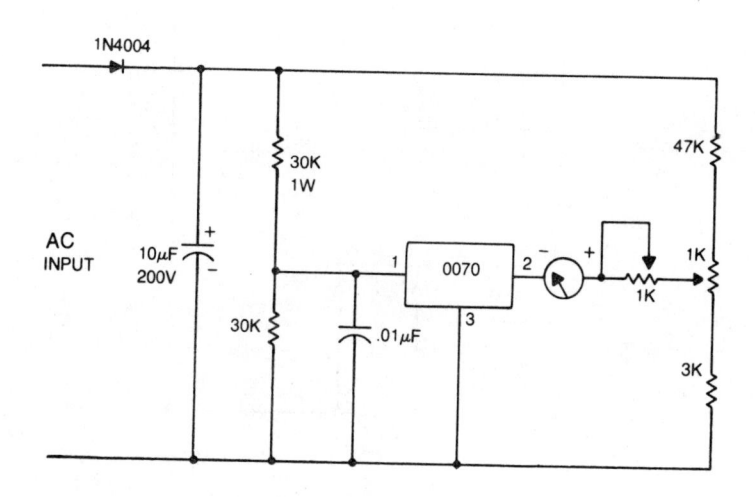

AC voltmeter

1468—DUAL 15 VOLT TRACKING REGULATOR

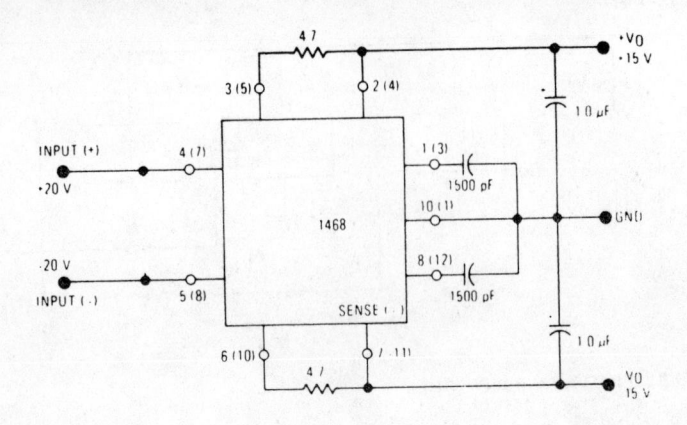

Basic 50-mA regulator

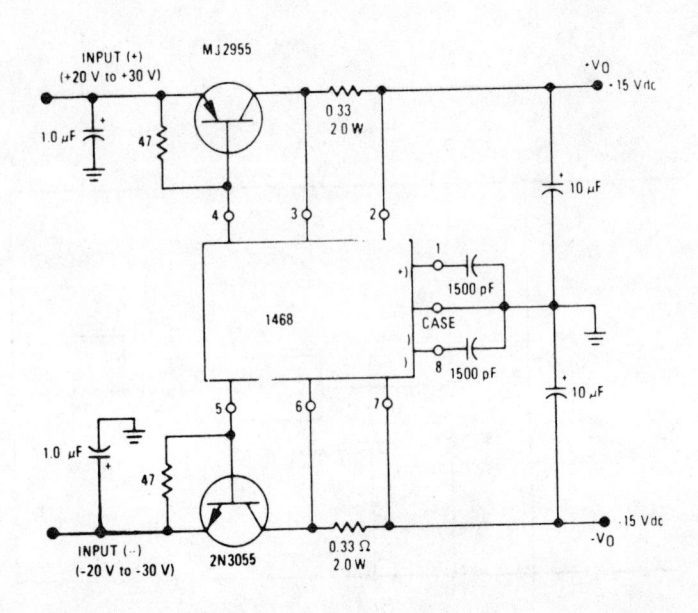

±1.5-ampere regulator

1469—FIVE VOLT HIGH CURRENT REGULATOR

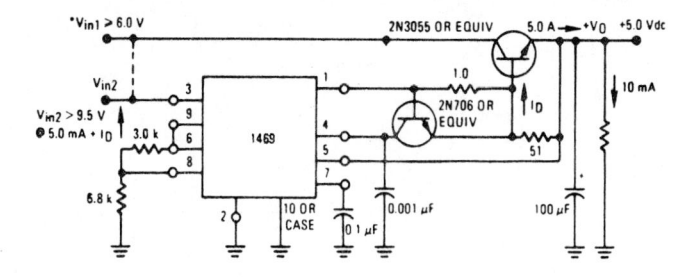

5 volt 5-ampere regulator

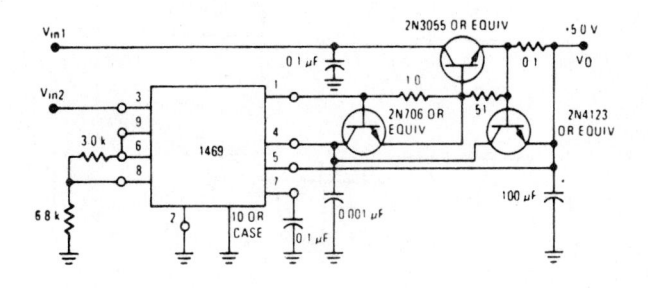

5-volt 5-ampere regulator

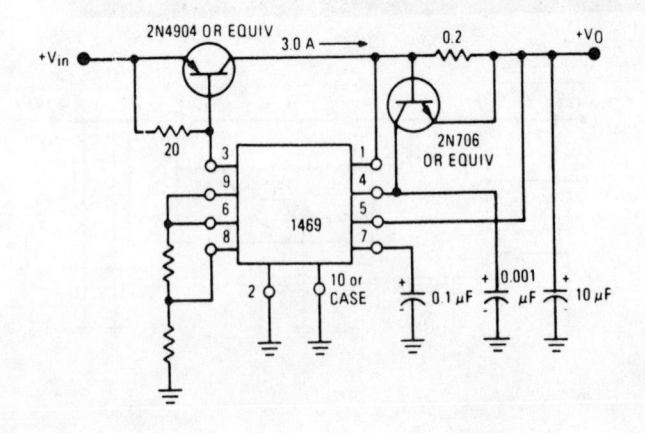

PNP current boost connection

1723— WIDE RANGE VOLTAGE REGULATOR

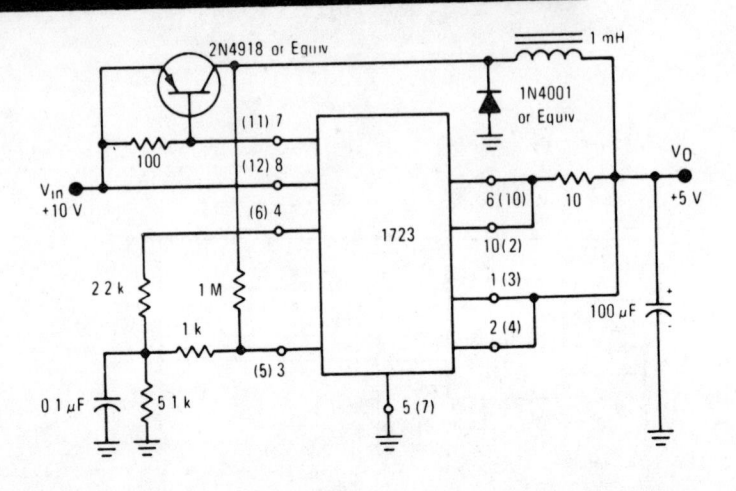

+5 V, 1-ampere switching regulator

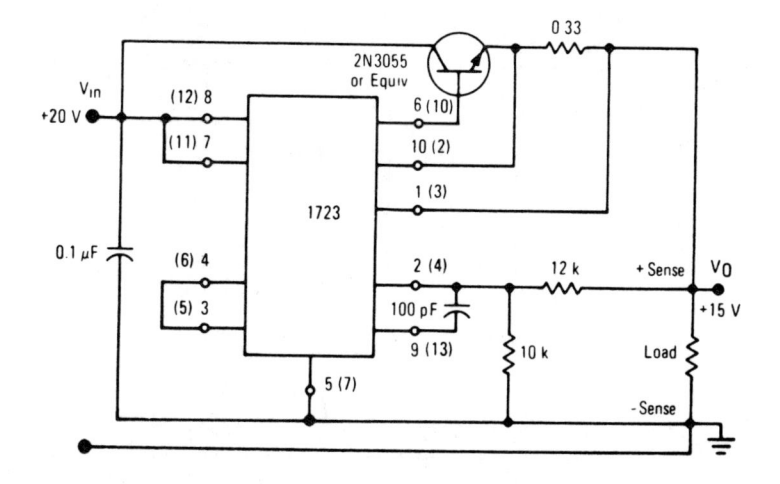

+15 V, 1-ampere regulator with remote sense

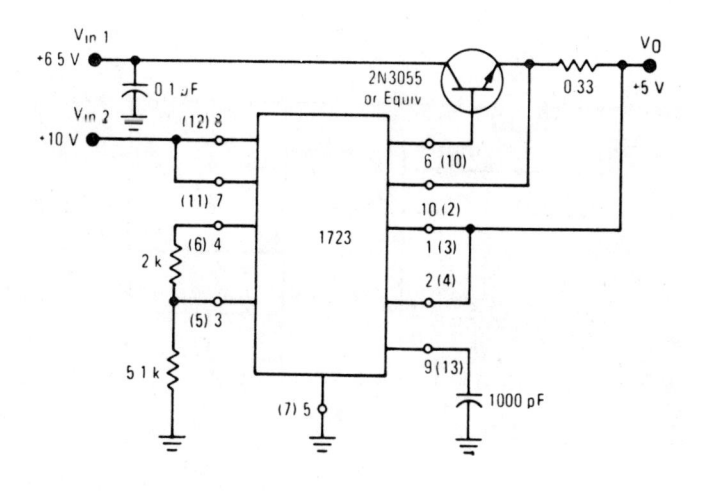

+5 V, 1-ampere high efficiency regulator

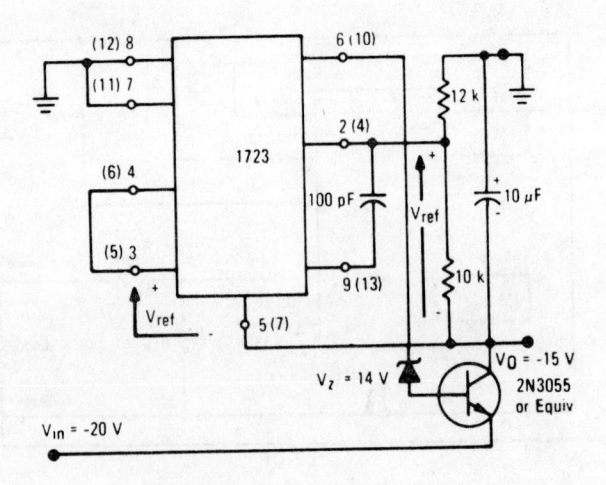

−15 V negative regulator

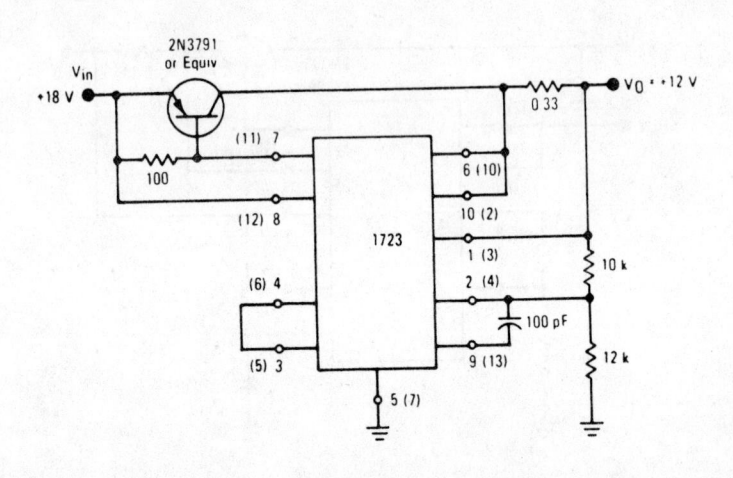

+12V, 1-ampere regulator using PNP current boost

4194 — DUAL TRACKING VOLTAGE REGULATOR

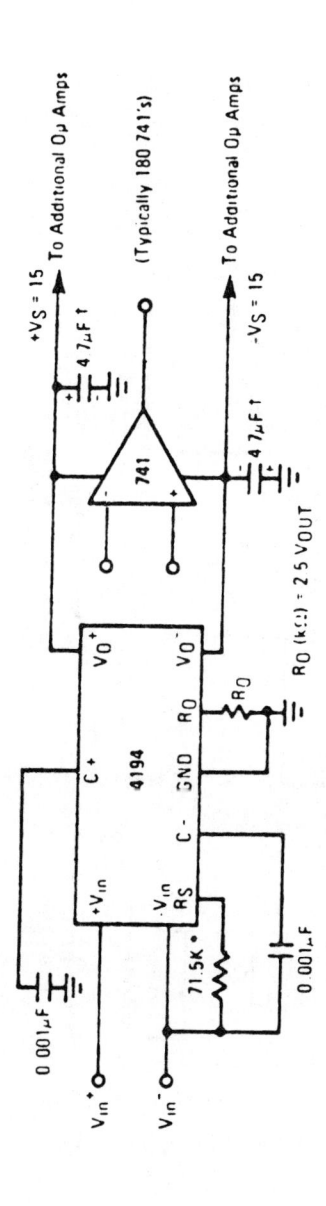

R_O (kΩ) = 2.5 V_{OUT}

(Typically 180 741's)

Balanced output voltage — op amp application

397

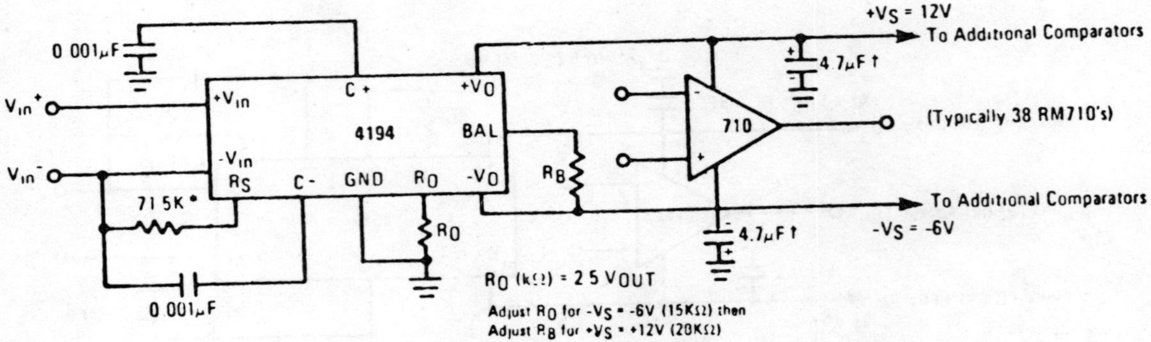

Unbalanced output voltage — comparator application

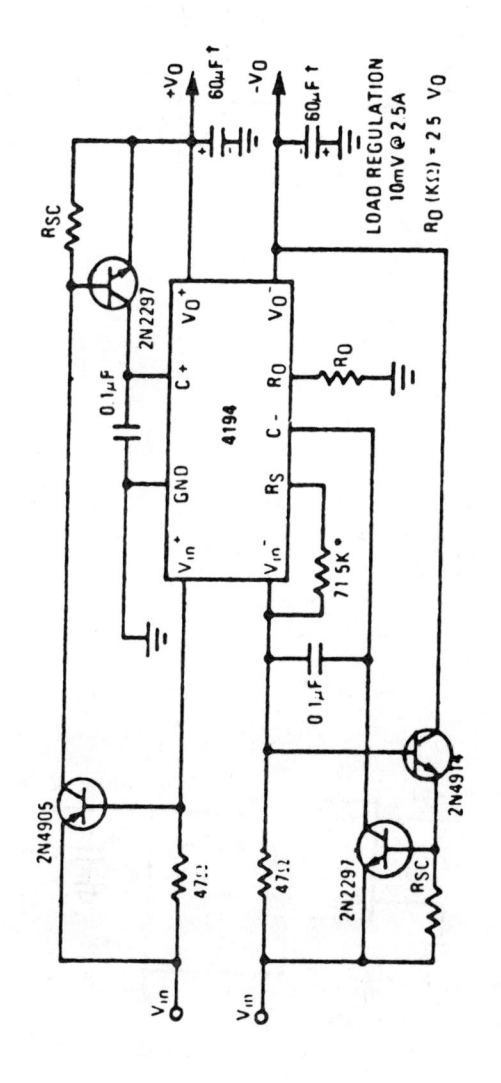

High output current application

4195—FIXED 15 VOLT DUAL TRACKING REGULATOR

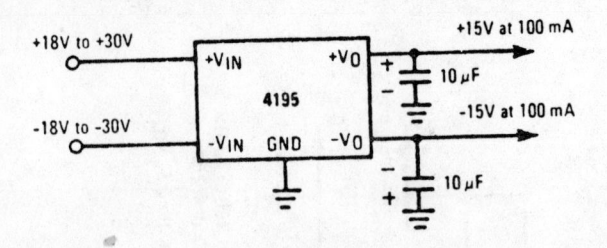

Balanced output ($V_o = \pm 15V$)

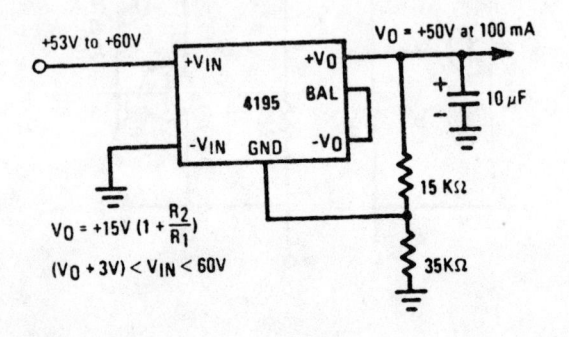

$V_O = +15V \left(1 + \frac{R_2}{R_1}\right)$

$(V_O + 3V) < V_{IN} < 60V$

Positive single supply ($+15V < V_o < +50V$)

400

High output current

7905 — NEGATIVE FIVE VOLT SERIES VOLTAGE REGULATOR

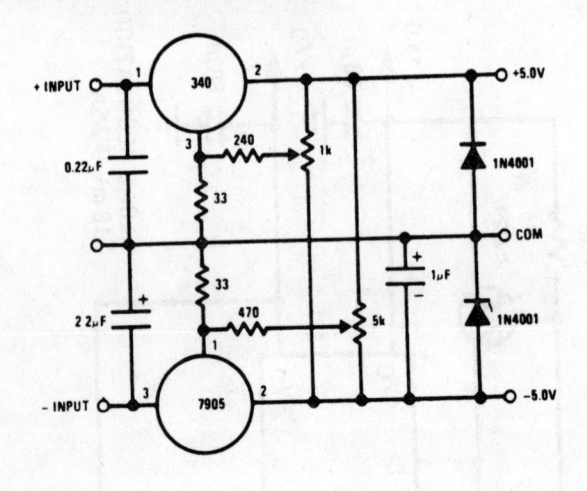

Dual trimmed supply

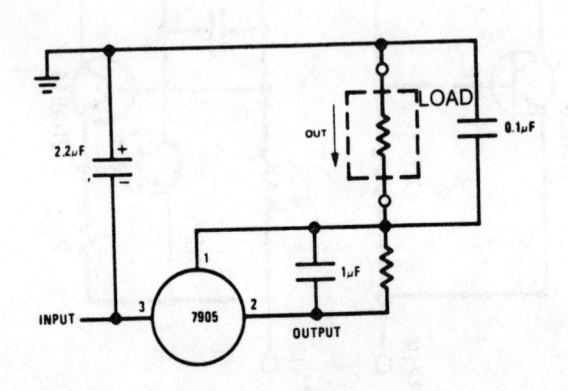

Current source

402

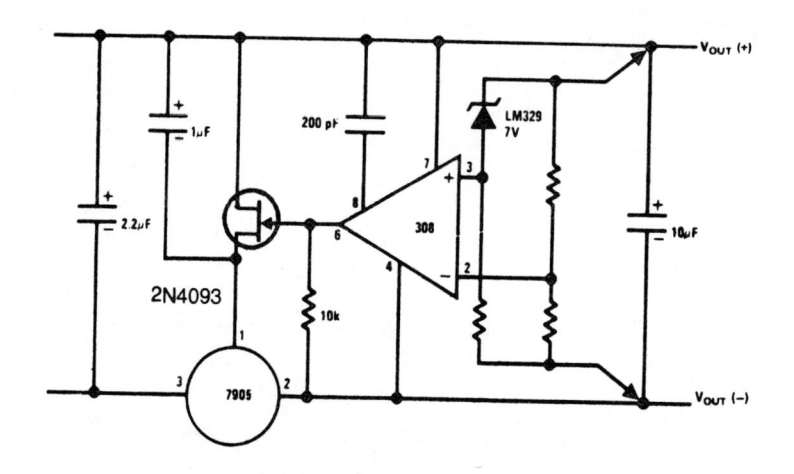

High stability 1 amp regulator

72309—FIVE VOLT INTEGRATED VOLTAGE REGULATOR

$V_O = 7\text{ V to }30\text{ V} \leq (V_I - 2\text{ V})$

Highly stable adjustable-output voltage regulator

78L05—FIVE VOLT INTEGRATED VOLTAGE REGULATOR

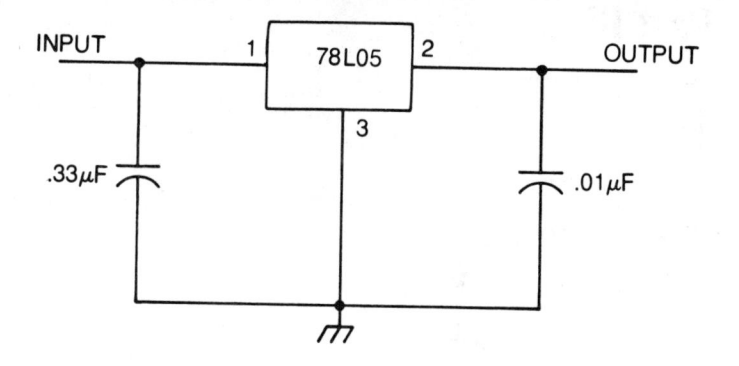

Fixed five volt output regulator

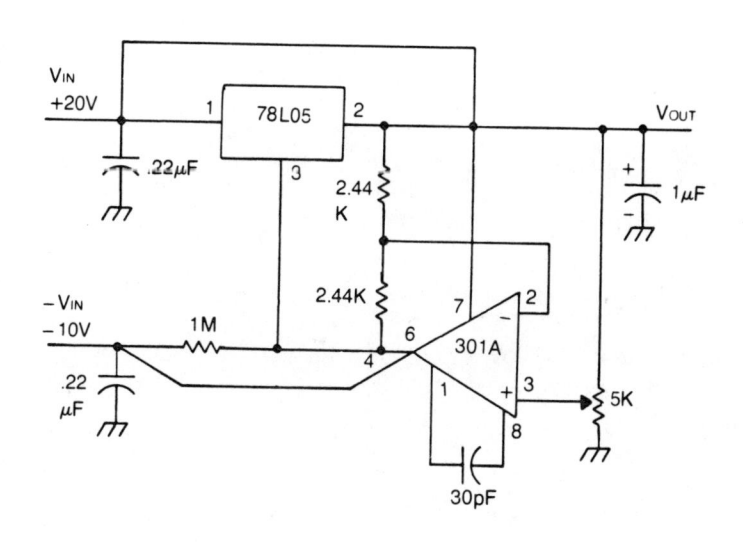

Variable output regulator

Part III

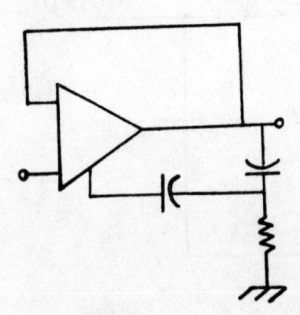

CMOS Integrated Circuits

CMOS digital ICs are the well-known "4000" series of devices. They contain more functions per chip than comparable TTL or LS ICs. Most can be used with a power supply ranging from positive 3 to 15 volts. Their only major drawback is their susceptibility to damage from static discharge.

4001 — CMOS QUAD NOR GATE

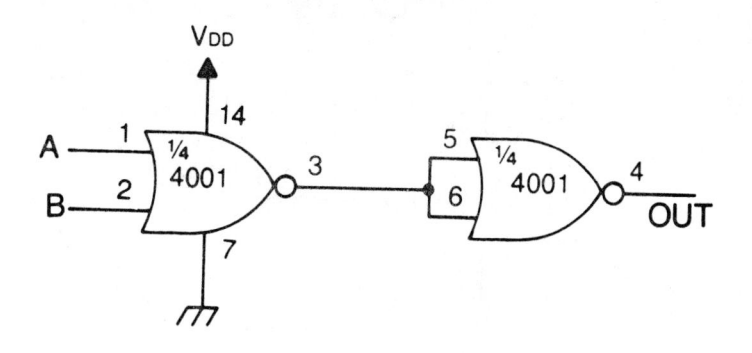

OR gate

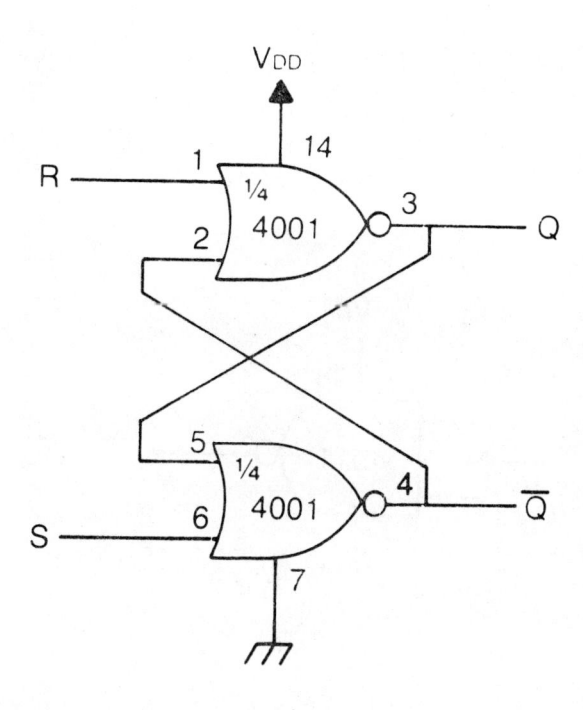

RS latch

4011 — CMOS QUAD NAND GATE

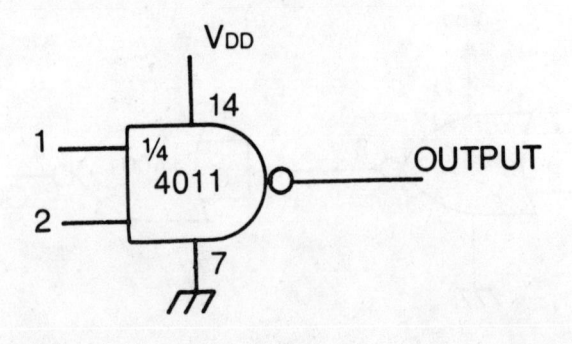

Basic NAND gate

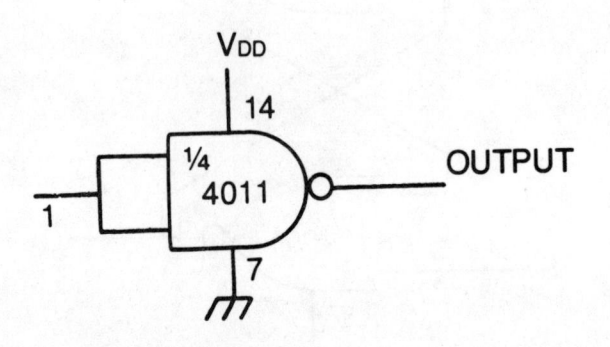

Inverter

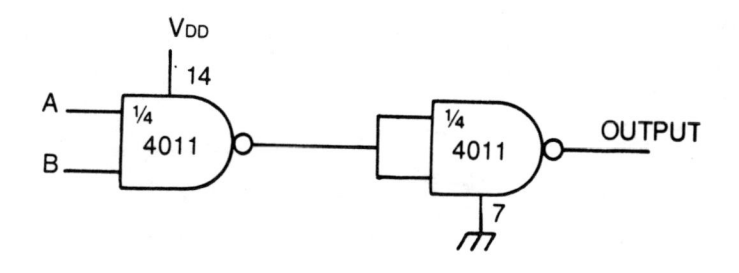

AND gate

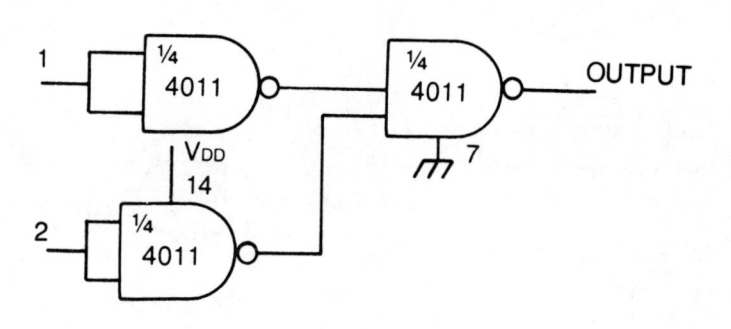

OR gate

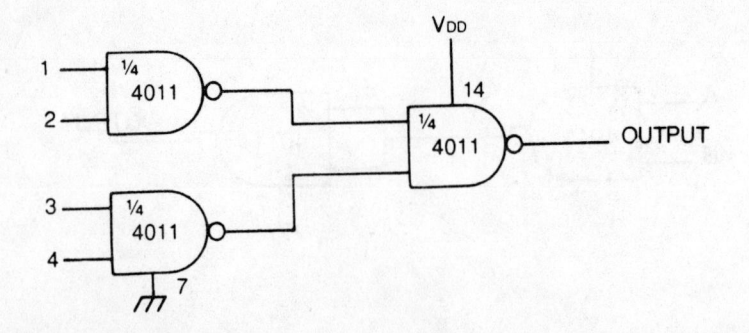

AND/OR gate

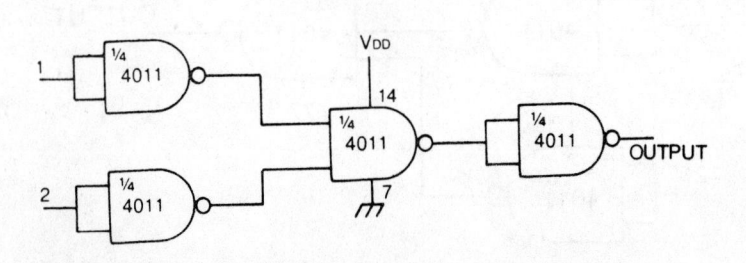

NOR gate

410

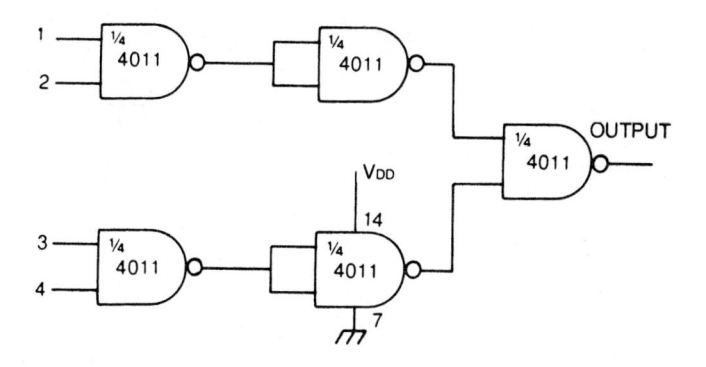

Quad input NAND gate

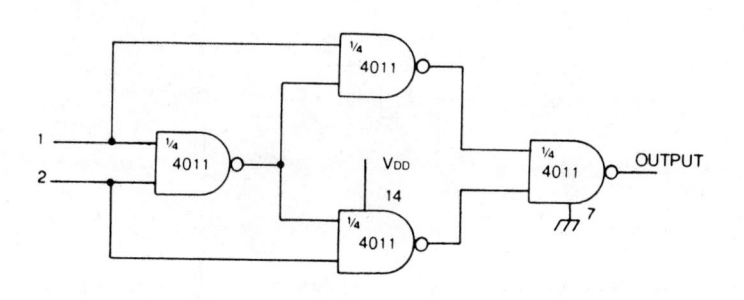

Exclusive OR gate

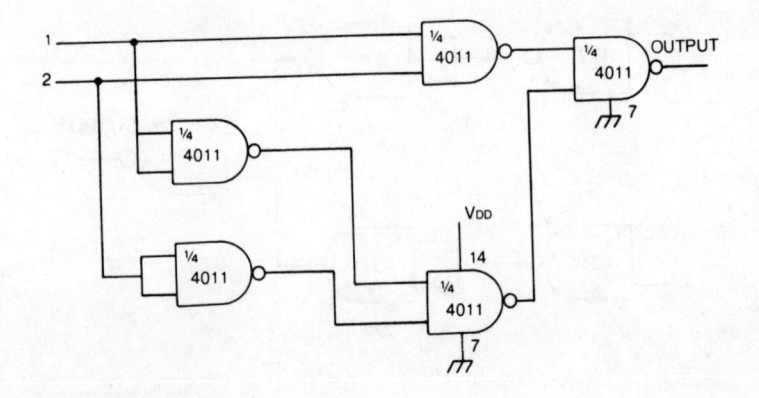

Exclusive NOR gate

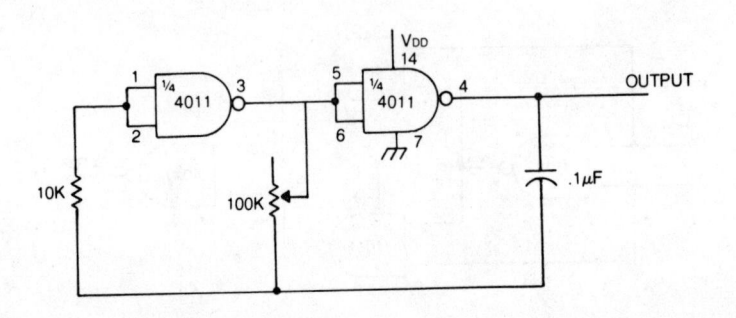

Clock pulse generator

412

4012 — CMOS DUAL FOUR INPUT NAND GATE

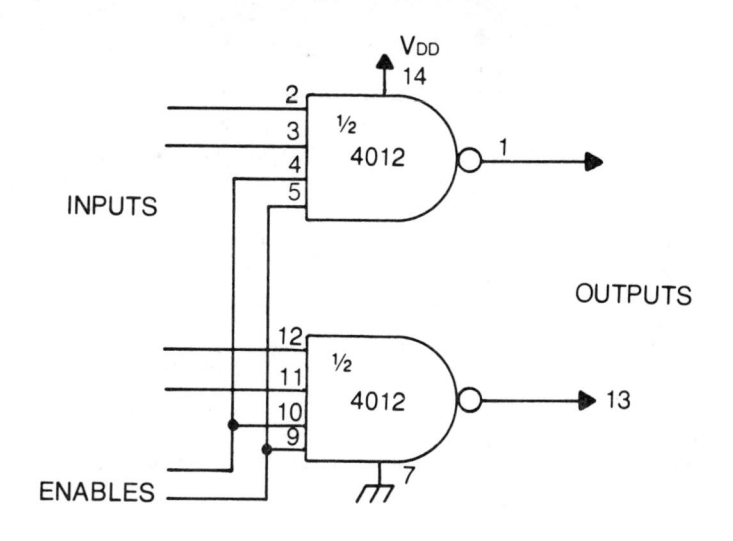

Enable input

4017 — CMOS DECADE COUNTER/DECODER

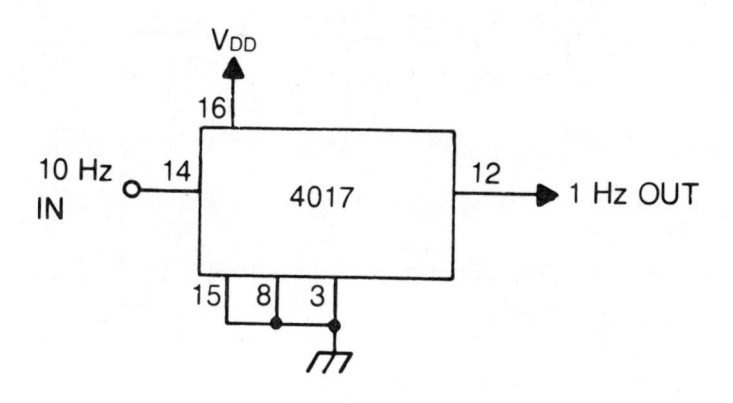

1-Hz timebase

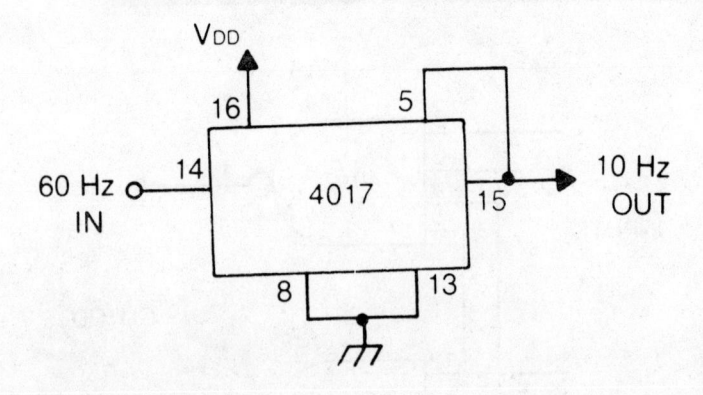

10-Hz timebase

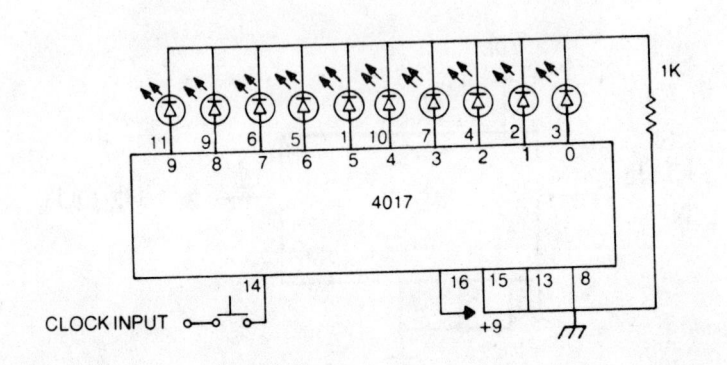

Random number generator

4023 — CMOS TRIPLE THREE INPUT NAND GATE

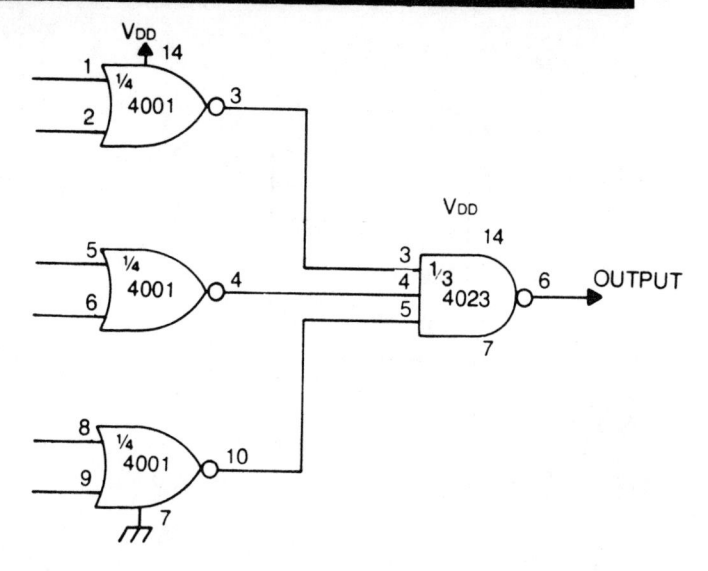

6-input OR gate

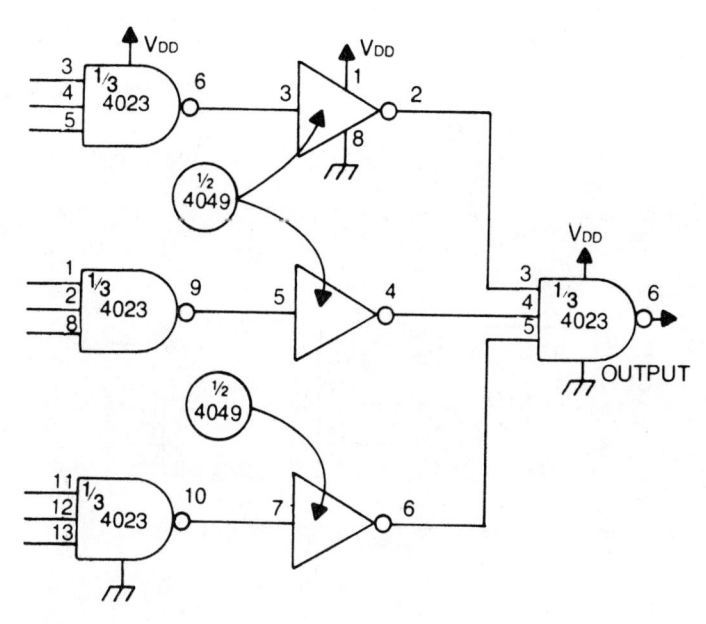

9-input NAND gate

'4027 — CMOS DUAL J-K FLIP-FLOP

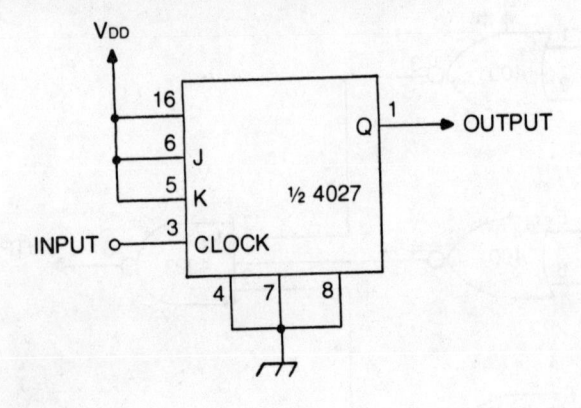

Divide-by-2 counter

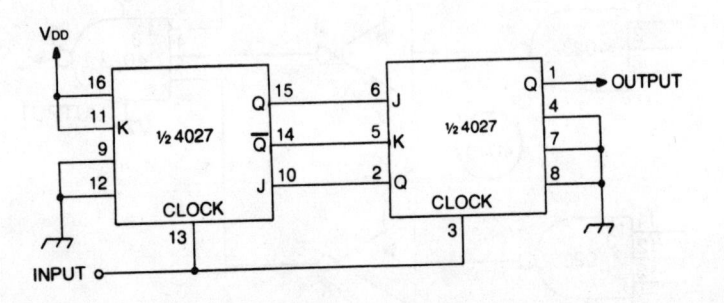

Divide-by-3 counter

Divide-by-4 counter

Divide-by-5 counter

4028 — CMOS BCD TO DECIMAL DECODER

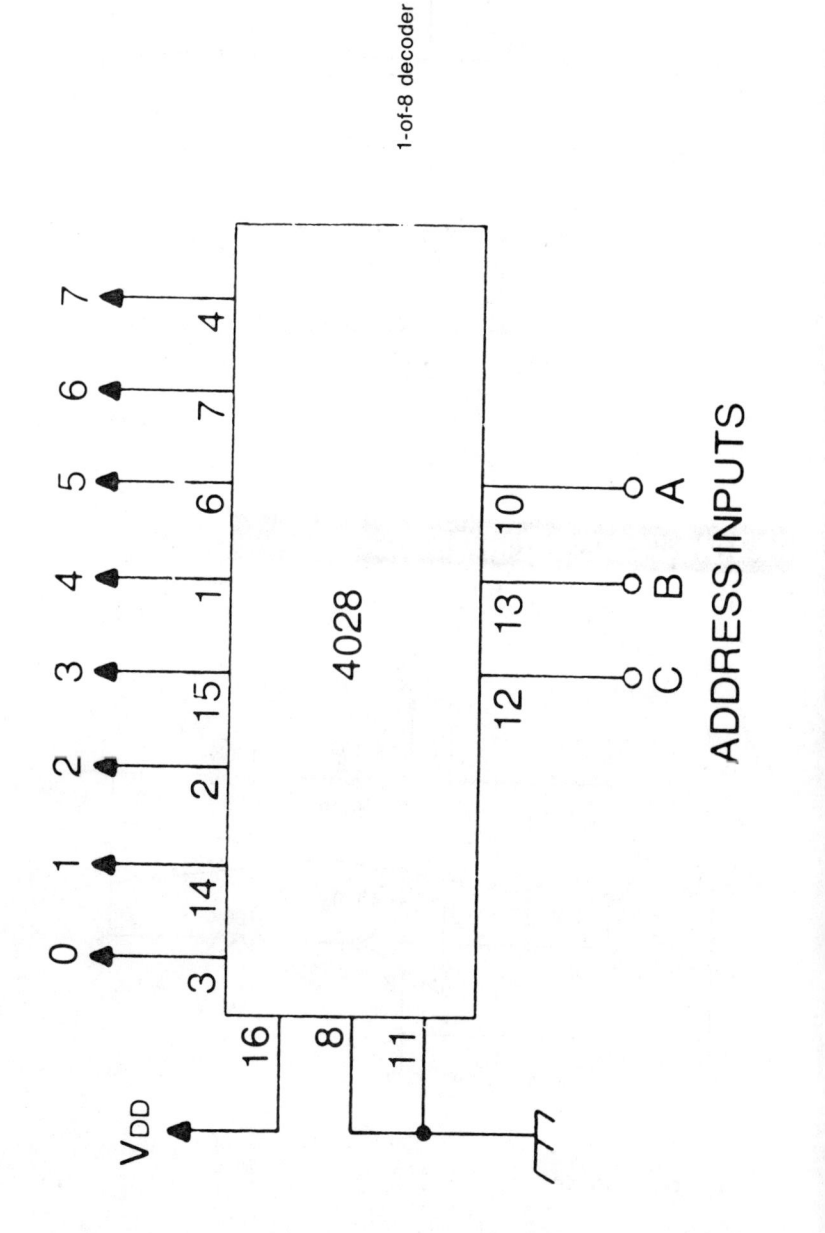

1-of-8 decoder

4046—CMOS PHASE LOCK LOOP INTEGRATED CIRCUIT

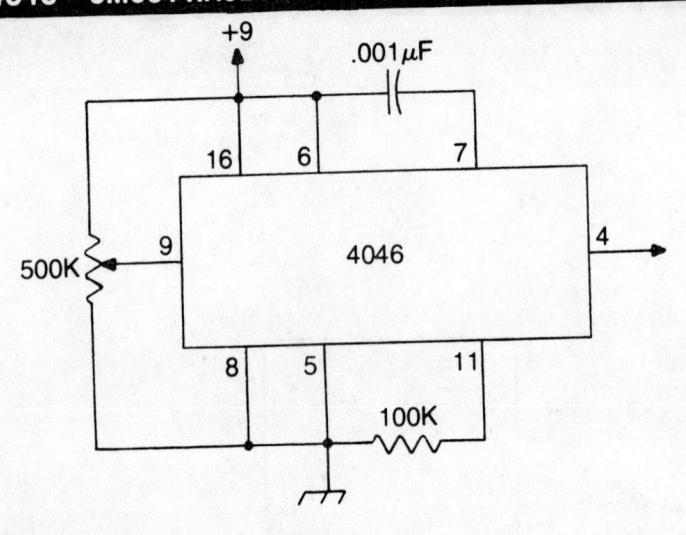

Tunable oscillator

4049—CMOS HEX INVERTING BUFFER

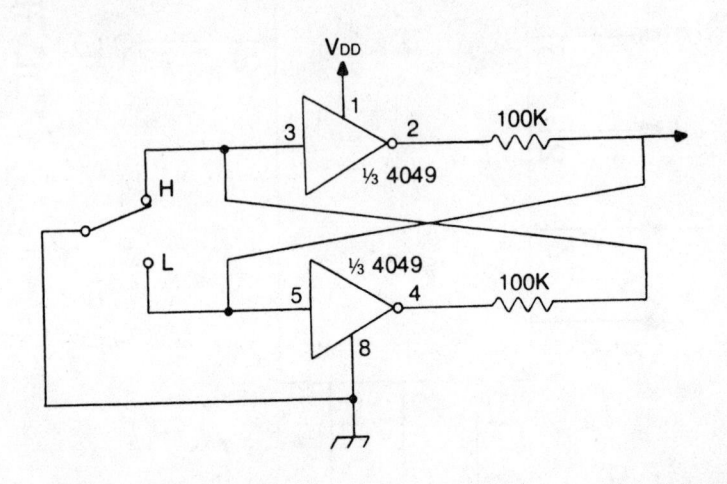

Bounceless switch

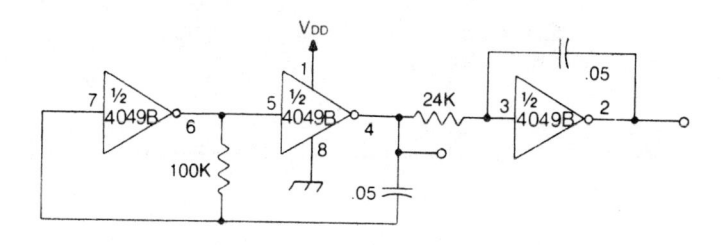

Triangle wave source

4051 — CMOS ANALOG MULTIPLEXER

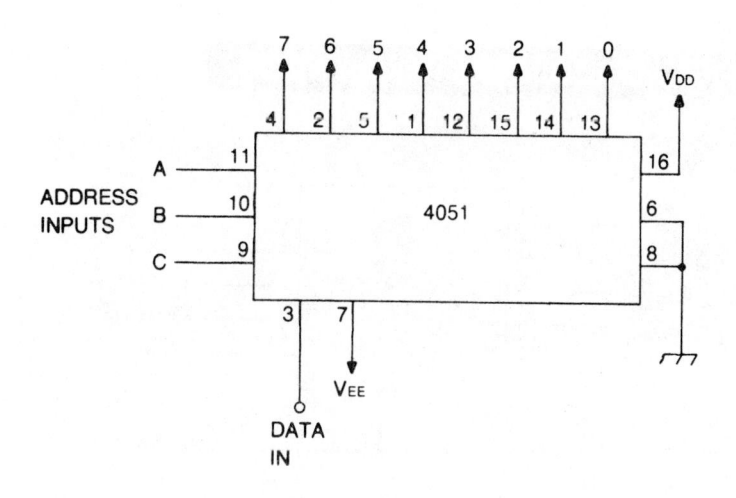

1-of-8 multiplexer

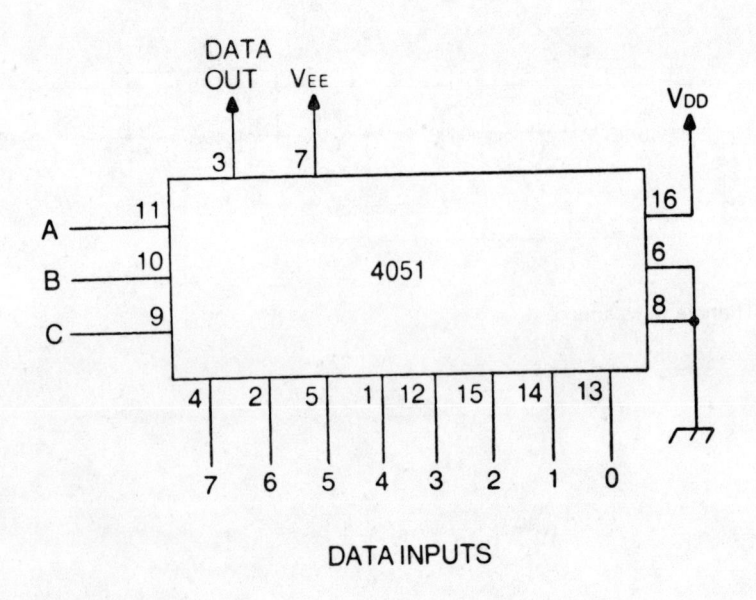

DATA INPUTS

1-of-8 demultiplexer

4066 — CMOS QUAD BILATERAL SWITCH

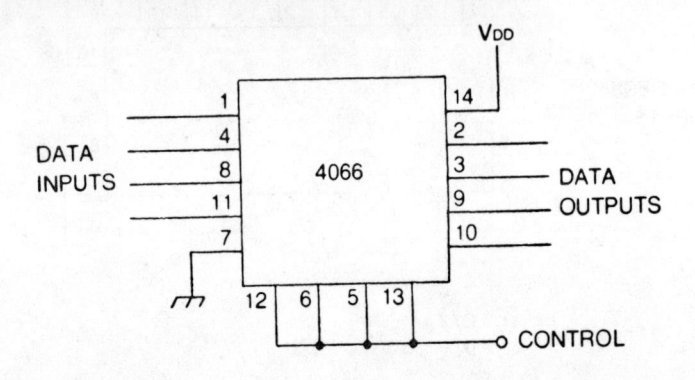

Data bus control

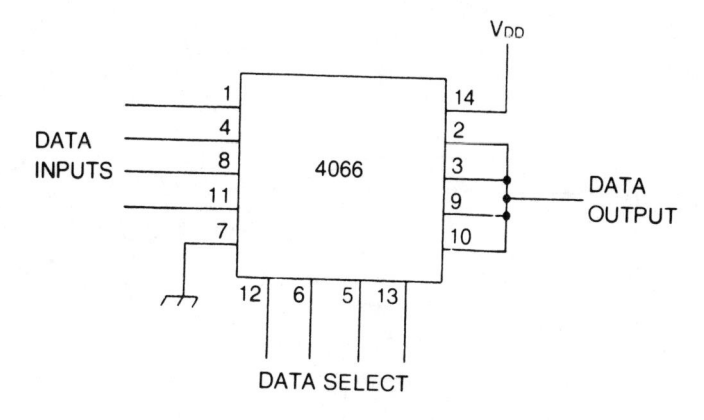

Data selector

4070—CMOS QUAD EXCLUSIVE OR GATE

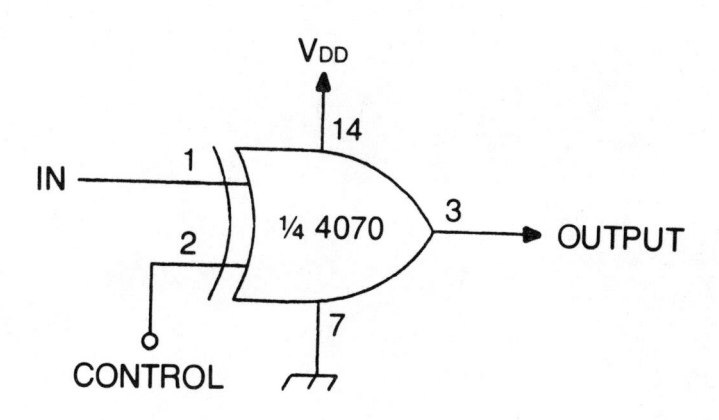

Controlled inverter

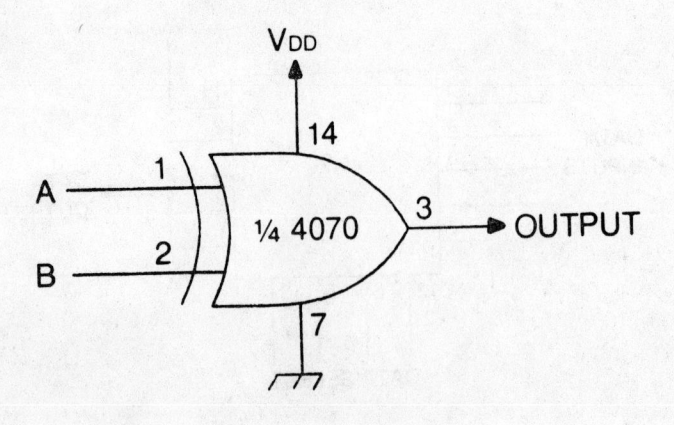

1-bit comparator

Part IV

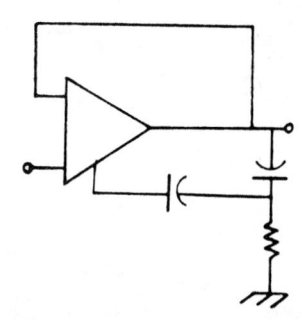

TTL/LS Integrated Circuits

TTL/LS digital ICs are the "7000" series of ICs. They are the most commonly used logic circuits and require a five volt power source. The LS designation stands for *low power Schottky*, which uses approximately 80% less power than a standard TTL chip.

7400—TTL QUAD NAND GATE

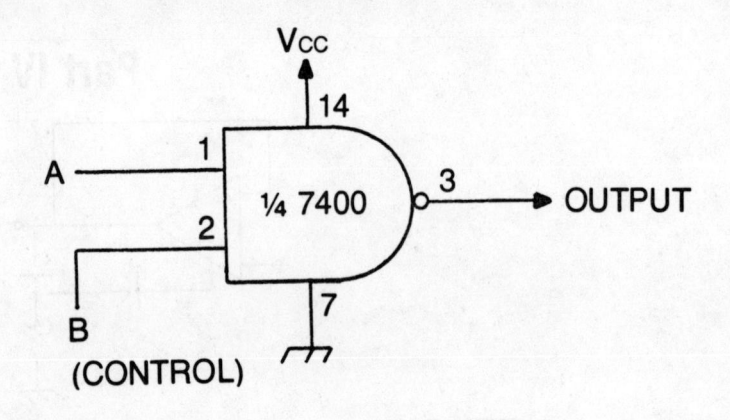

Control gate

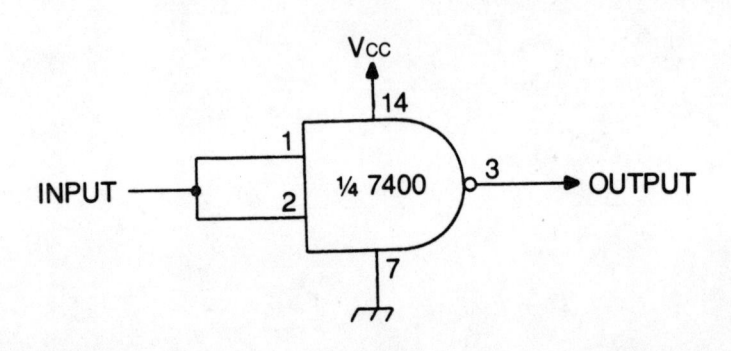

Inverter

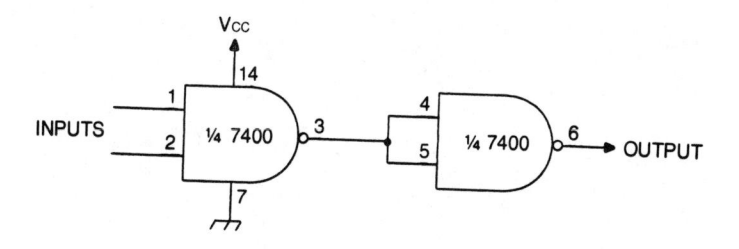

AND gate

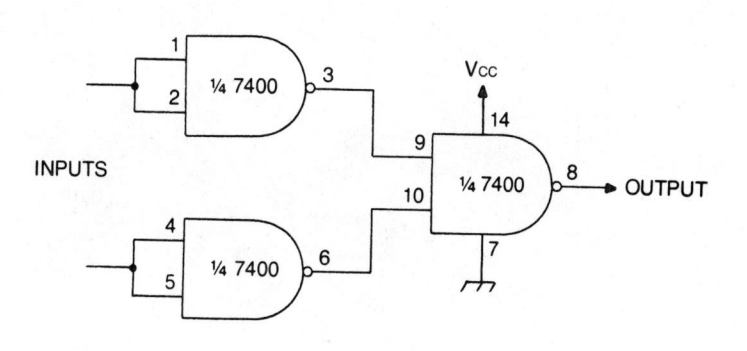

OR gate

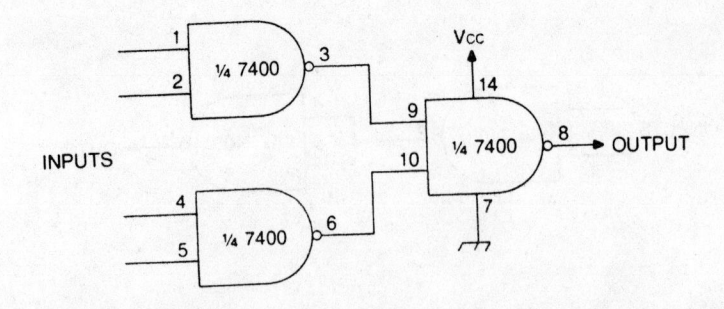

AND-OR gate

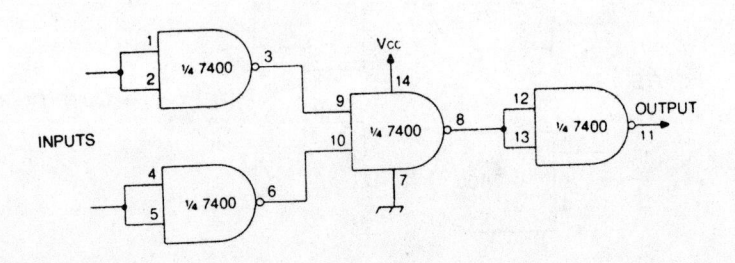

NOR gate

428

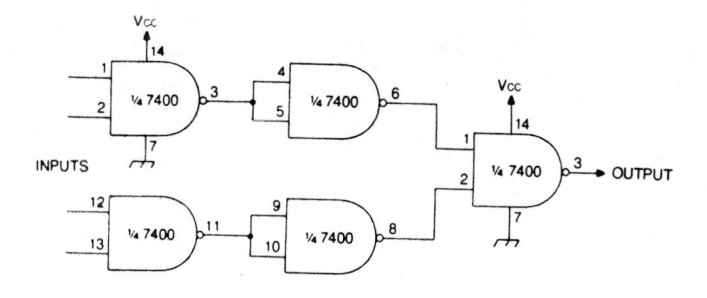

4-input NAND gate

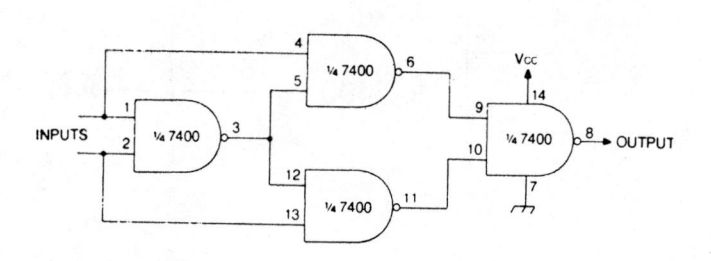

Exclusive-OR gate

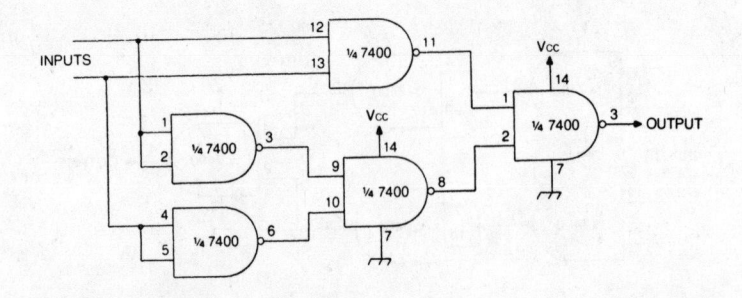

Exclusive-NOR gate

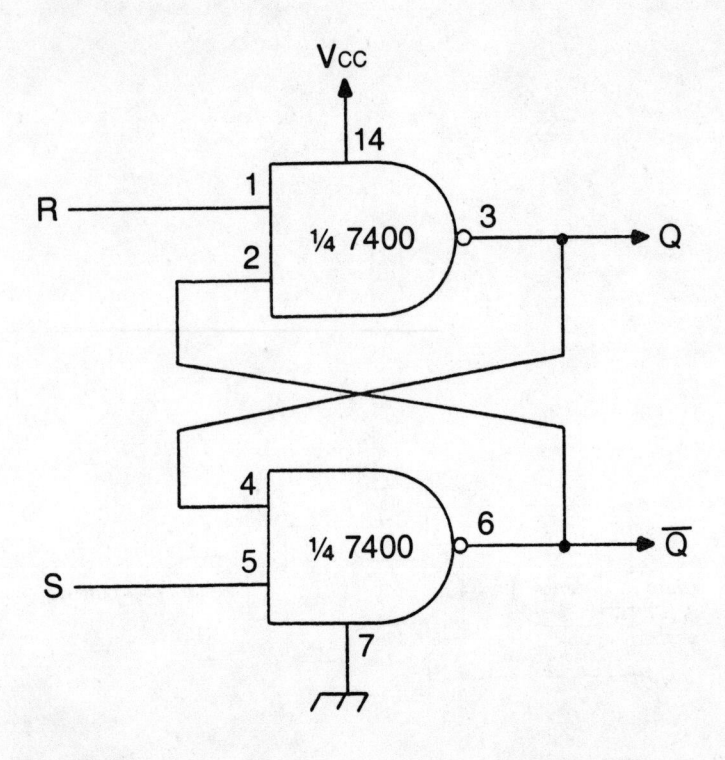

RS latch

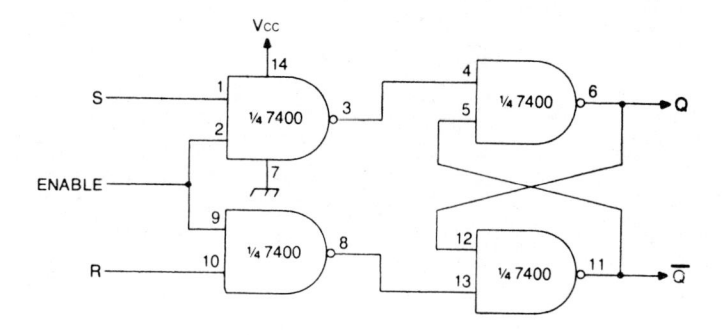

Gated RS latch

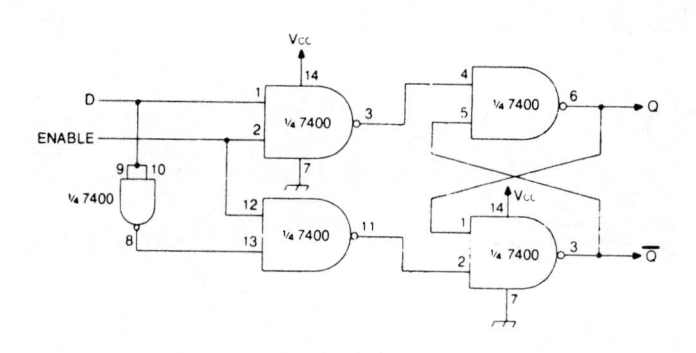

D flip-flop

431

7402—TTL QUAD NOR GATE

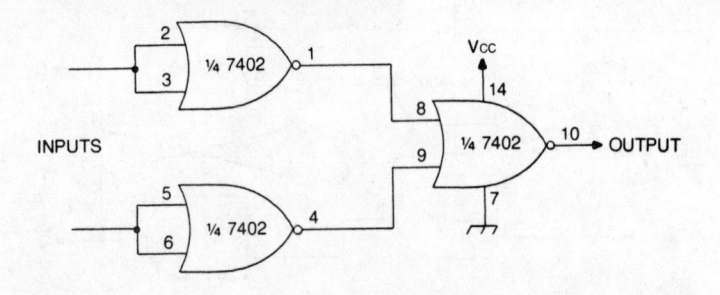

AND gate

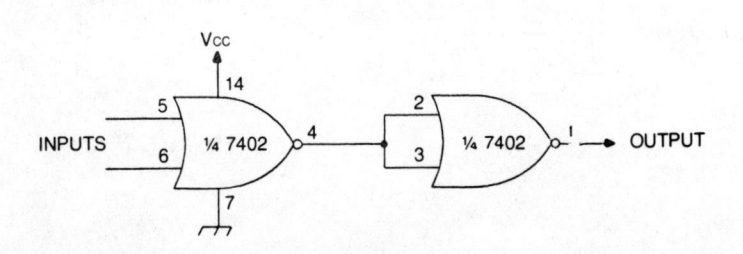

OR gate

432

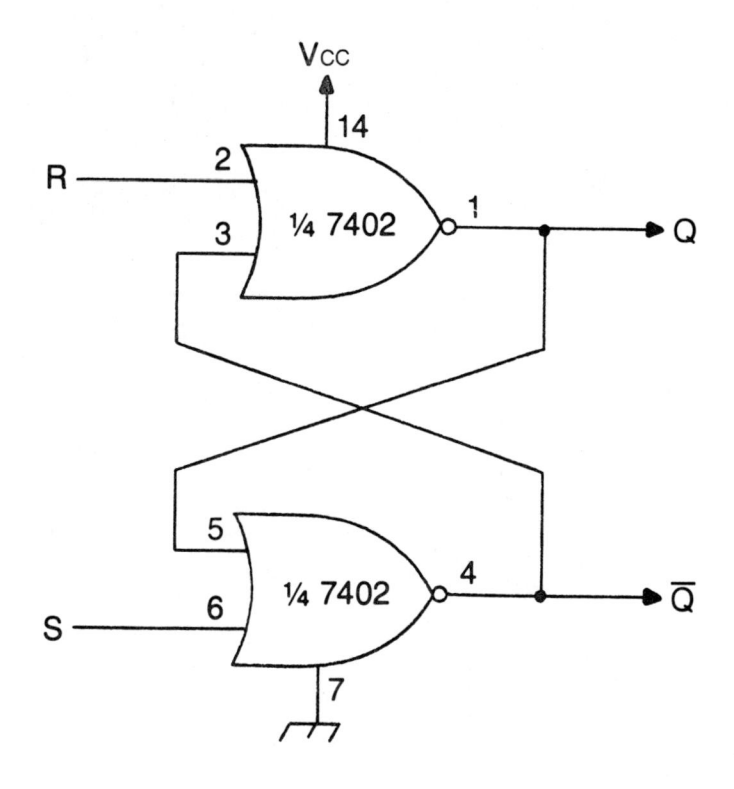

RS latch

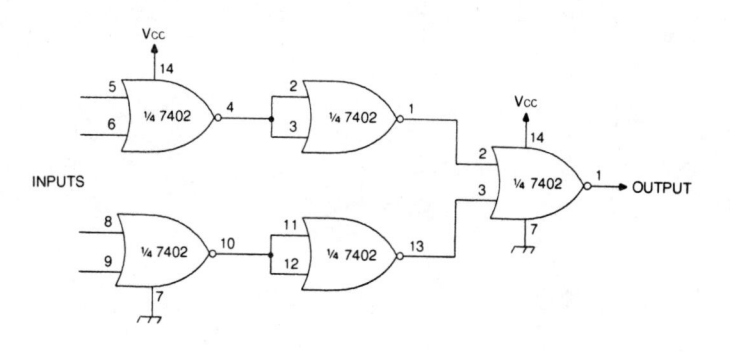

4-input NOR gate

433

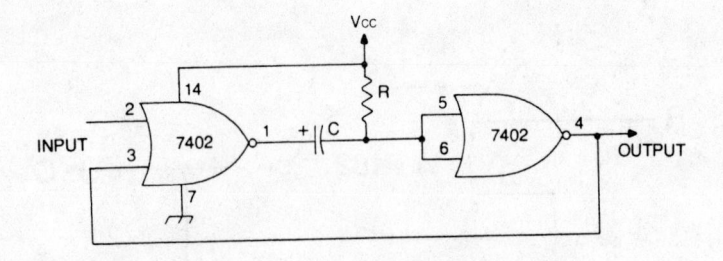

One-shot

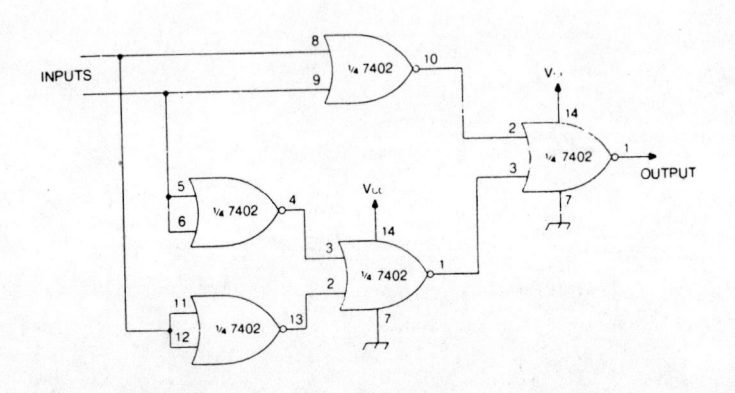

Exclusive-OR gate

434

7408 — TTL QUAD AND GATE

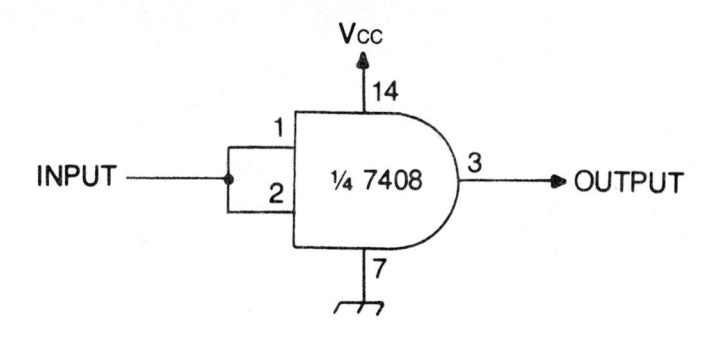

AND gate buffer

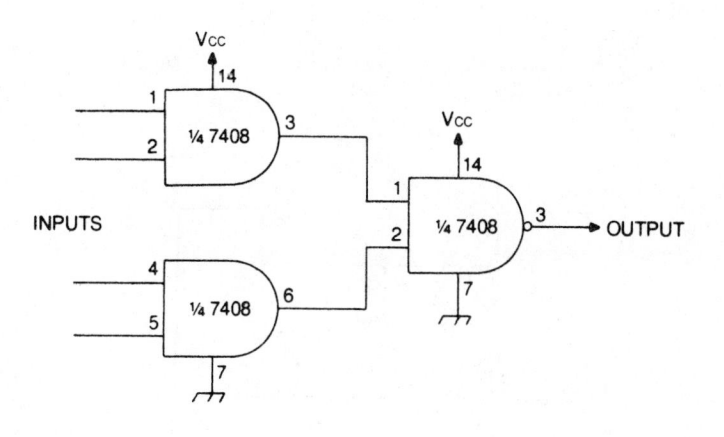

4-input NAND gate

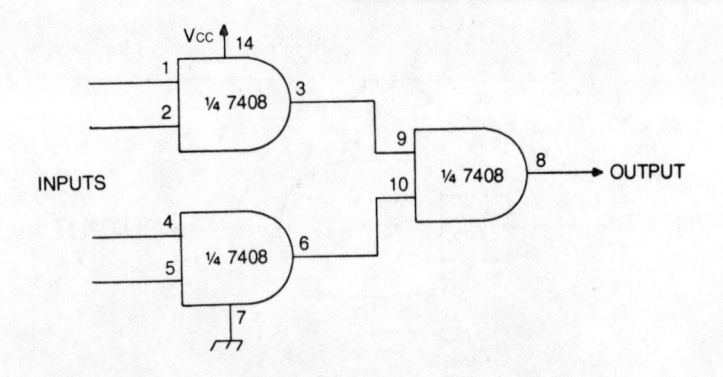

4-input AND gate

7474—TTL DUAL D FLIP FLOP

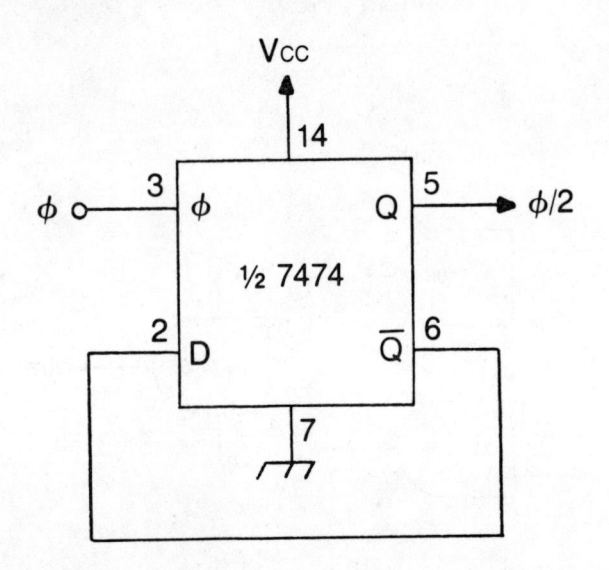

Divide-by-two counter

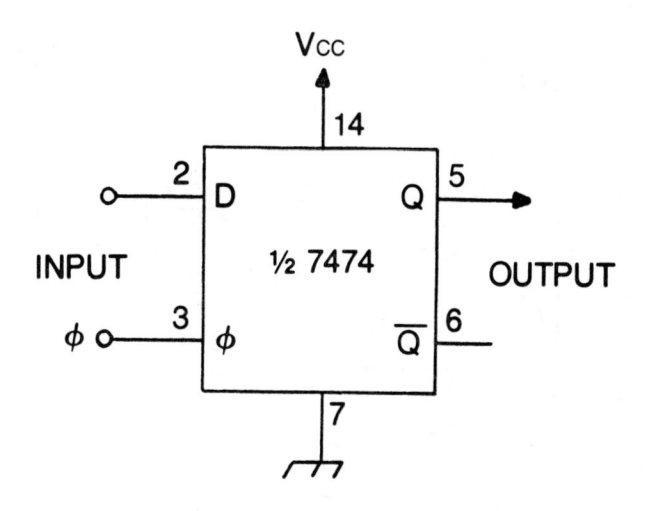

Wave shaper

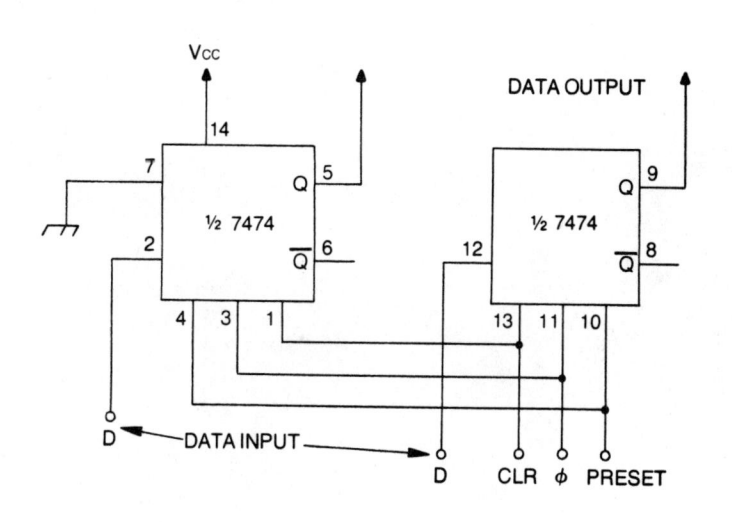

2-bit storage register

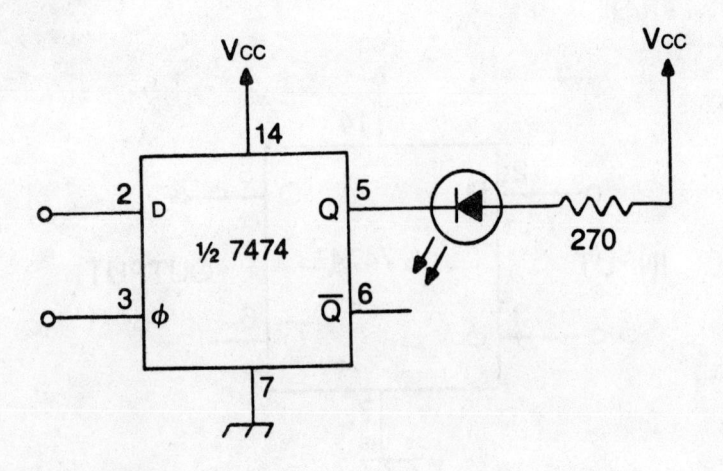

Phase detector

7476—TTL DUAL J-K FLIP FLOP

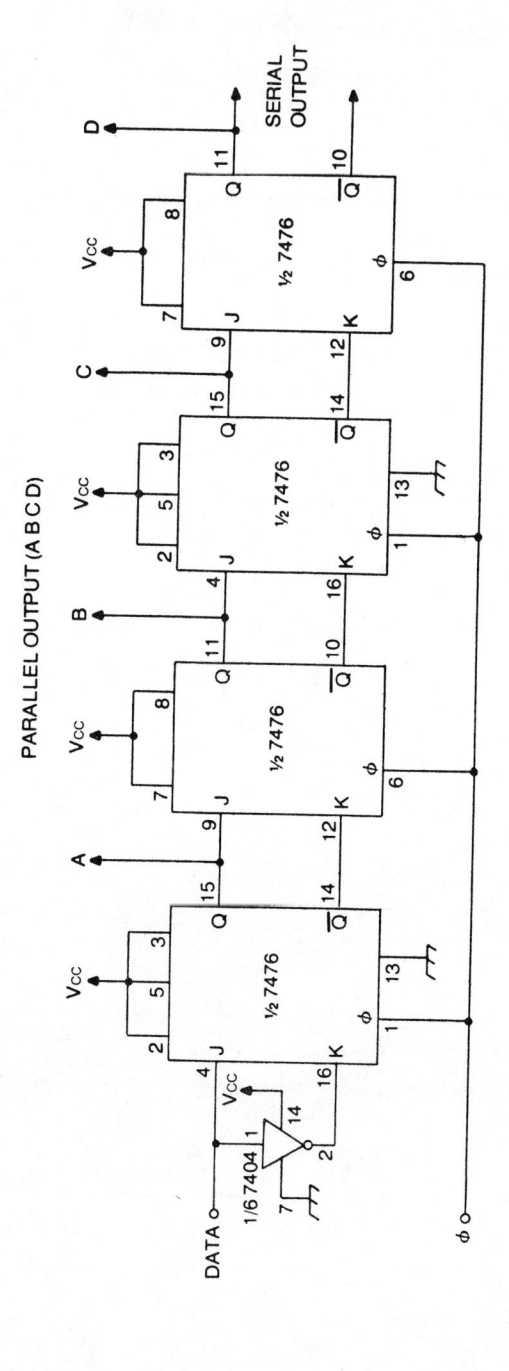

4-bit serial shift register

7490—TTL BCD DECADE COUNTER

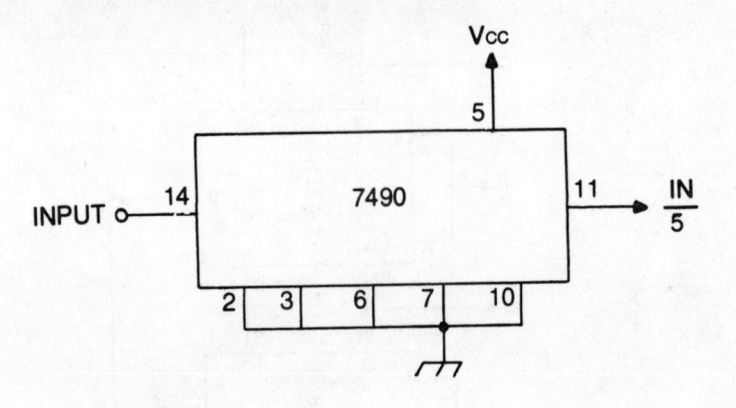

Divide-by-5 counter

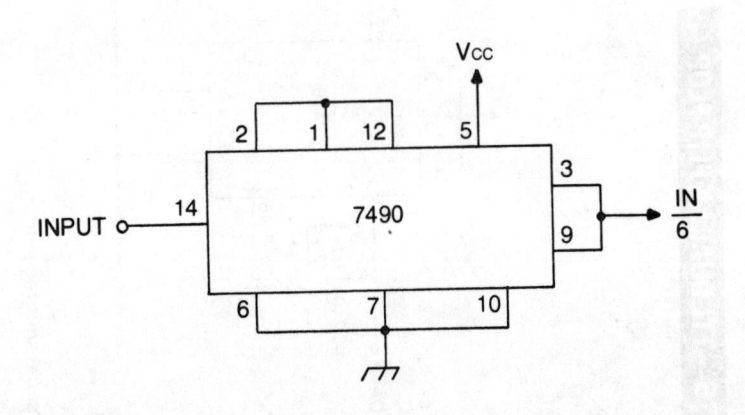

Divide-by-6 counter

440

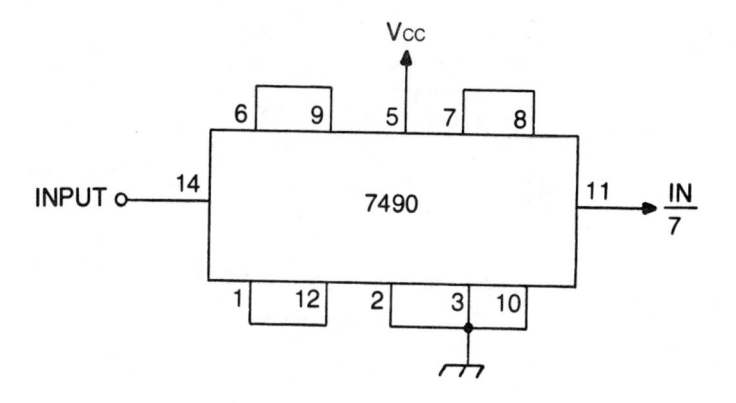

Divide-by-7 counter

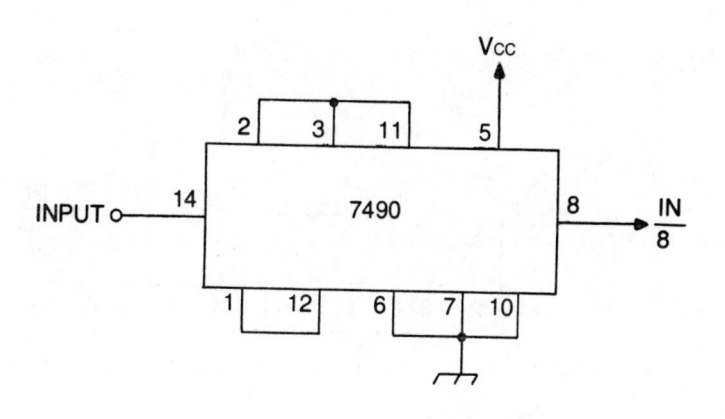

Divide-by-8 counter

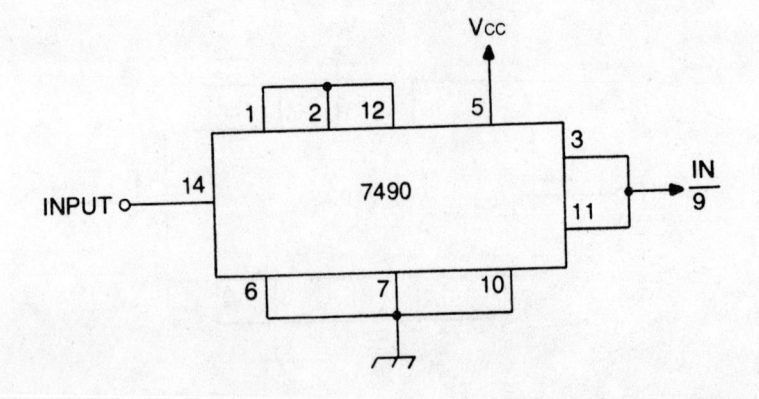

Divide-by-9 counter

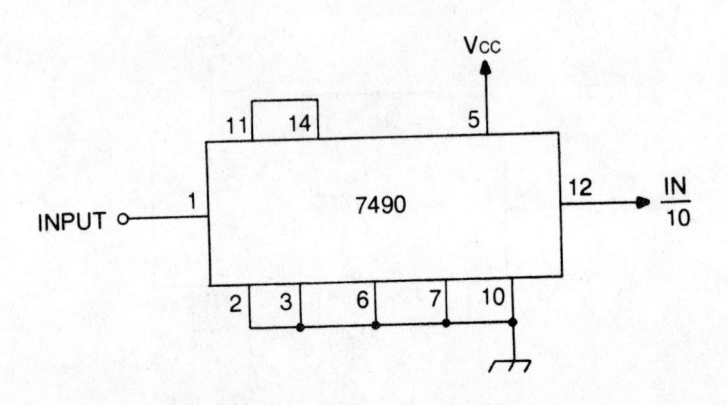

Divide-by-10 counter

7492—TTL DIVIDE BY 12 BINARY COUNTER

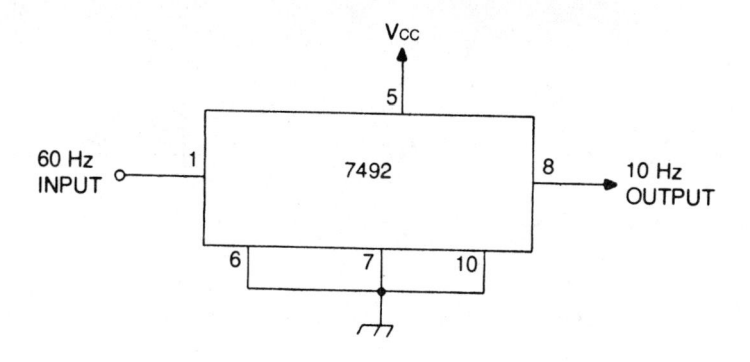

10-Hz pulse source

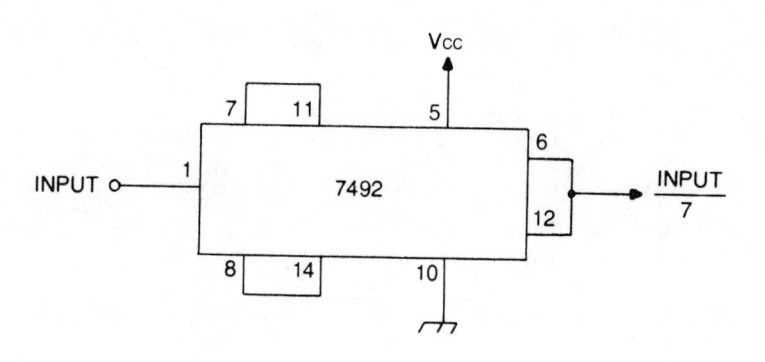

Divide-by-7 counter

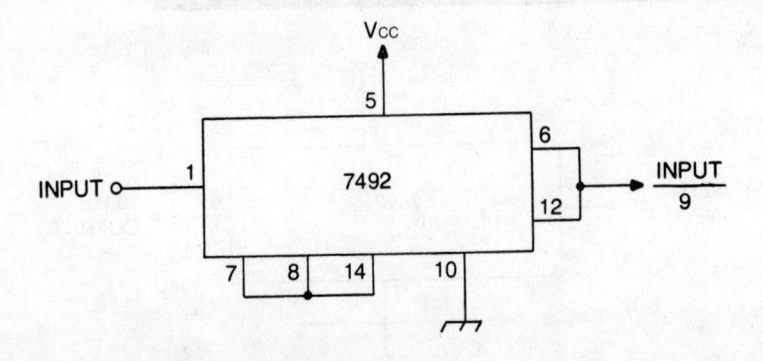

Divide-by-9 counter

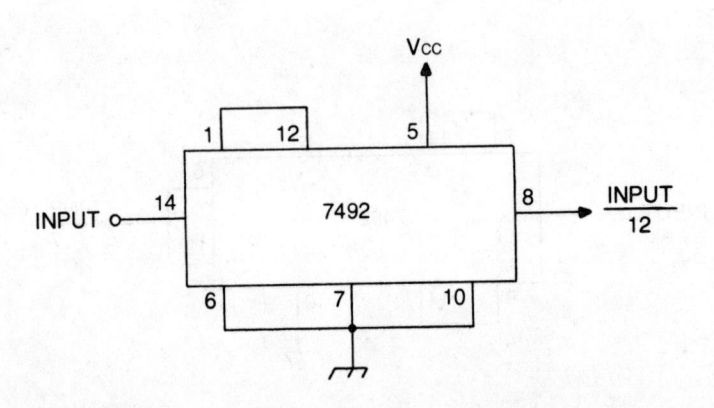

Divide-by-12 counter

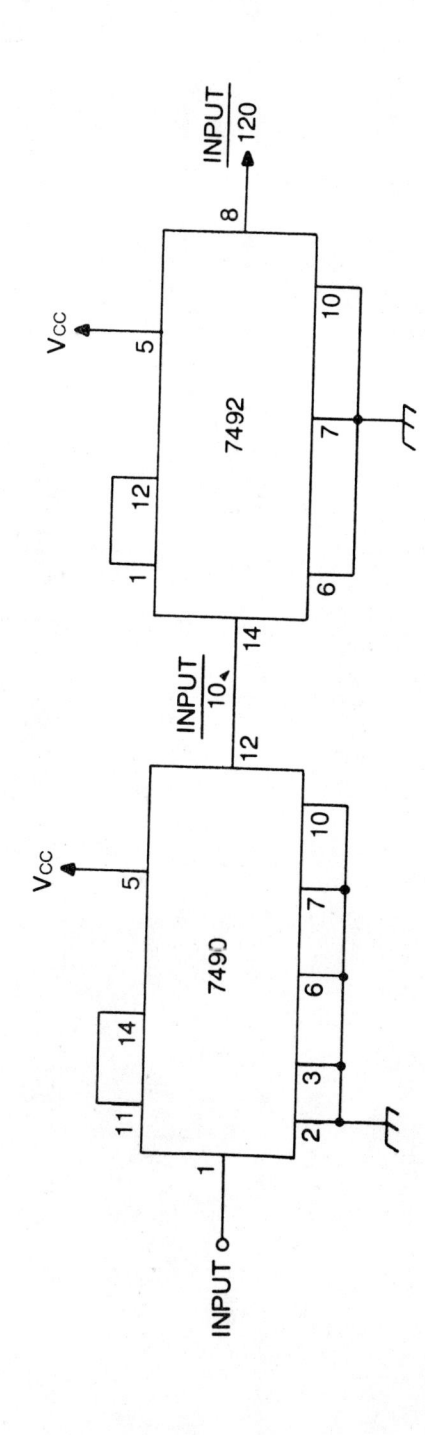

Divide-by-120 counter

74154 — TTL FOUR TO SIXTEEN LINE DECODER

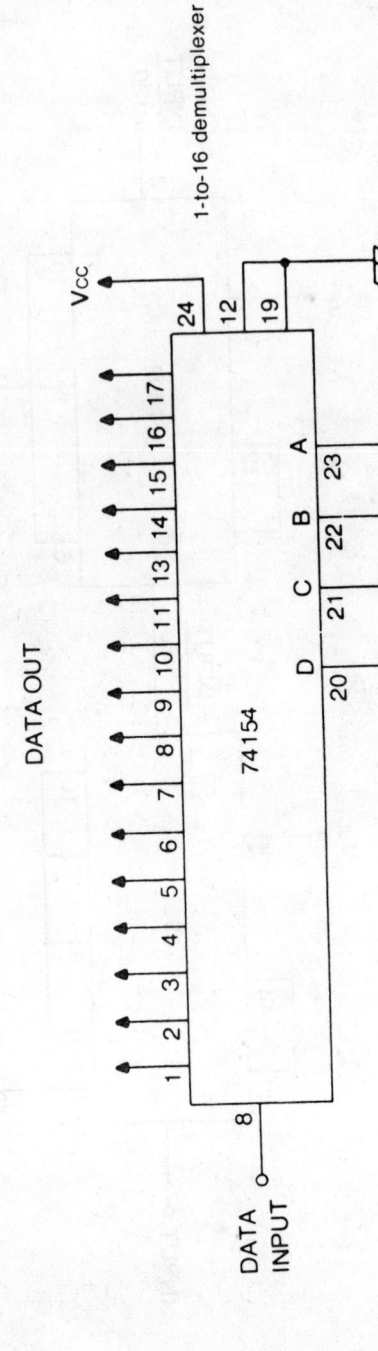

1-to-16 demultiplexer

74192—TTL BCD UP/DOWN COUNTER

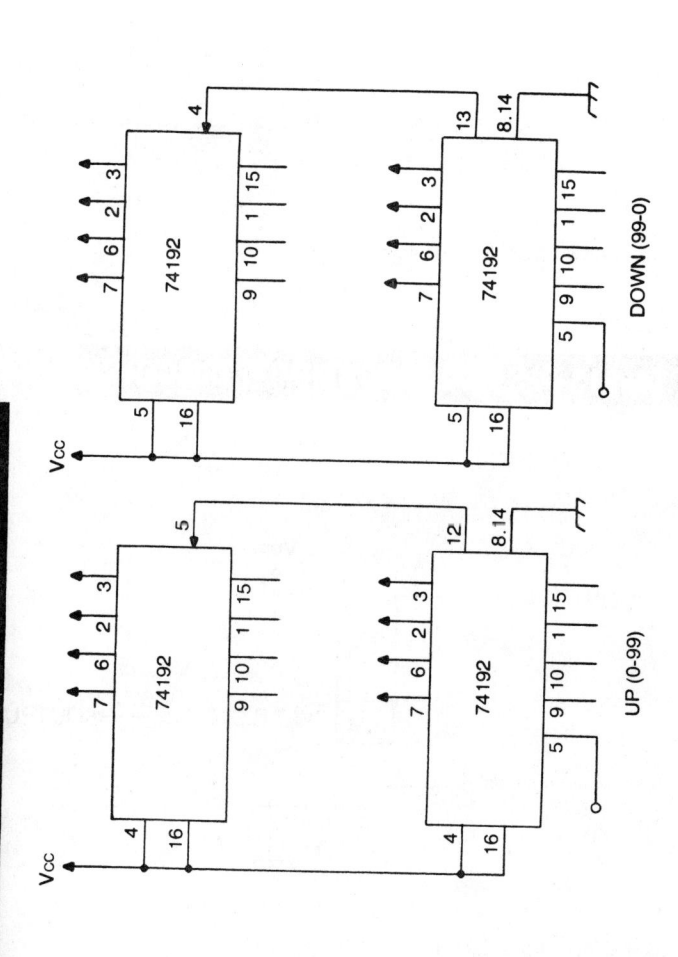

Cascaded counters

74193—TTL FOUR BIT UP/DOWN COUNTER

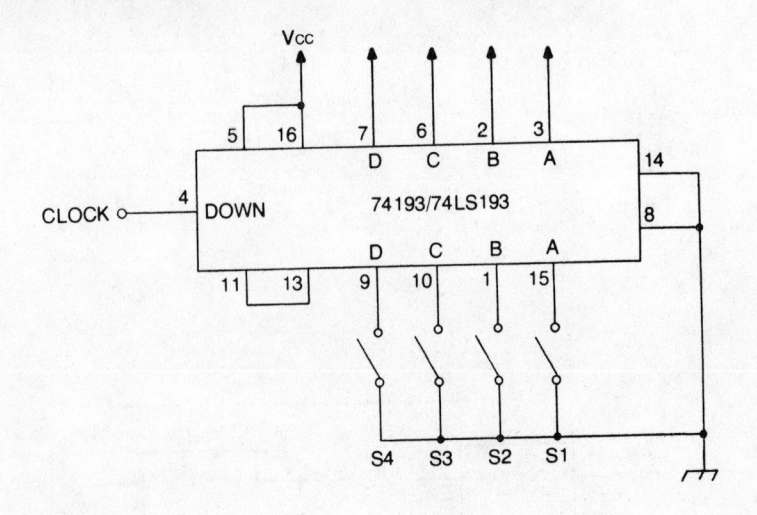

Count down from N and recycle

74LS13—LS DUAL NAND SCHMITT TRIGGER

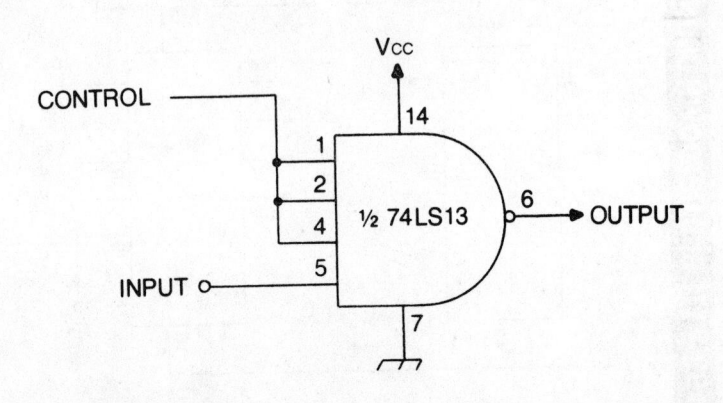

Gated threshold detector

448

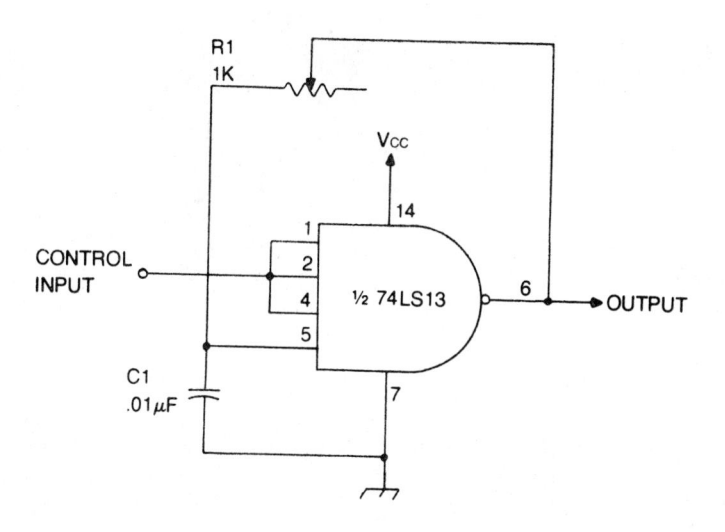

Gated oscillator

74LS30—LS EIGHT INPUT NAND GATE

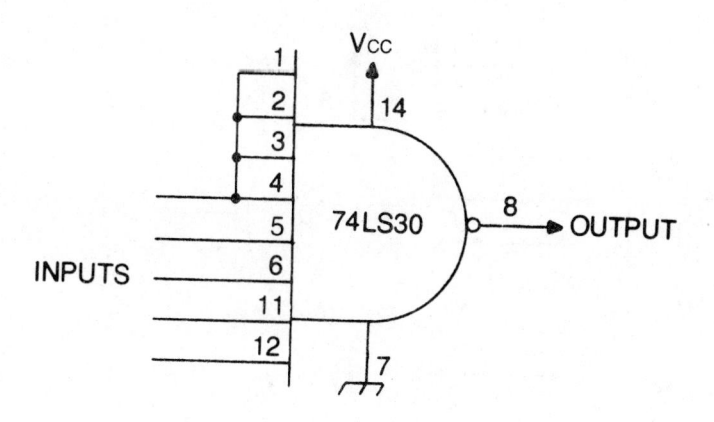

5-input NAND

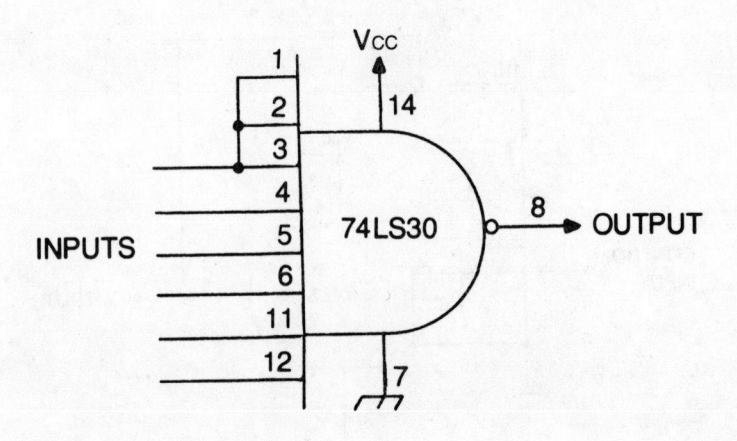

6-input NAND

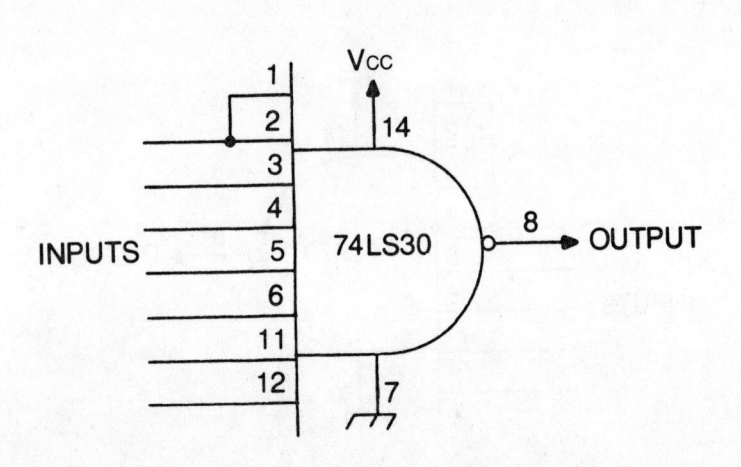

7-input NAND

450

74LS85—LS FOUR BIT MAGNITUDE COMPARATOR

8-bit comparator

74LS132—LS QUAD NAND SCHMITT TRIGGER

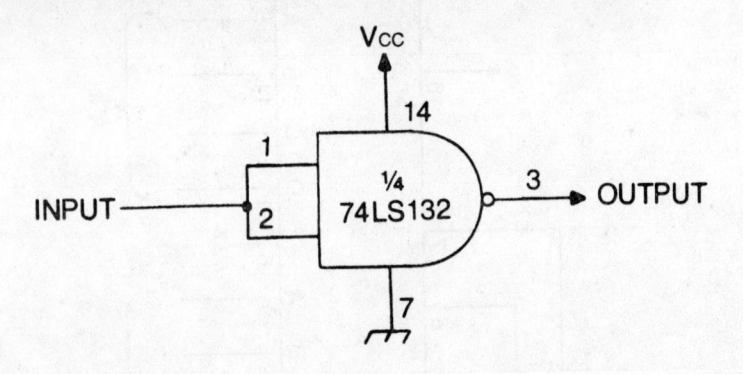

Wave shaper

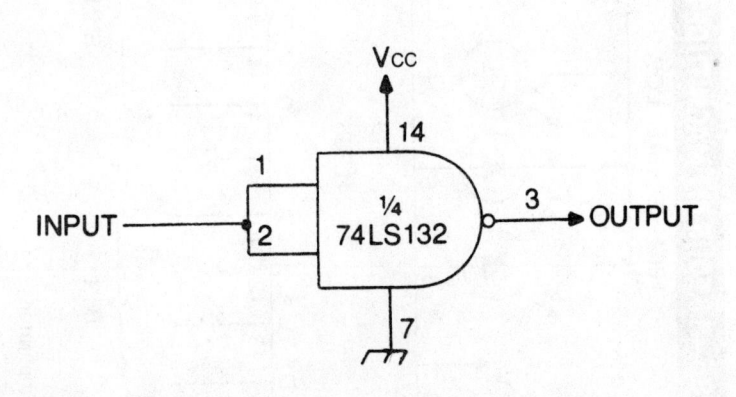

Pulse restorer

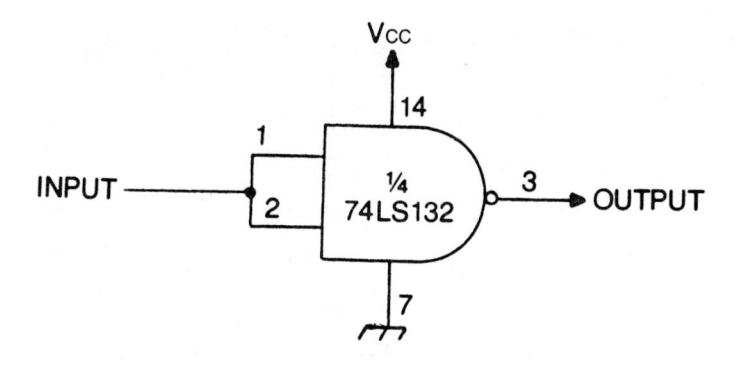

Noise eliminator

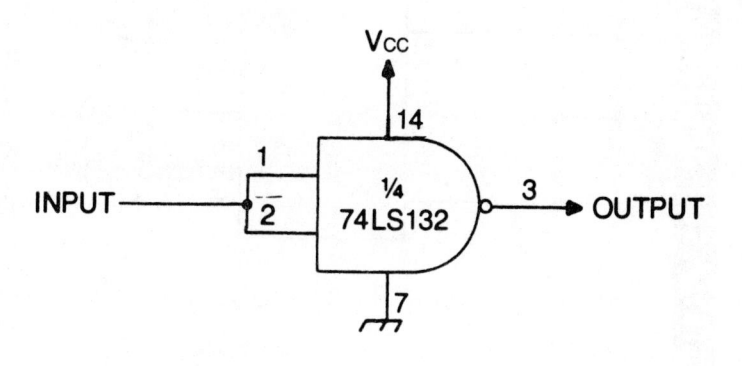

Threshold detector

74LS151 — LS ONE OF EIGHT DATA SELECTOR

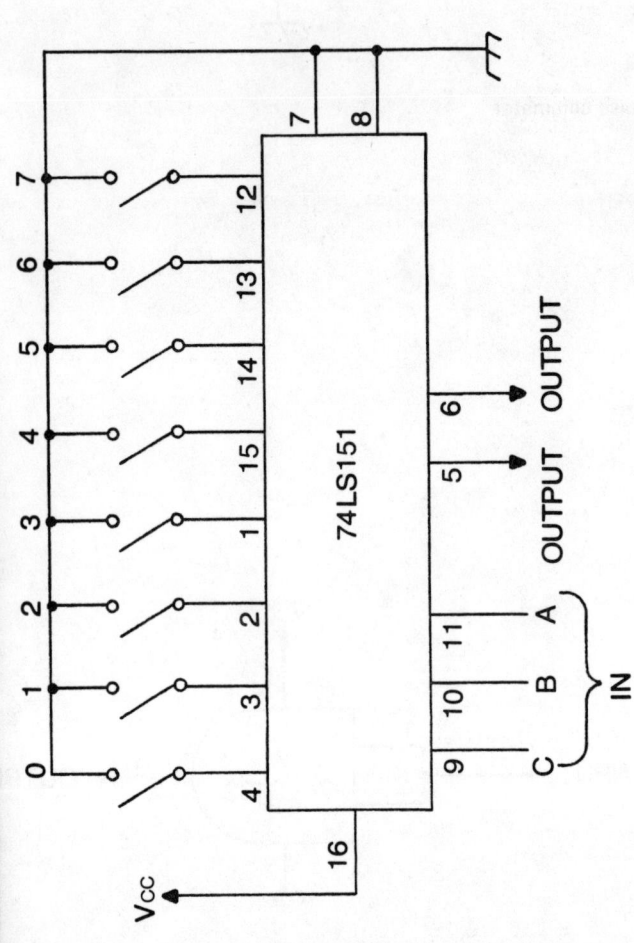

74LS161—LS FOUR BIT UP COUNTER

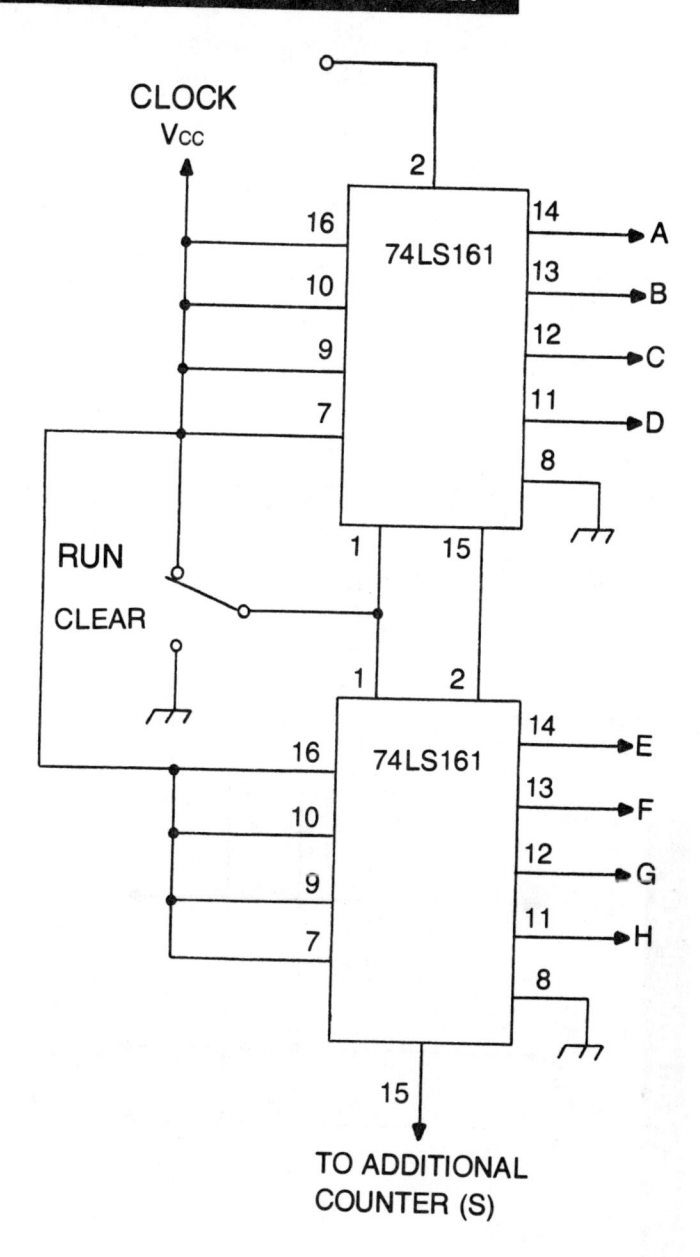

8-bit counter

74LS175 — LS QUAD D FLIP-FLOP

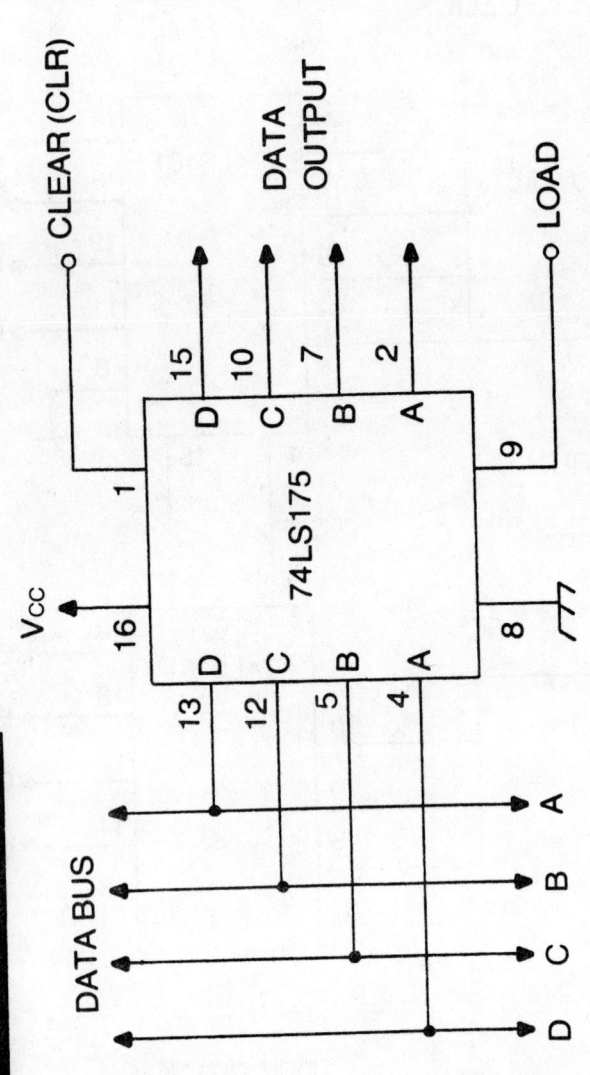

4-bit data register

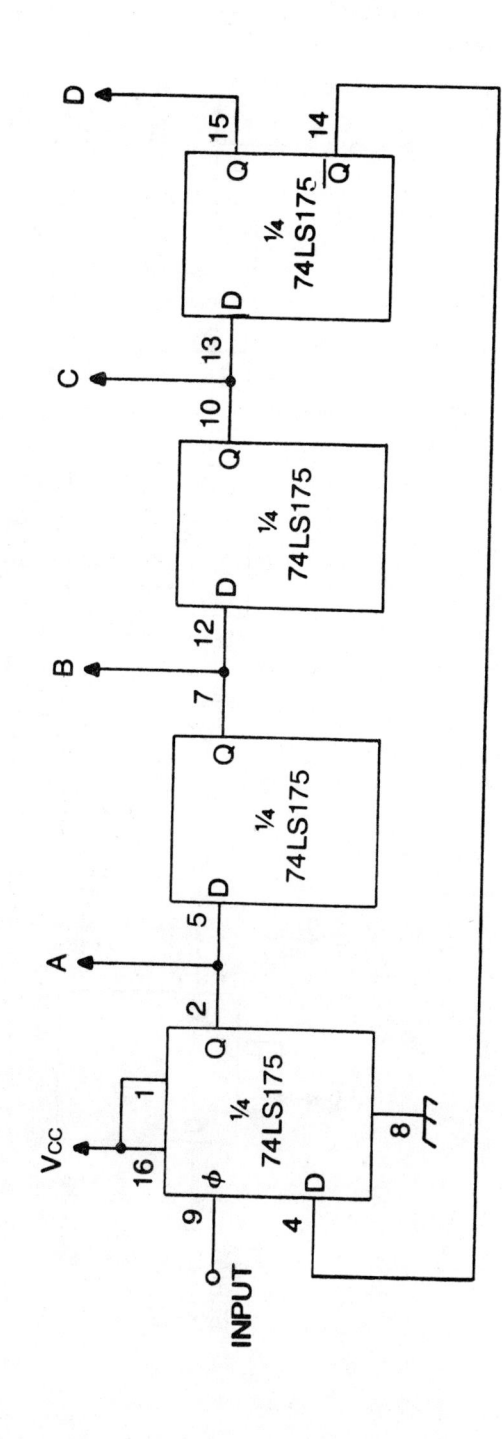

Modulo-8 counter

Serial in/out, parallel out shift register

74LS196 — LS BCD DECADE COUNTER

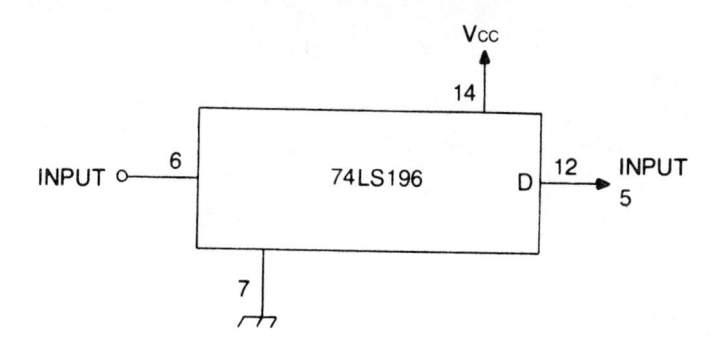

Divide-by-5 counter

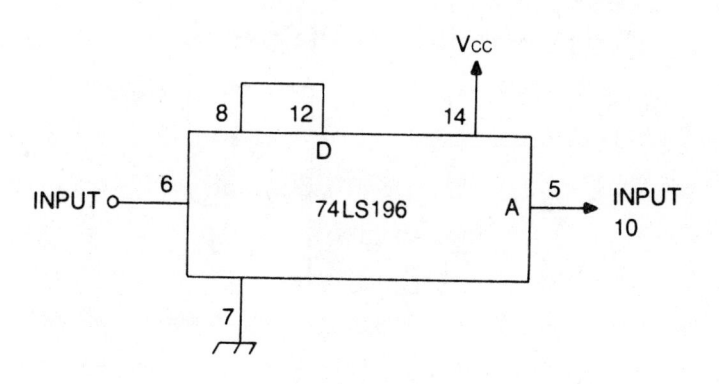

Divide-by 10 counter

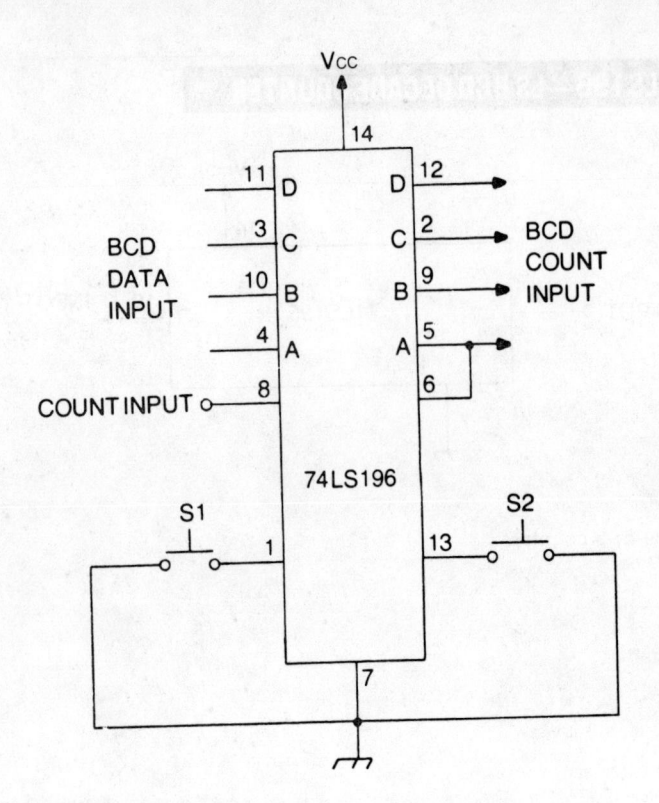

Decade counter

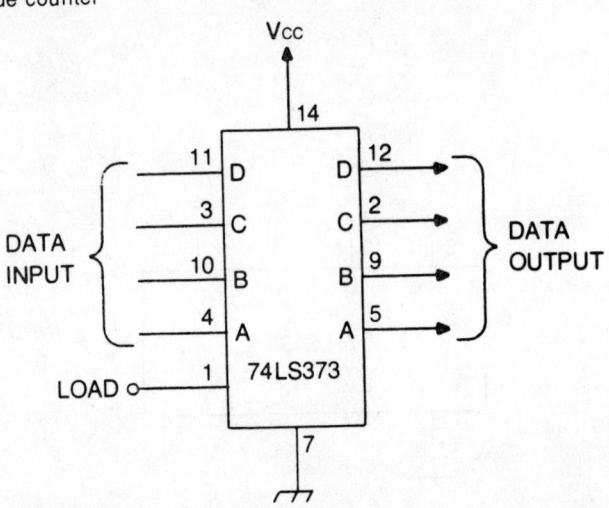

4-bit latch

460

74LS373—LS OCTAL D TYPE LATCH

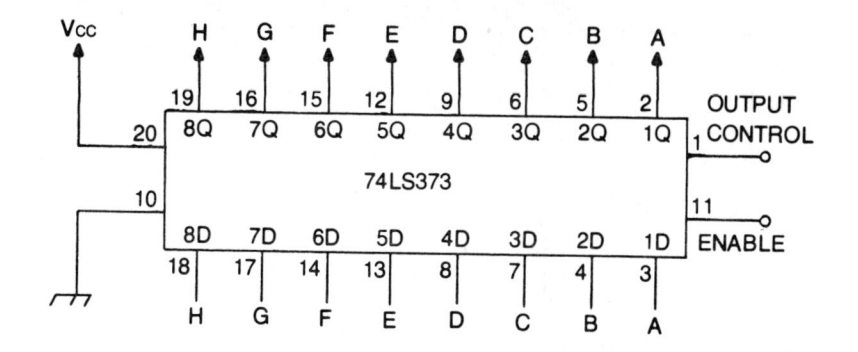

Three state eight bit storage register

74LS374—LS OCTAL D EDGE-TRIGGERED FLIP-FLOP

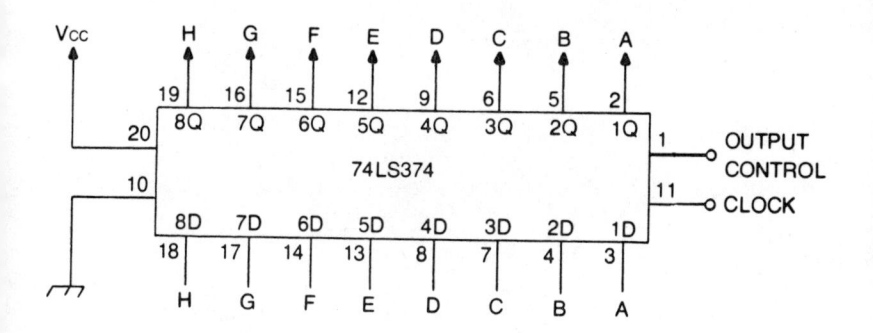

Clocked 3-state register

Part V

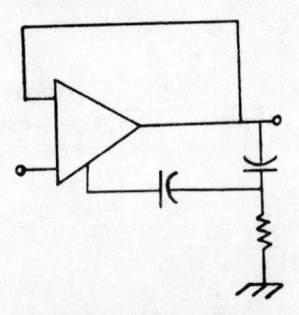

Radio And Television Integrated Circuits

Not all chips wind up in audio amplifiers or logic circuits. Numerous ICs have been developed for use in radio and television applications. Many of these are unknown to the casual electronics experimenter. However, they are now becoming available on the surplus electronics market. This section will look at how some of these ICs can be used.

172—INTEGRATED AM IF STRIP

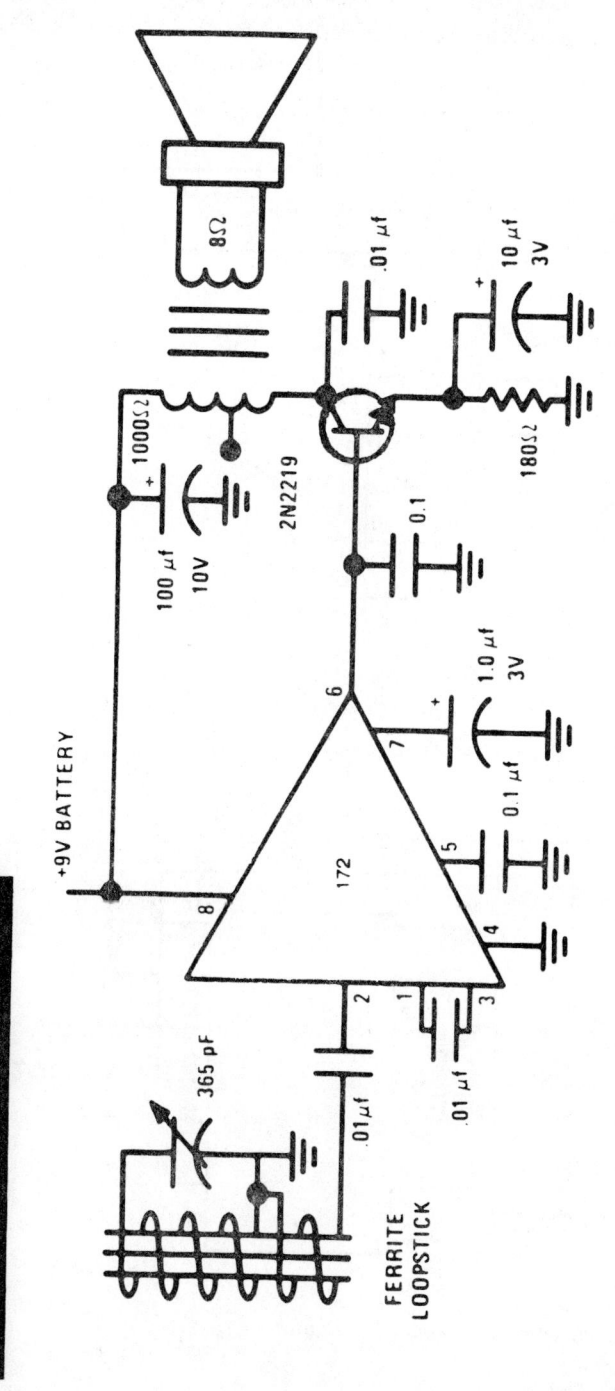

T.R.F. broadcast receiver

175 — MONOLITHIC DIFFERENTIAL PAIR OSCILLATOR

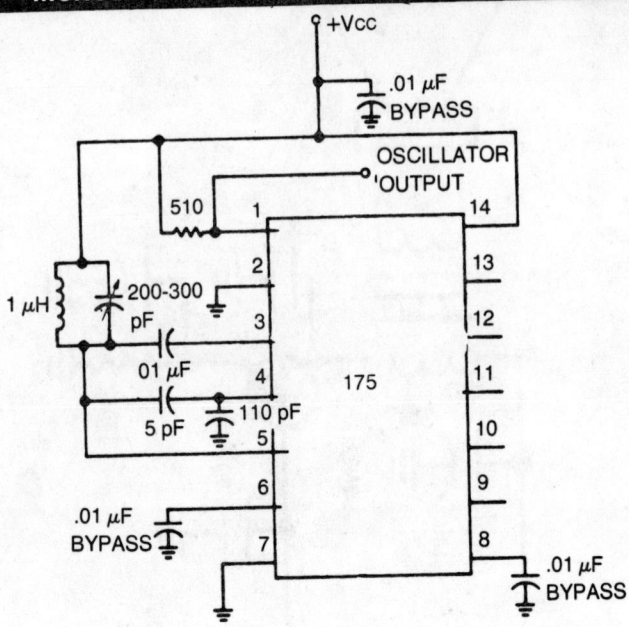

10 MHz L-C sine wave oscillator

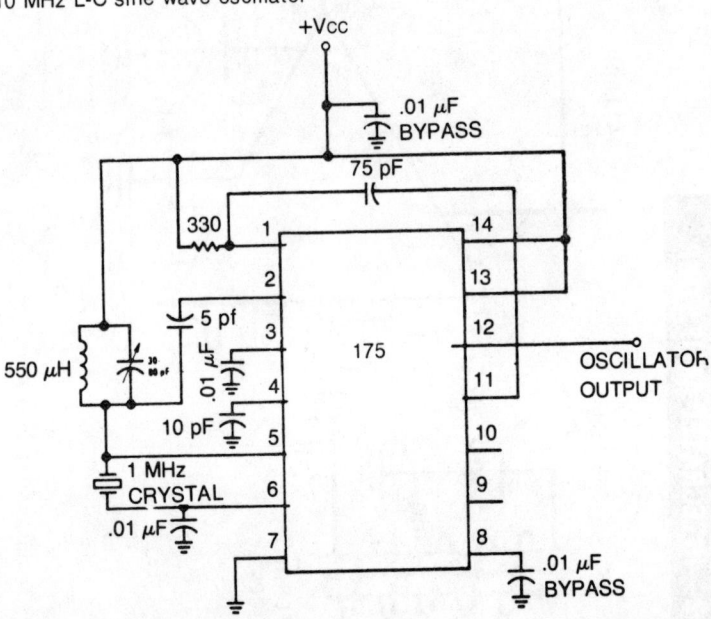

1 MHz crystal oscillator with TTL output

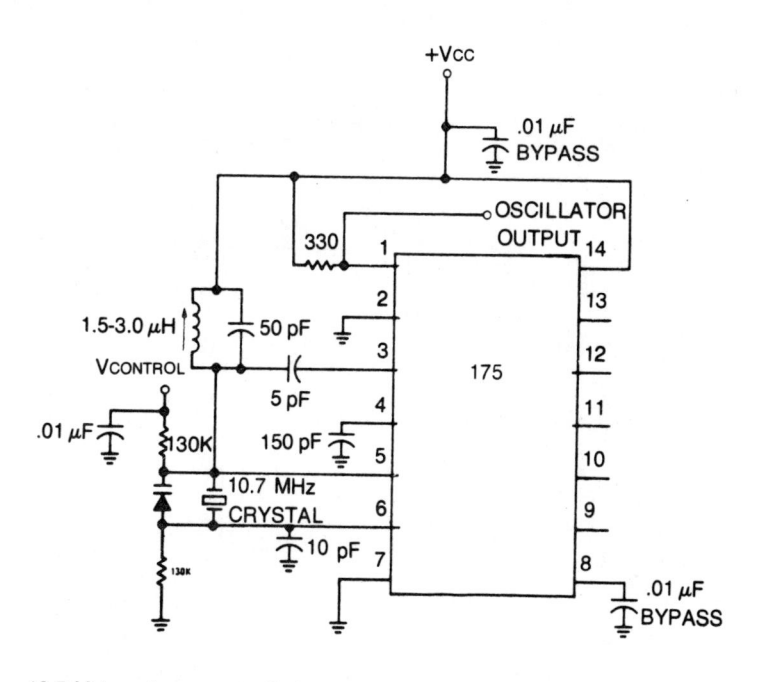

10.7 MHz voltage-controlled crystal oscillator parallel tuning

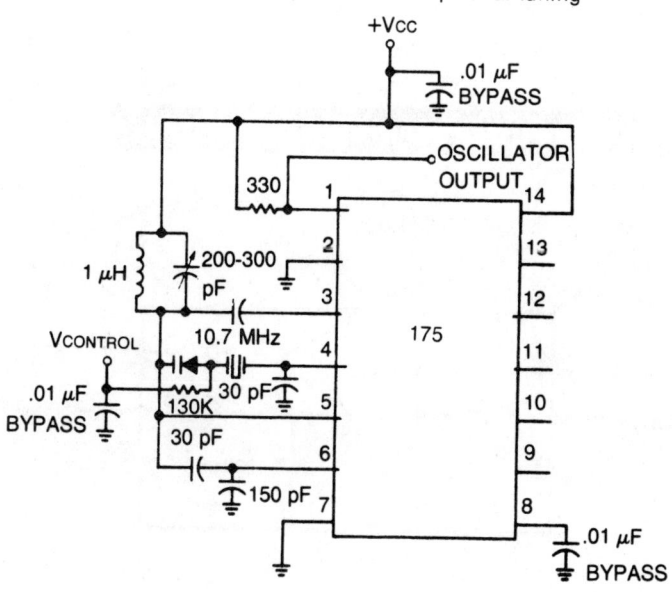

10.7 MHz voltage controlled crystal oscillator series tuning

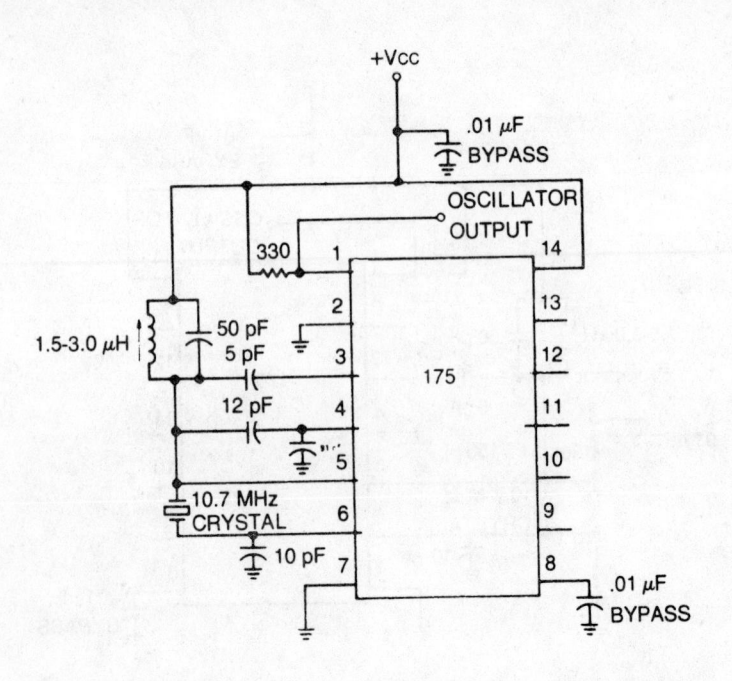

10.7 MHz parallel resonant crystal oscillator

703—LOW POWER DRAIN RF/IF AMPLIFIER

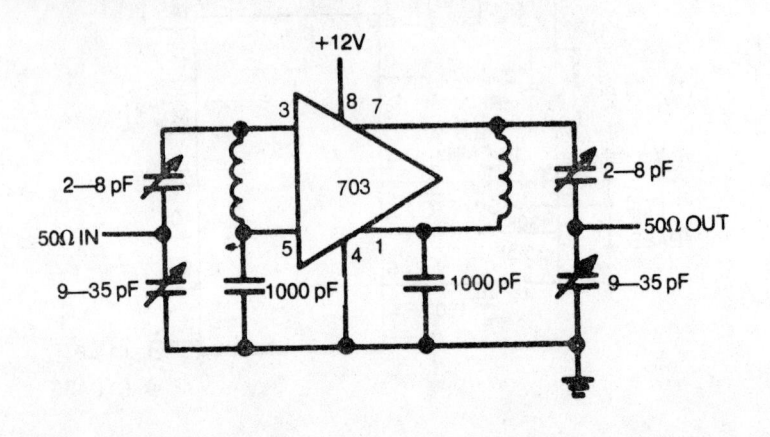

100 MHz narrow band amplifier

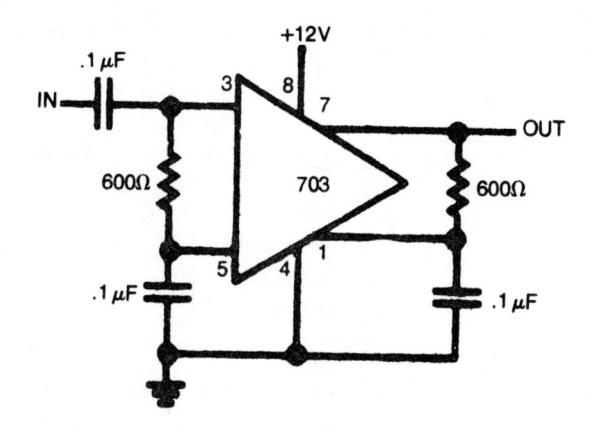

RC coupled video amplifier

720 — AM RADIO SYSTEM INTEGRATED CIRCUIT

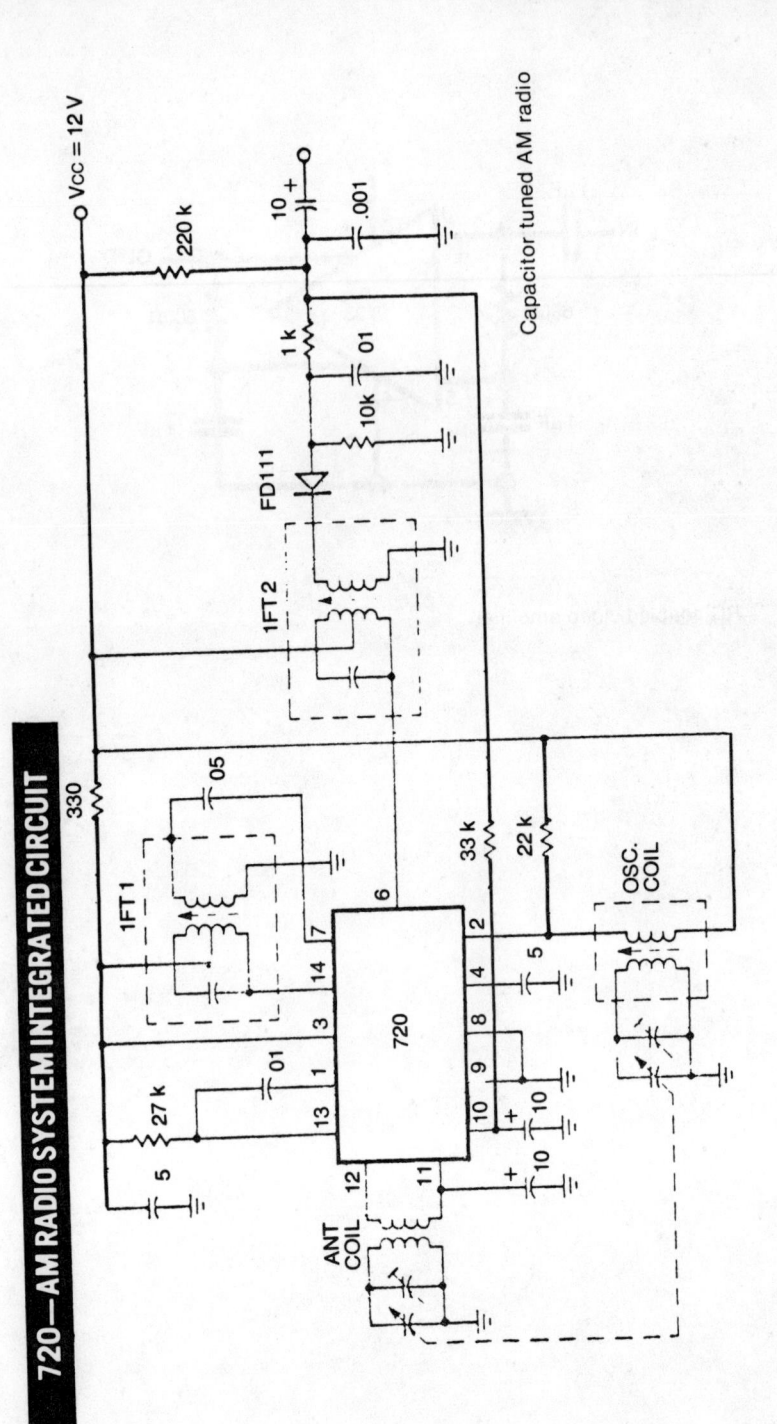

Capacitor tuned AM radio

732—FM STEREO MULTIPLEX DECODER

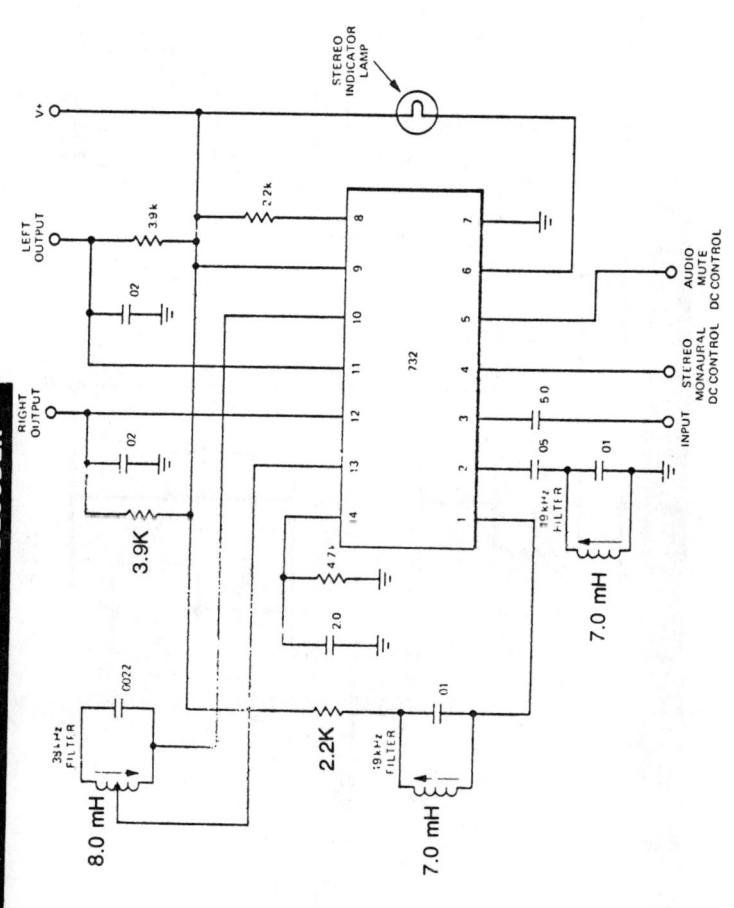

Fm stereo multiplex decoder circuit

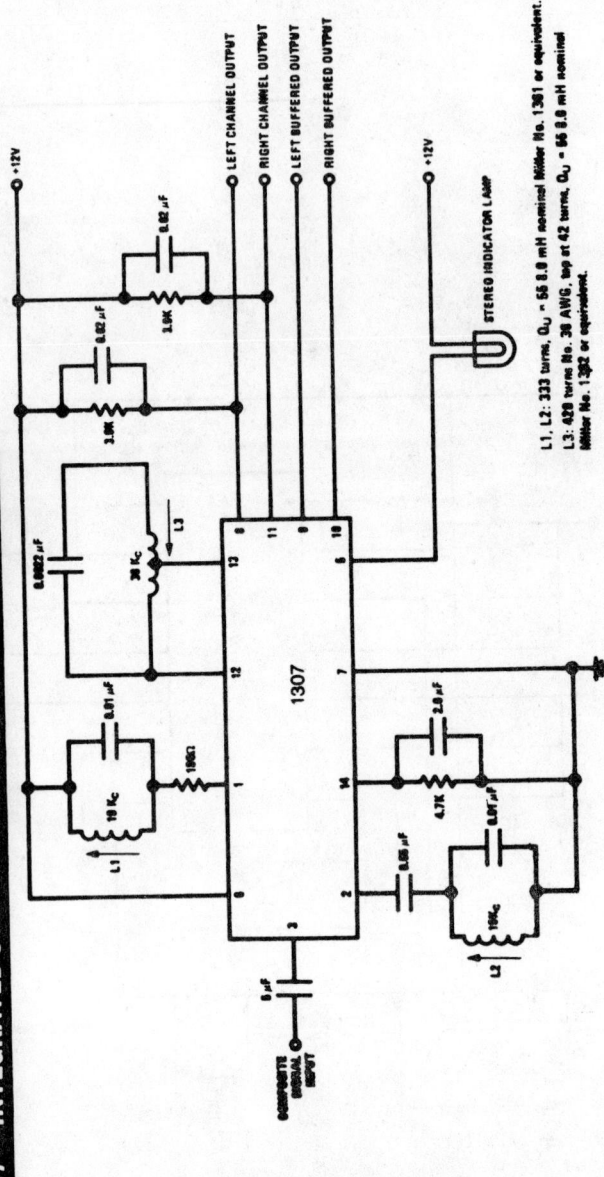

1307 — INTEGRATED STEREO MULTIPLEX DEMODULATOR

L1, L2: 333 turns, $Q_u \approx 56$ 8.0 mH nominal Miller No. 1361 or equivalent.
L3: 420 turns No. 36 AWG, tap at 42 turns, $Q_u \approx 56$ 8.0 mH nominal Miller No. 1362 or equivalent.

Stereo multiplex demodulator with beacon lamp

1310 — PHASE LOCKED LOOP FM STEREO DEMODULATOR

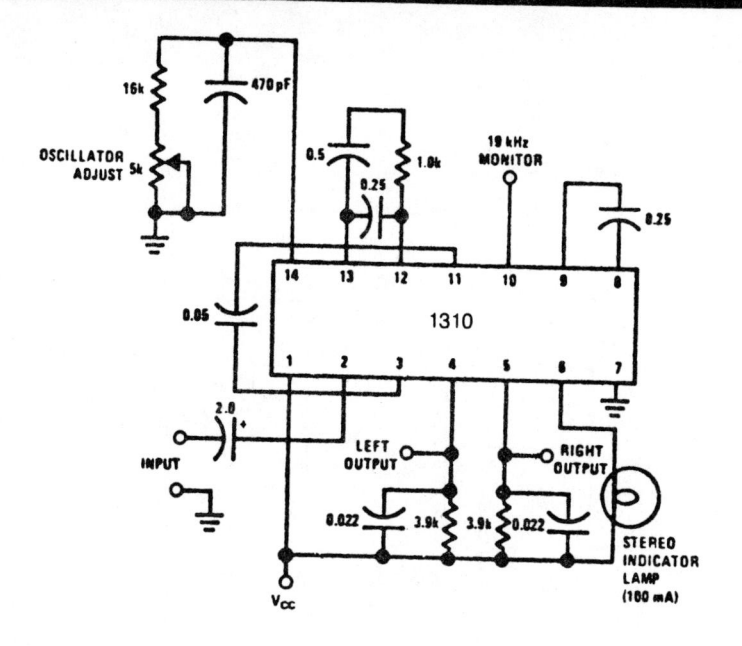

Fm stereo demodulator with indicating lamp

1349—INTEGRATED IF AMPLIFIER WITH AGC

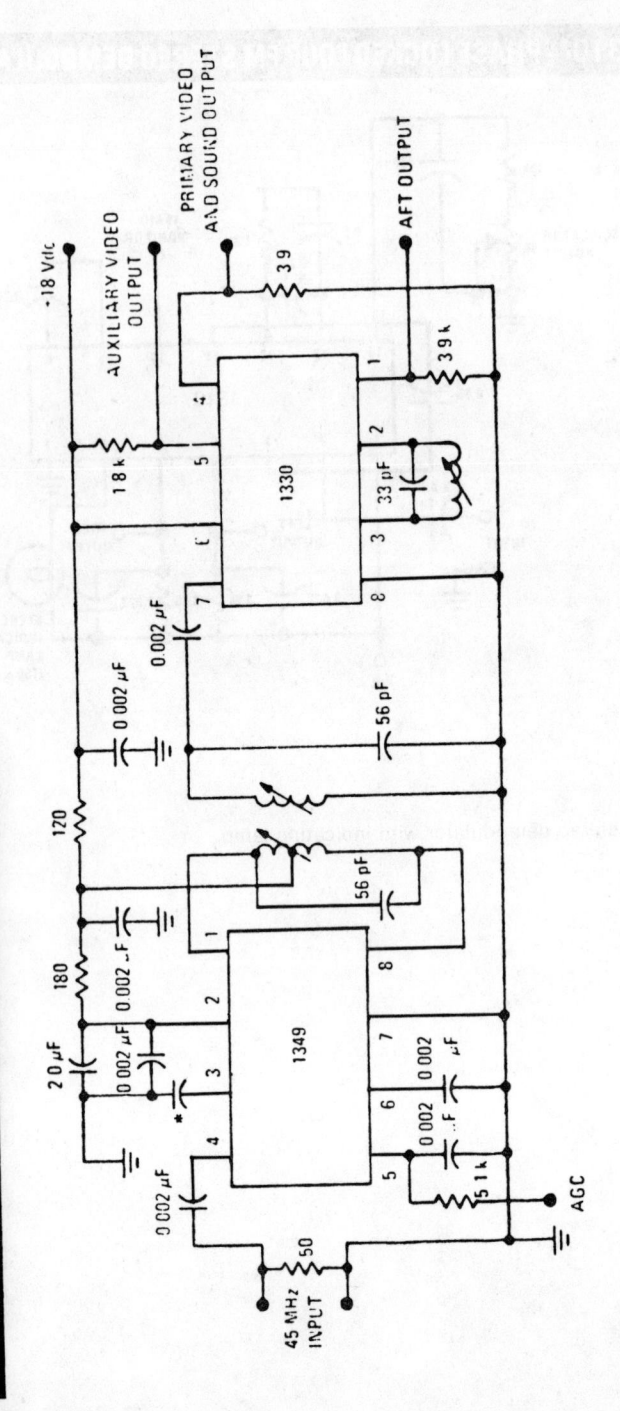

Video IF amplifier

1350—INTEGRATED IF AMPLIFIER

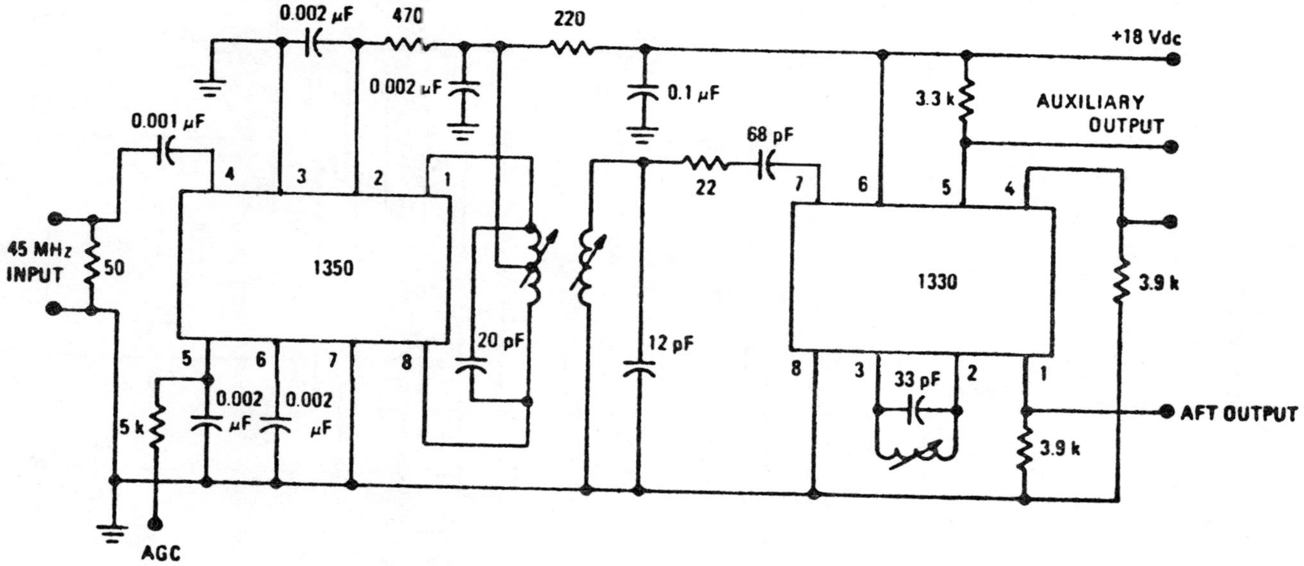

Video IF amplifier

1351 — INTEGRATED TV SOUND CIRCUIT

4.5 MHz typical application

1352—TV VIDEO AMPLIFIER WITH AGC

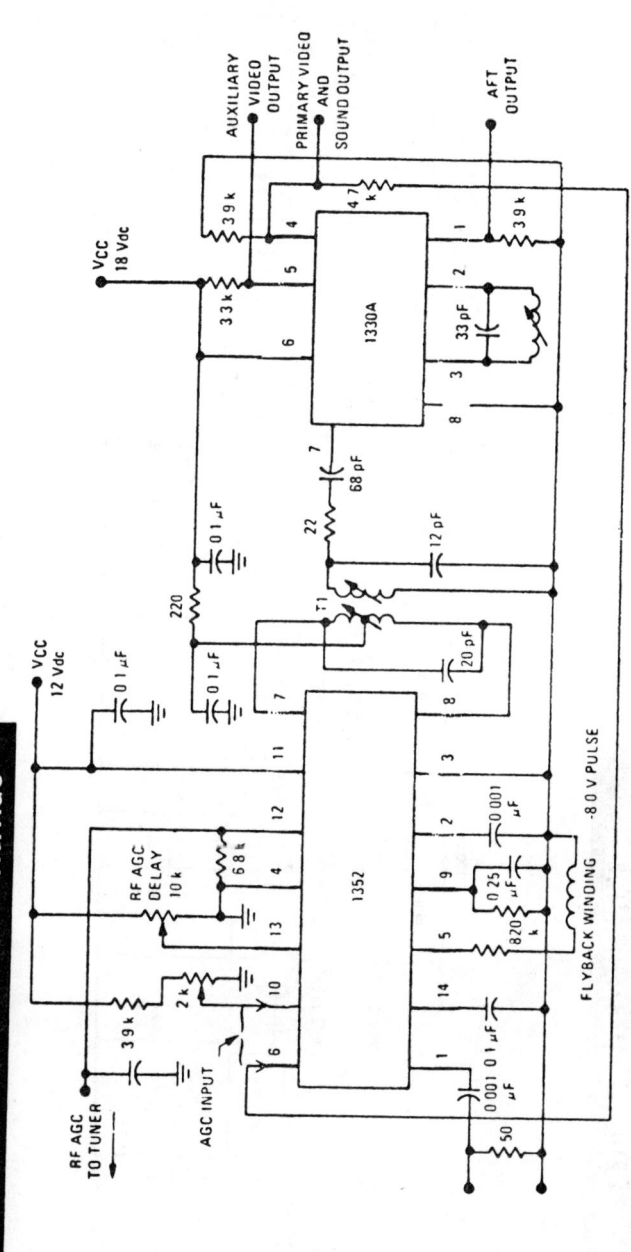

Typical video IF amplifier application

1355 — BALANCED FOUR STAGE HIGH GAIN FM/IF AMPLIFIER

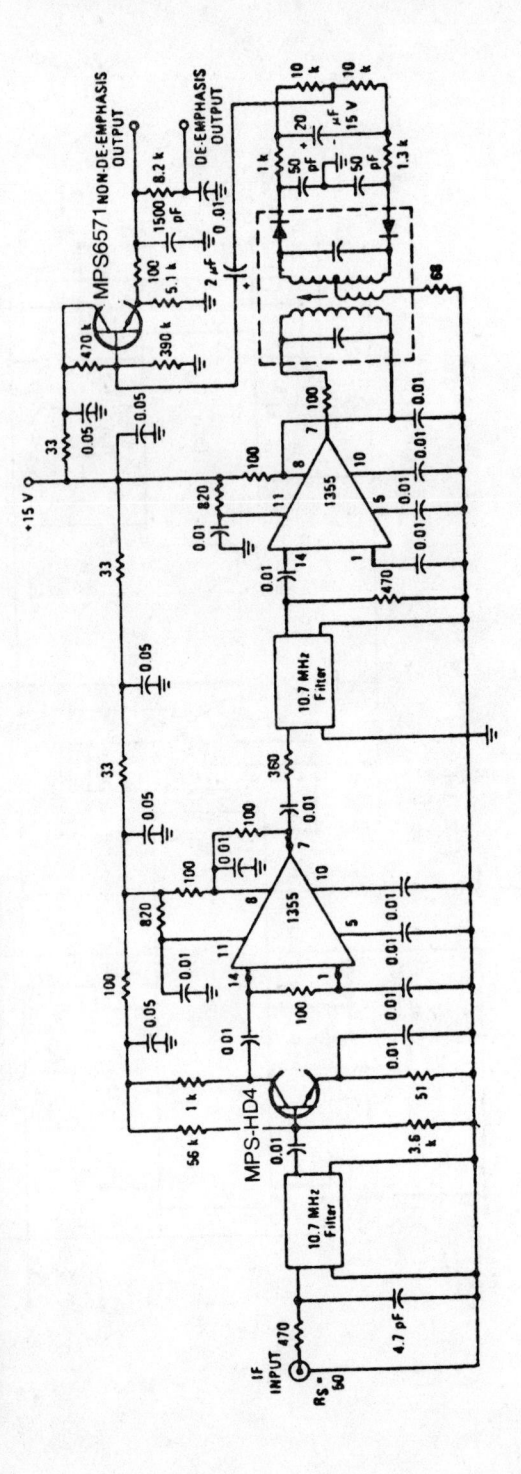

Dual 1355 FM IF application

1357 — INTEGRATED IF AMPLIFIER AND QUADRATURE DETECTOR

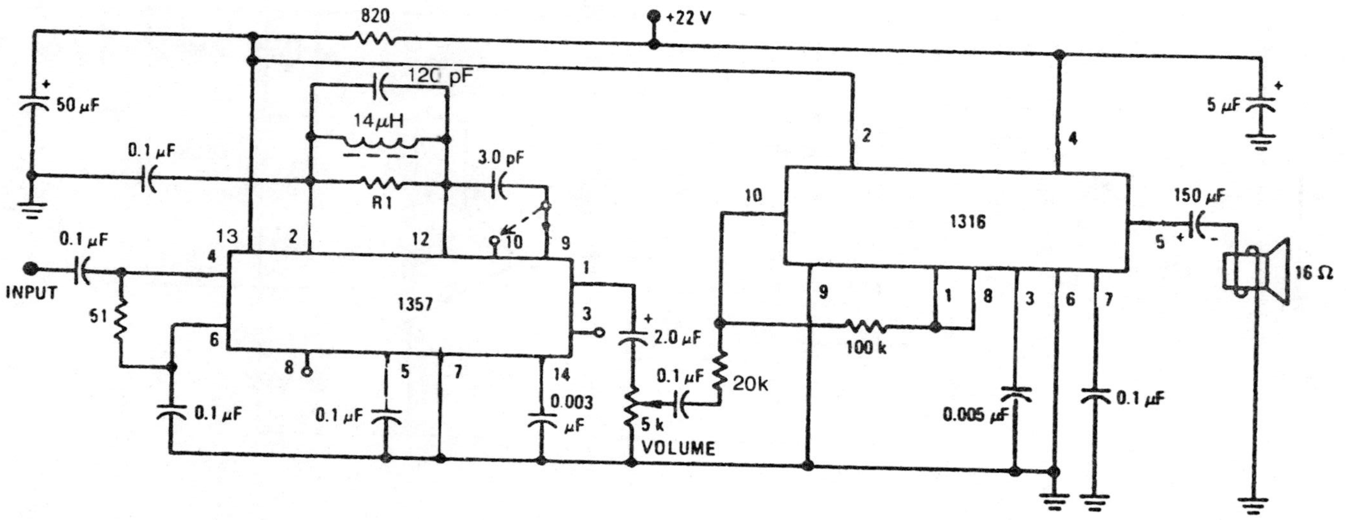

TV typical application circuit

477

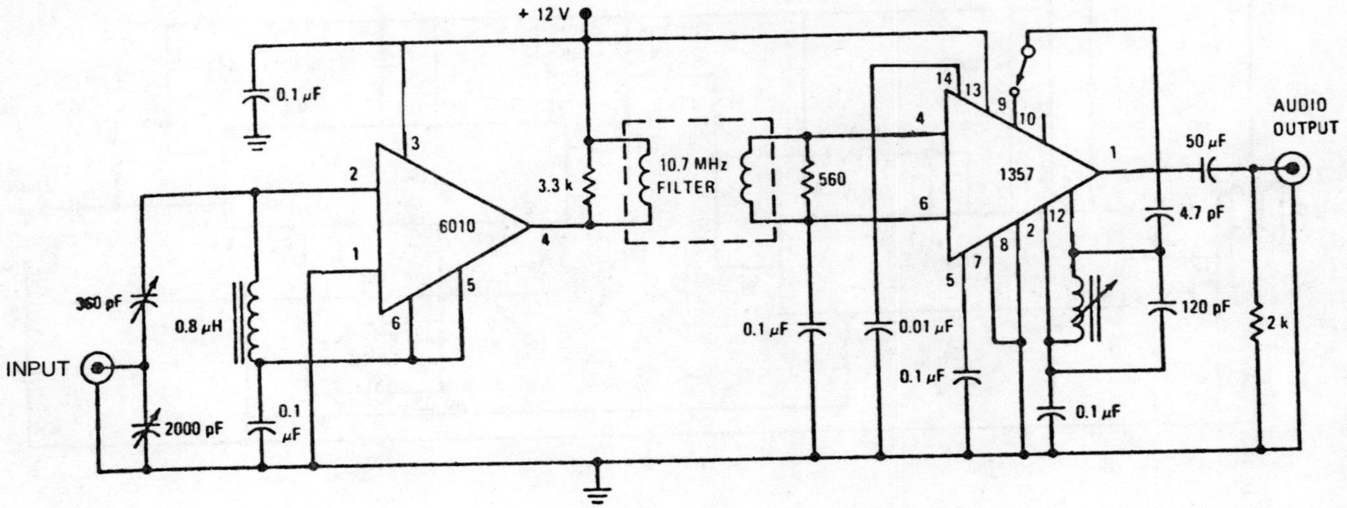

Fm radio typical application circuit

1358 — INTEGRATED TELEVISION SOUND IF AMPLIFIER

TV application circuit

$$R_S = \frac{V^+ - 11}{0.033} \, (\Omega)$$

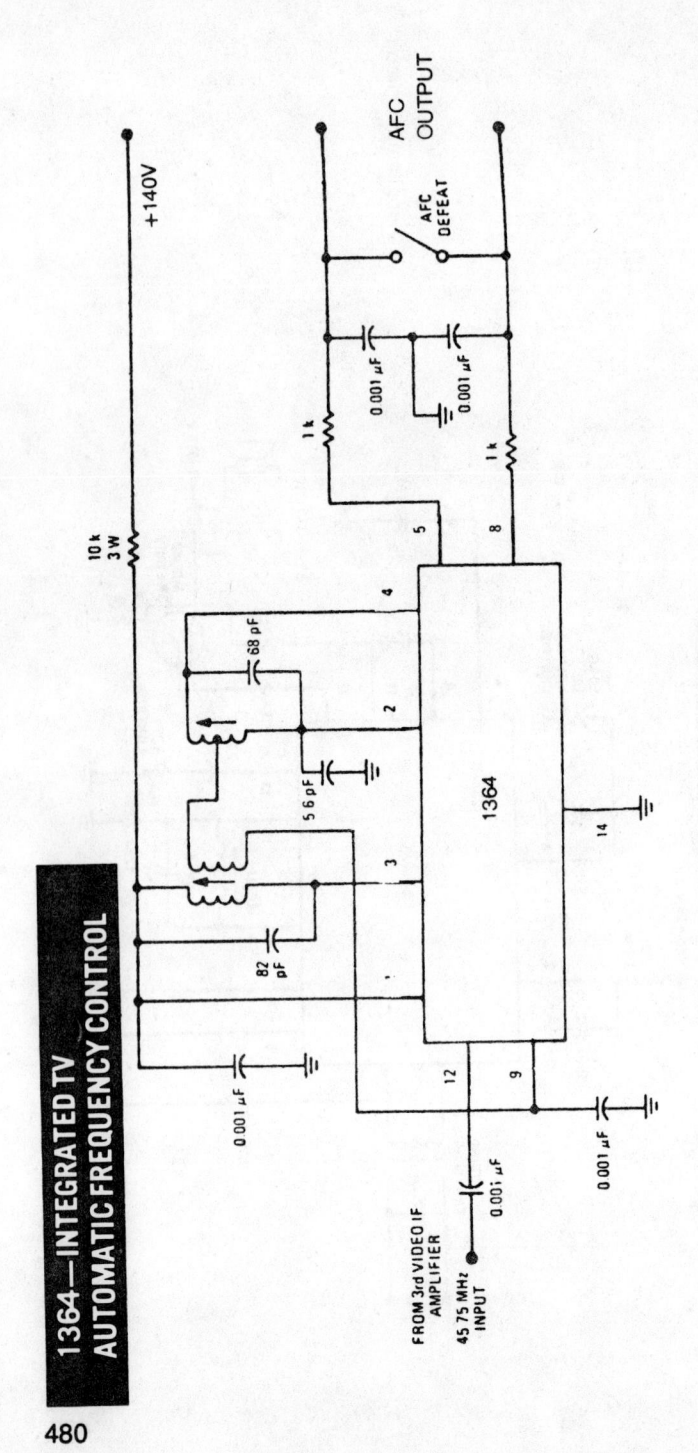

1364 — INTEGRATED TV AUTOMATIC FREQUENCY CONTROL

Typical AFC application

1391—INTEGRATED PHASE LOCK LOOP

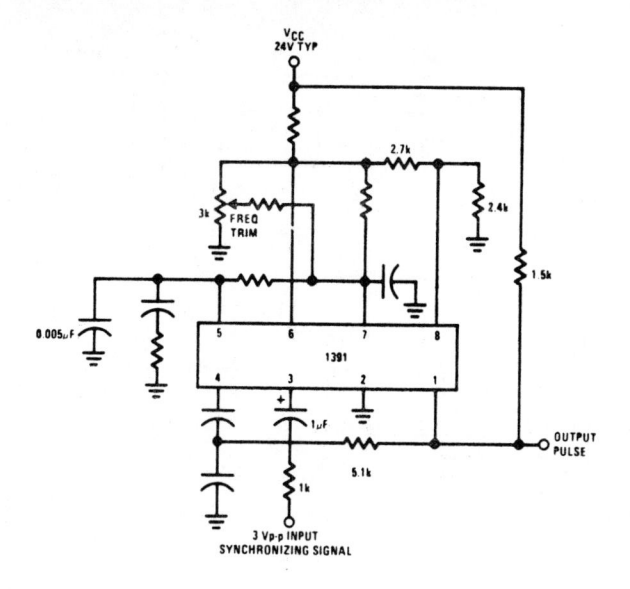

General purpose phase-lock loop

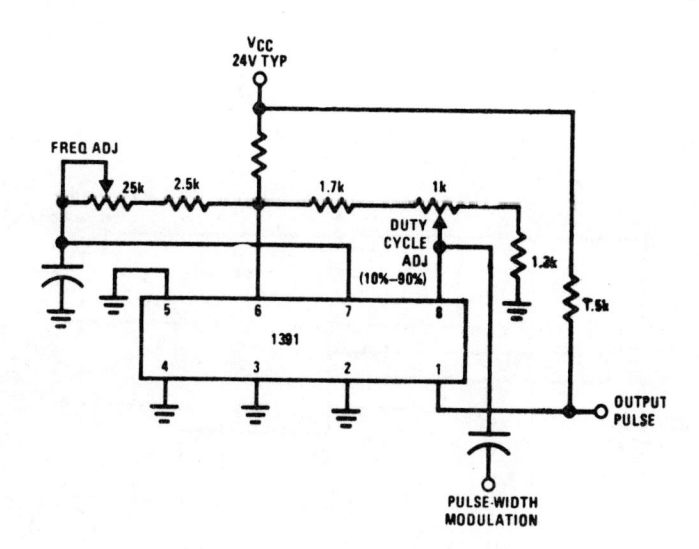

Variable duty cycle oscillator

1496 — BALANCED MODULATOR/DEMODULATOR

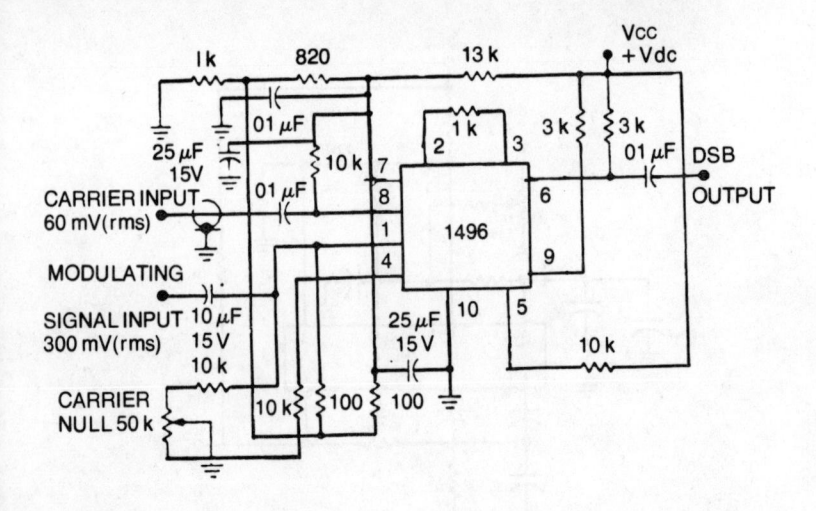

Balanced modulator (+12 Vdc single supply)

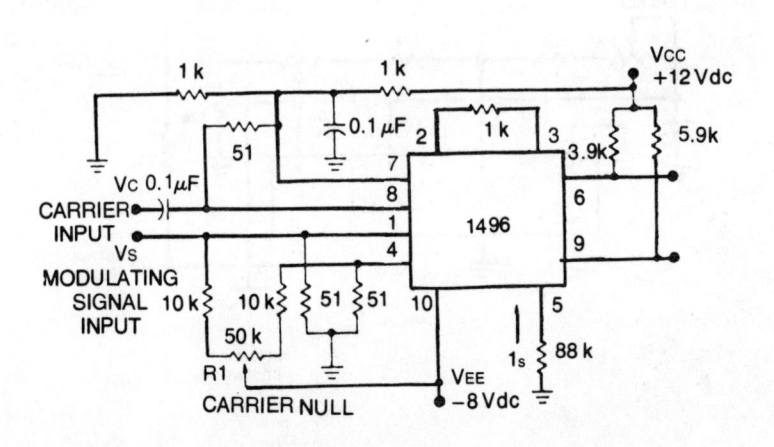

Balanced modulator-demodulator

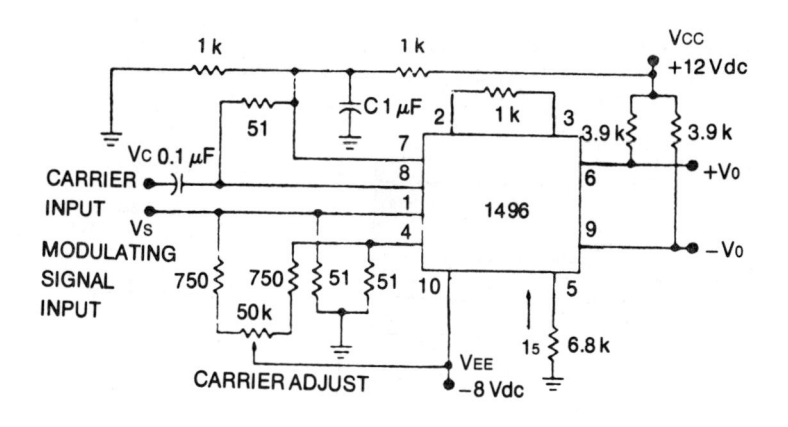

AM modulator circuit

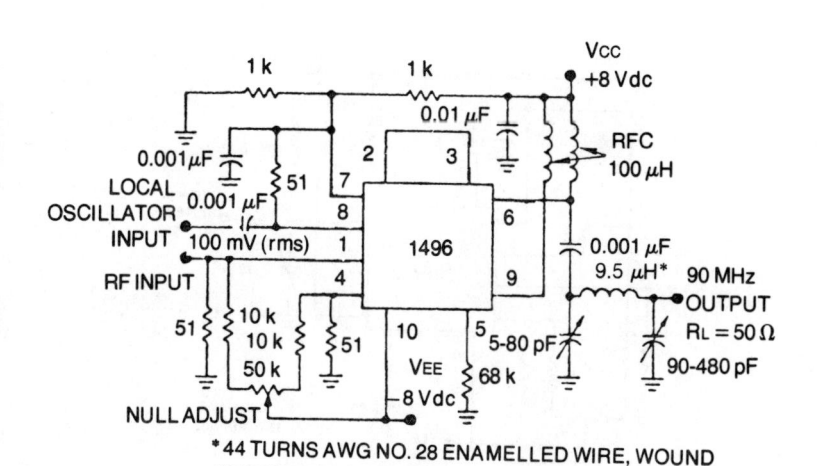

Doubly balanced mixer (broadband inputs, 9.0 MHz tuned output)

*44 TURNS AWG NO. 28 ENAMELLED WIRE, WOUND ON MICROMETALS TYPE 44.6 TOROID CORE

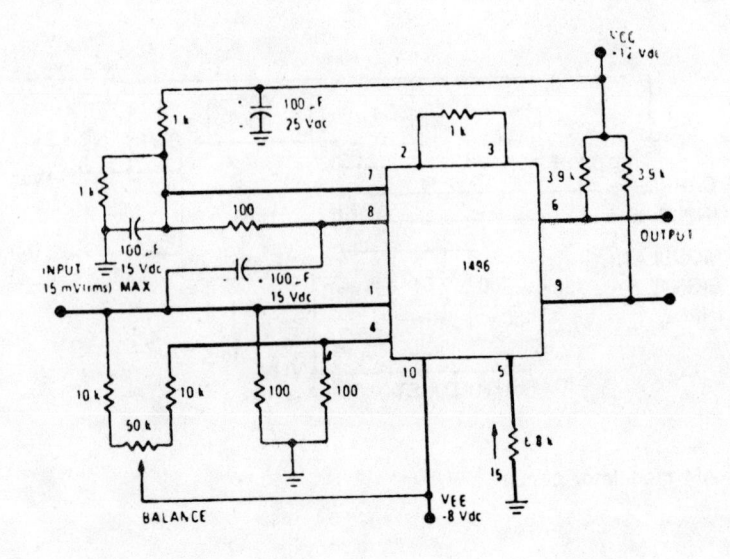

Low-frequency doubler

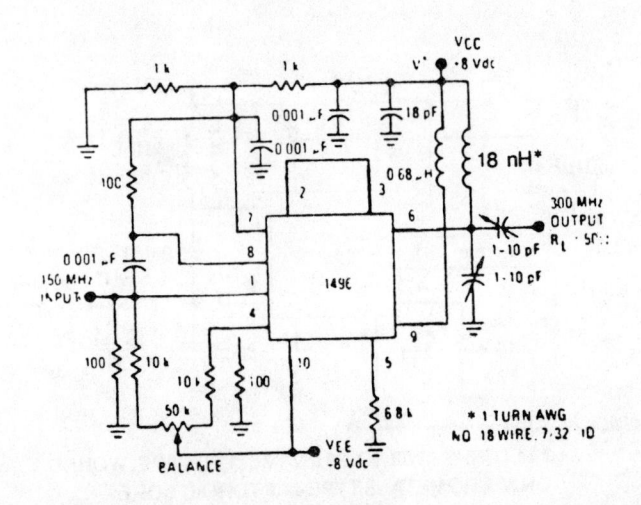

150 to 300 MHz doubler

SSB product detector

Broadband frequency doubler

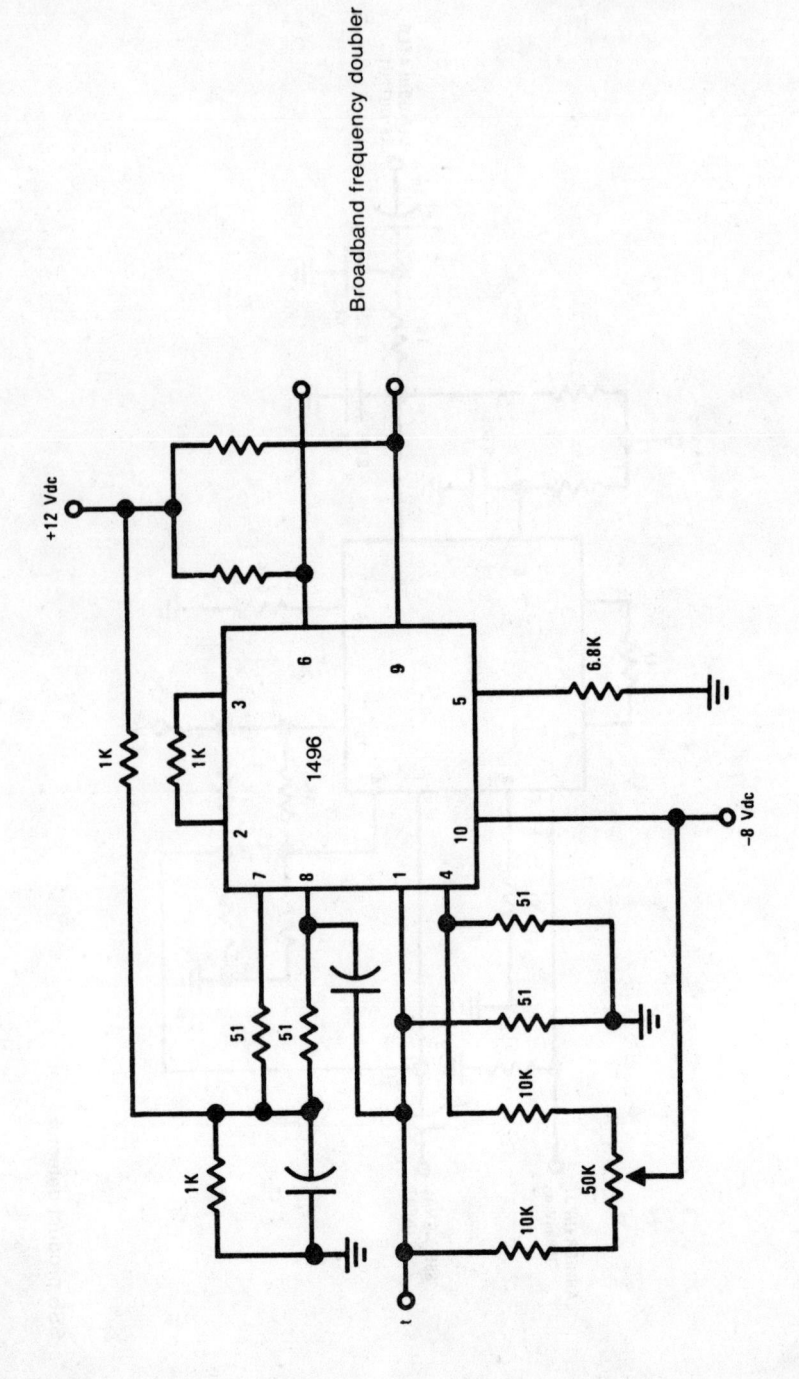

1590 — WIDEBAND INTEGRATED AMPLIFIER WITH A.G.C.

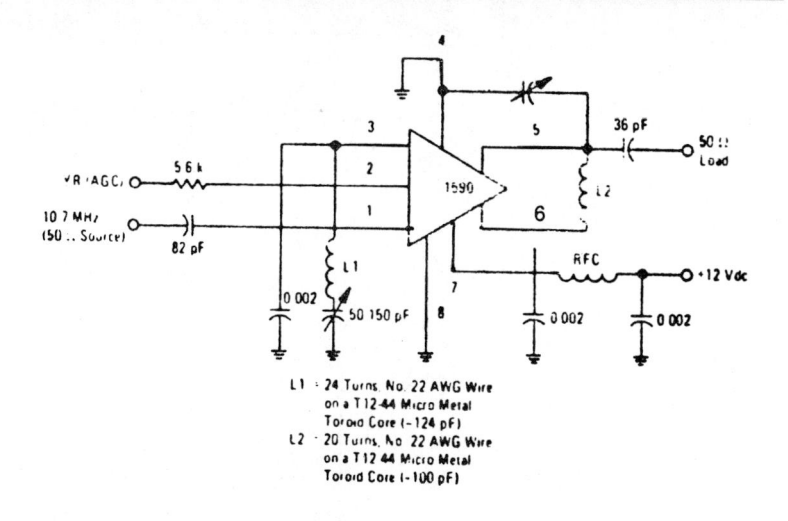

L1 = 24 Turns, No. 22 AWG Wire on a T12-44 Micro Metal Toroid Core (~124 pF)
L2 = 20 Turns, No. 22 AWG Wire on a T12-44 Micro Metal Toroid Core (~100 pF)

10.7-MHz amplifier gain ≃ 55 dB, BW ≃ 100 kHz

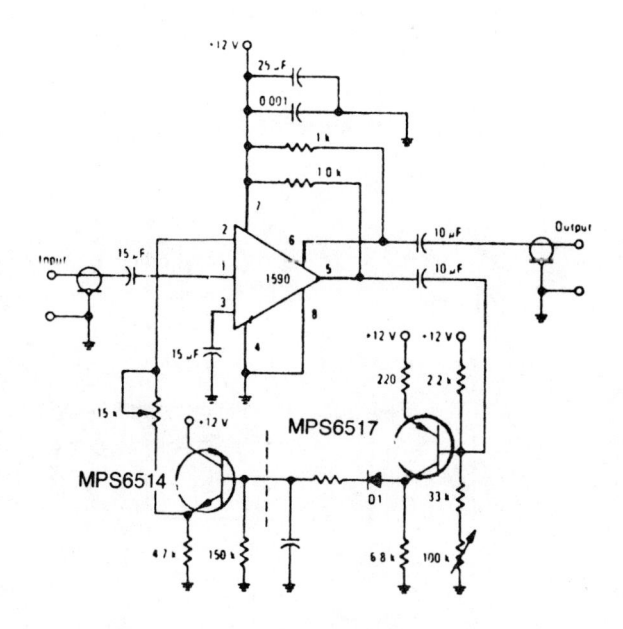

Speech compressor

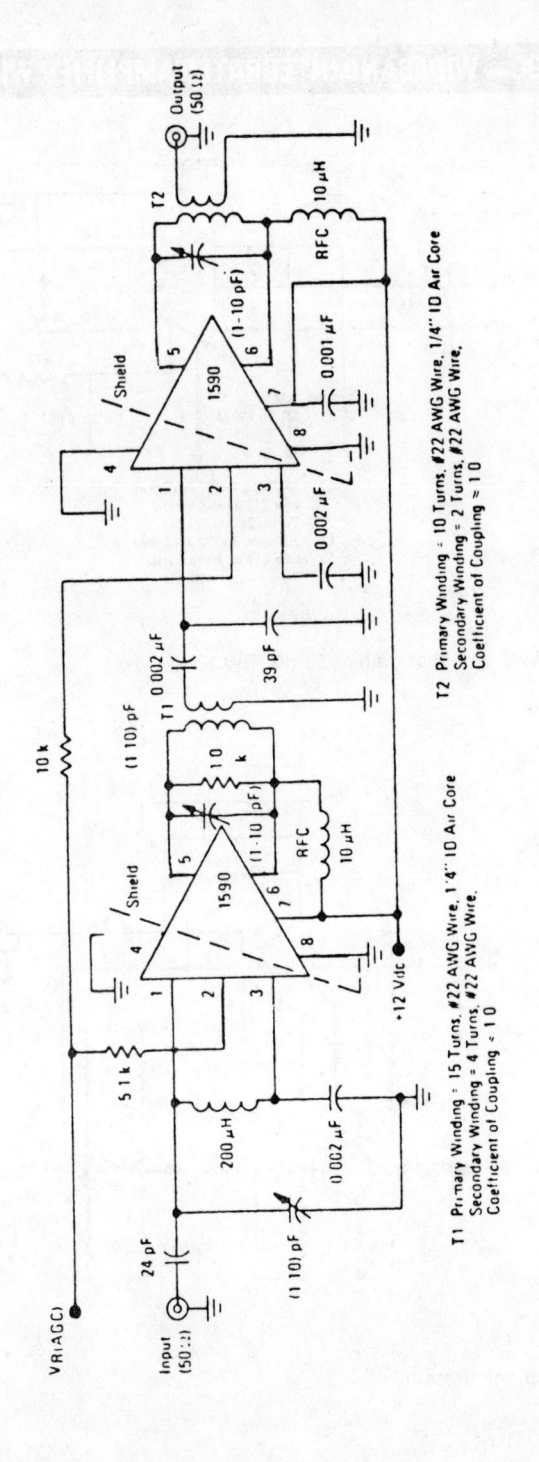

Two stage 60 MHz IF amplifier (Power gain ≈ 80 dB, BW ≈ 1.5 MHz)

T1 Primary Winding = 15 Turns, #22 AWG Wire, 1/4" ID Air Core
Secondary Winding = 4 Turns, #22 AWG Wire.
Coefficient of Coupling ≈ 1.0

T2 Primary Winding = 10 Turns, #22 AWG Wire, 1/4" ID Air Core
Secondary Winding = 2 Turns, #22 AWG Wire.
Coefficient of Coupling ≈ 1.0

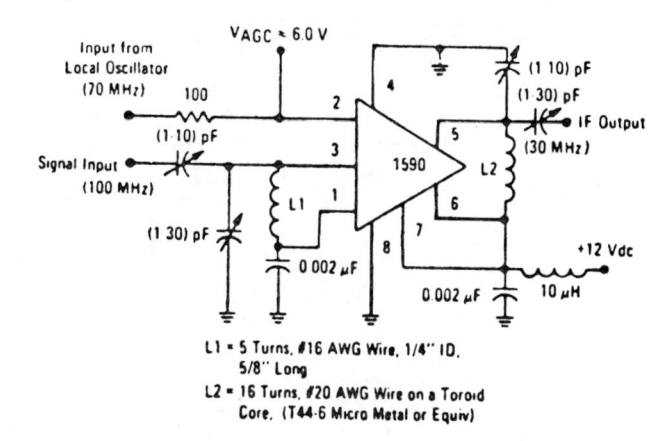

100 MHz mixer

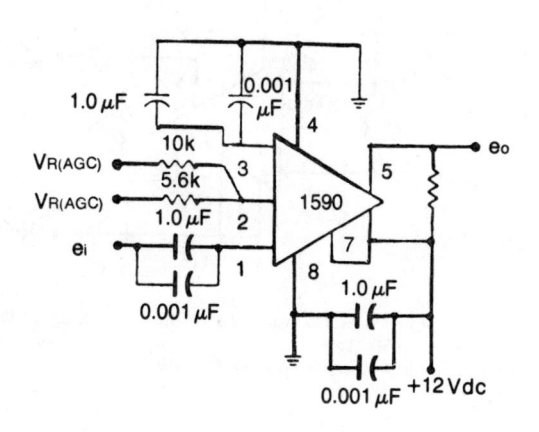

Video amplifier

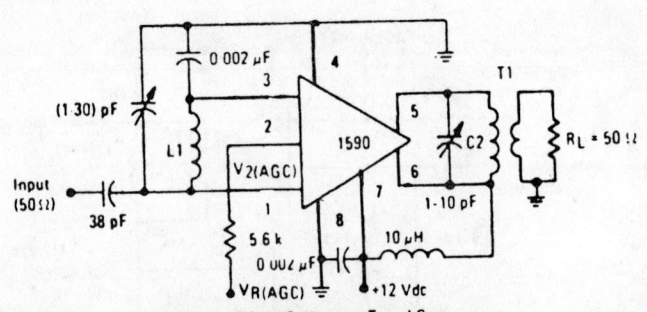

L1 = 12 Turns #22 AWG Wire on a Toroid Core.
(T37-6 Micro Metal or Equiv)
T1 Primary = 17 Turns #20 AWG Wire on a Toroid Core.
(T44-6 Micro Metal or Equiv)
Secondary = 2 Turns #20 AWG Wire

30-MHz amplifier (Power Gain = 50 dB, BW ≈ 1.0 Mhz)

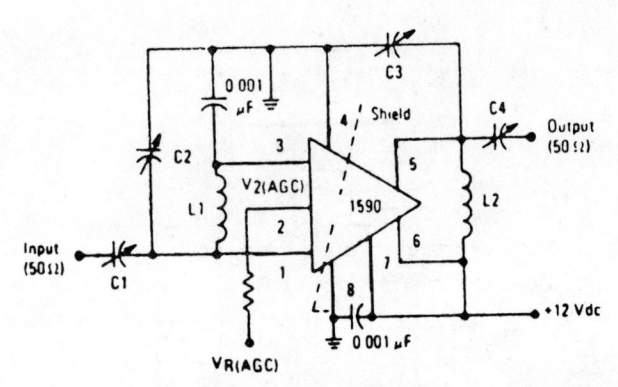

L1 = 7 Turns, #20 AWG Wire, 5/16" Dia.,
5/8" Long
L2 = 6 Turns, #14 AWG Wire, 9/16" Dia.,
3/4" Long

C1,C2,C3 = (1-30) pF
C4 = (1-10) pF

60 Mhz power gain test circuit

490

1733—DIFFERENTIAL VIDEO WIDEBAND AMPLIFIER

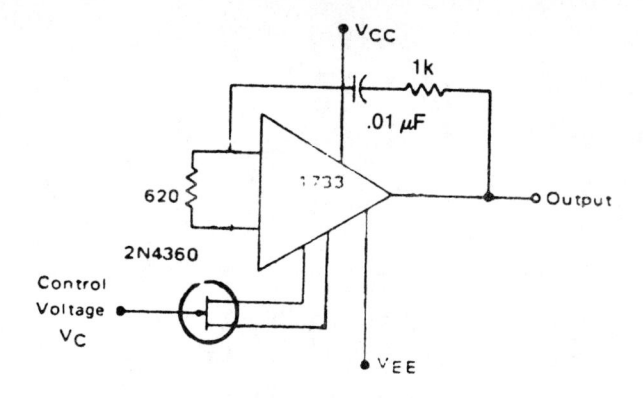

Voltage controlled oscillator

1800 — PHASE LOCKED LOOP FM STEREO DEMODULATOR

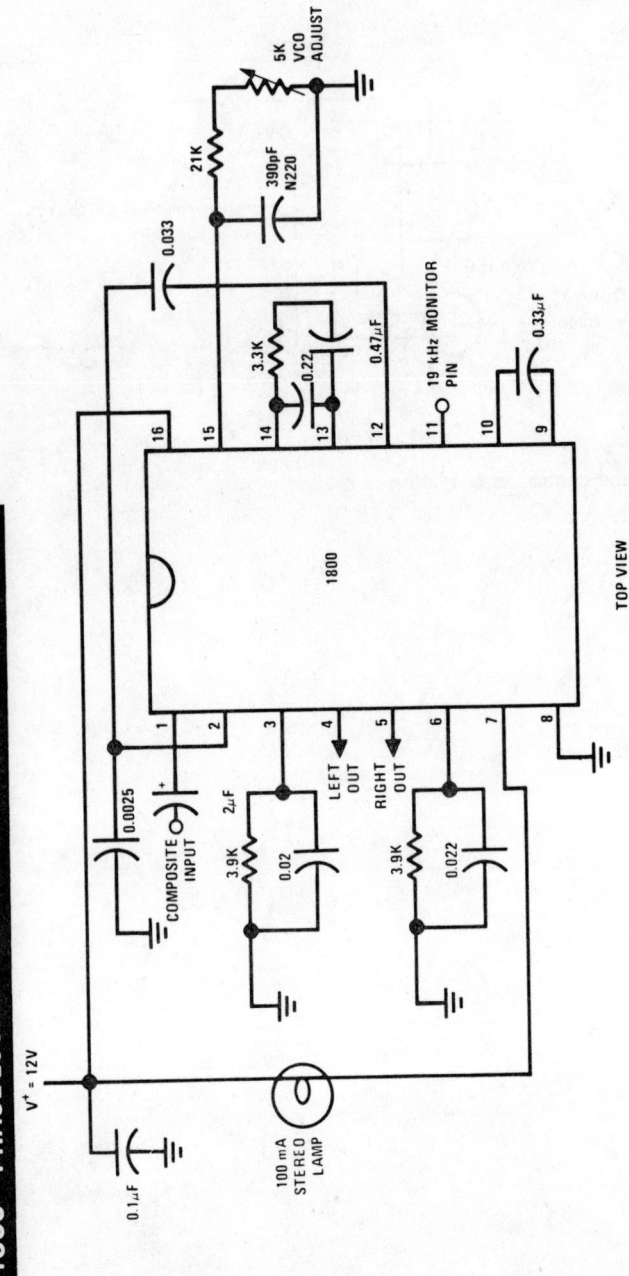

Fm stereo demodulator with beacon

1808 — MONOLITHIC TELEVISION SOUND SYSTEM

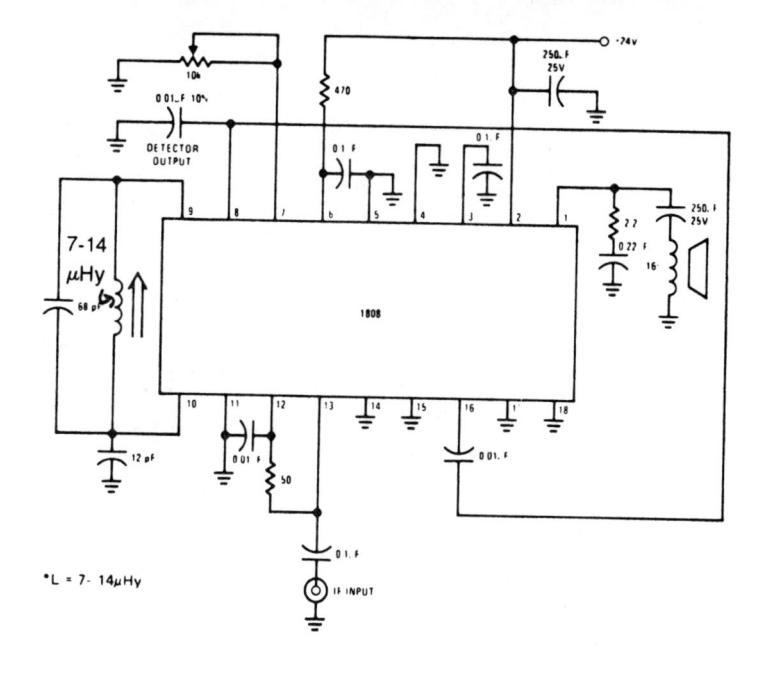

Television sound system

1889 — INTEGRATED VIDEO MODULATOR

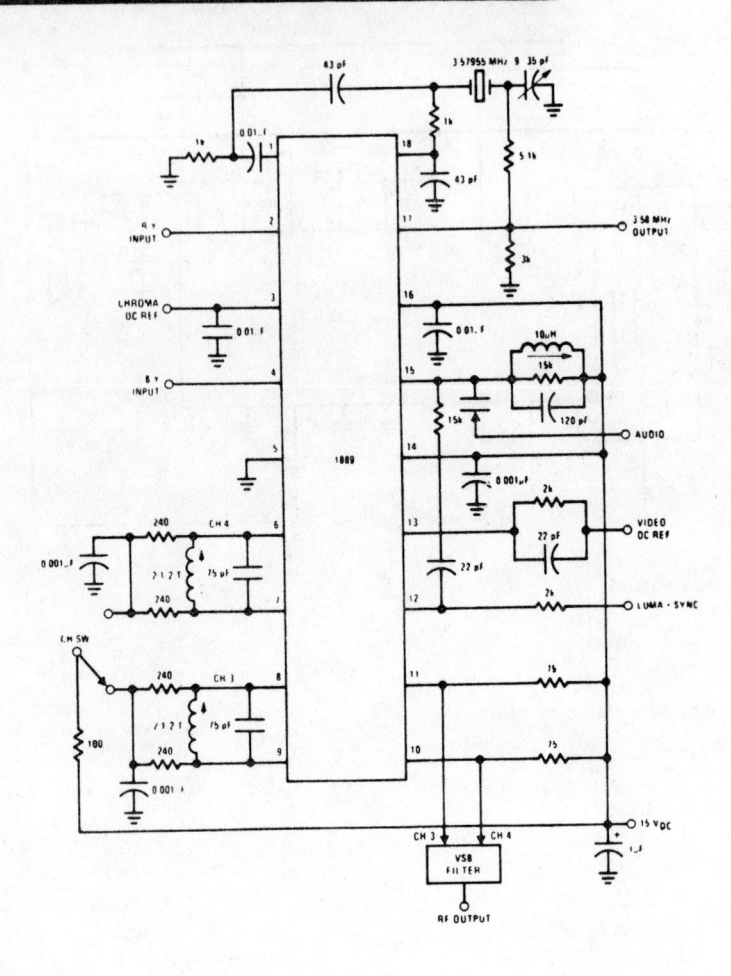

AC test circuit

2111—INTEGRATED FM DETECTOR AND LIMITER

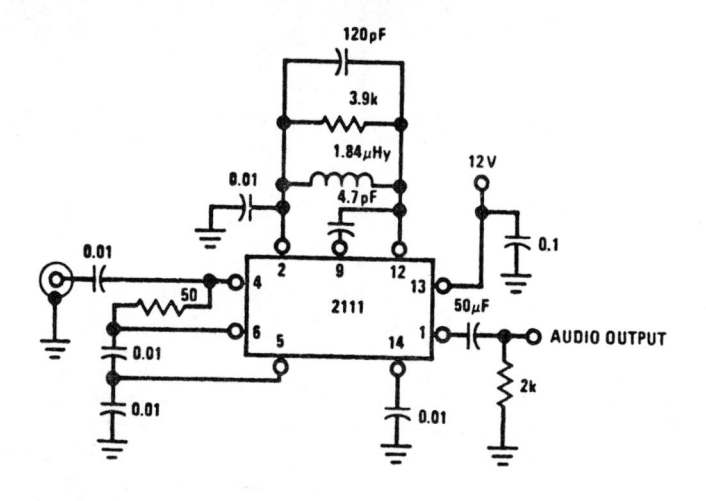

FM IF amplifier

3310 — WIDE BAND AMPLIFIER

FM/IF amplifier

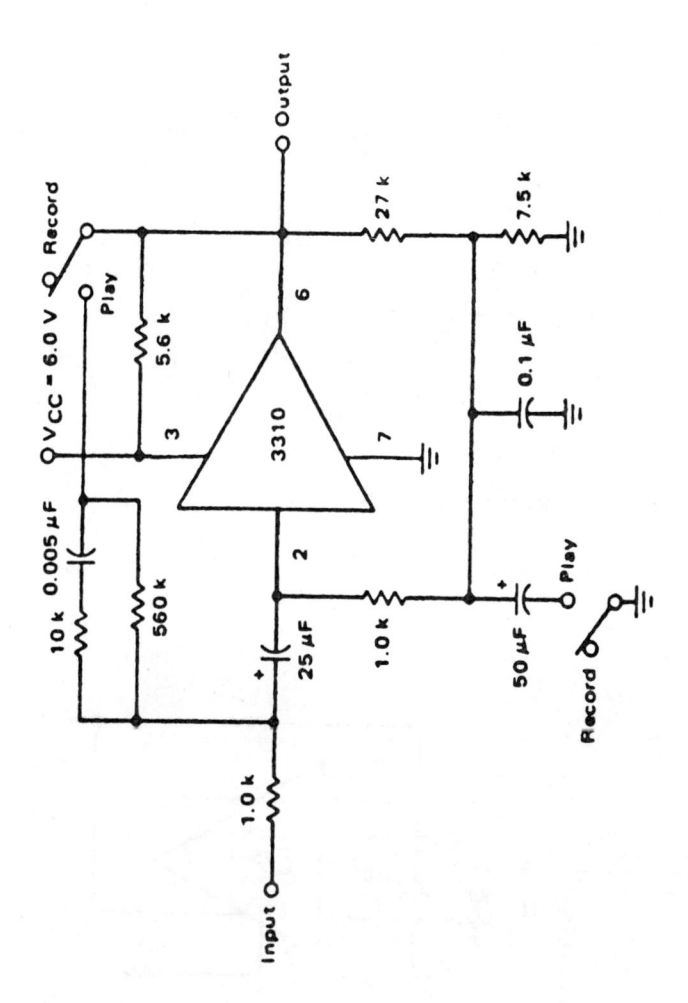

Record/Play preamplifier for cassette and portable tape recorders.

7511 — VIDEO OPERATIONAL AMPLIFIER

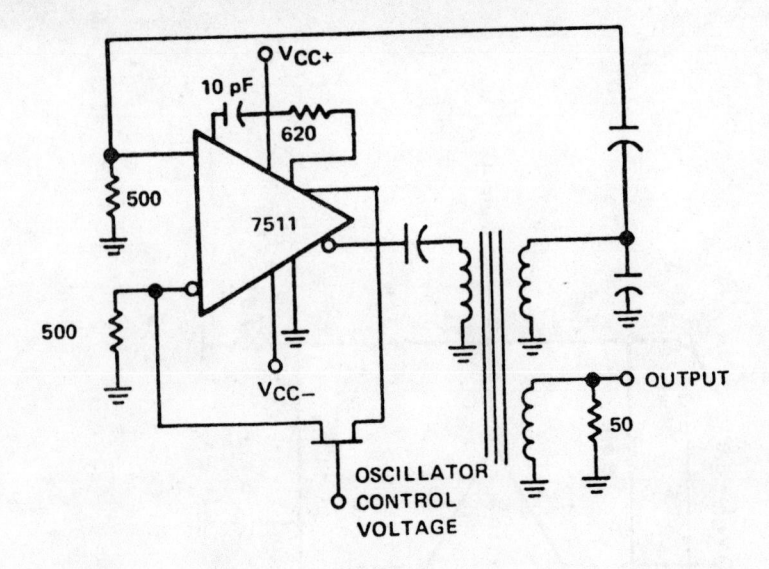

Basic transformer-feedback oscillator

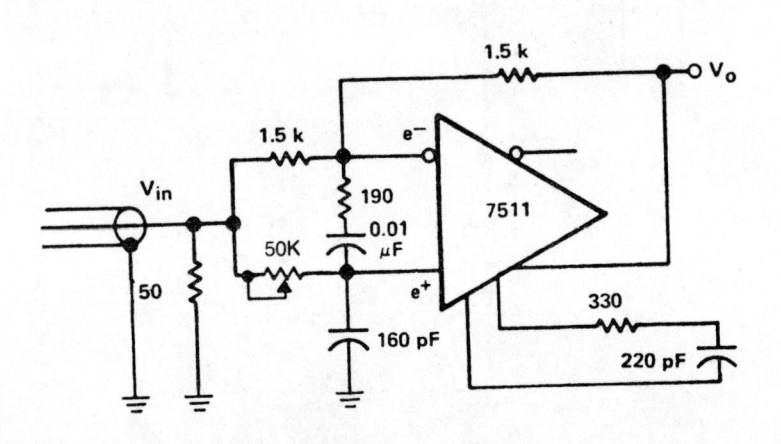

Variable-phase-shift circuit

Part VI

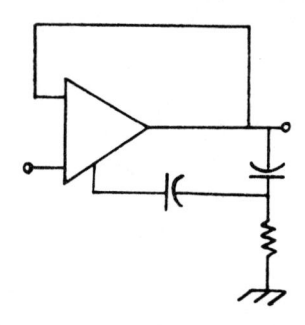

Special Purpose Devices

Sometimes you can't help but wonder if some circuits aren't designed more to amuse the design engineers than for a serious purpose! Whatever the reason, several interesting and novel ICs are now available. They can be used to produce many interesting and unusual projects.

567 — TONE DECODER

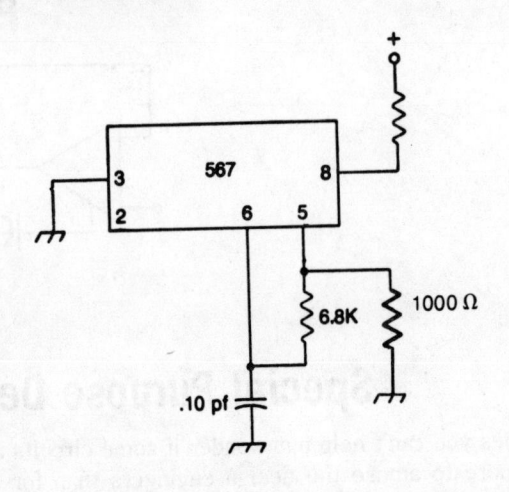

Oscillator with quadrature output

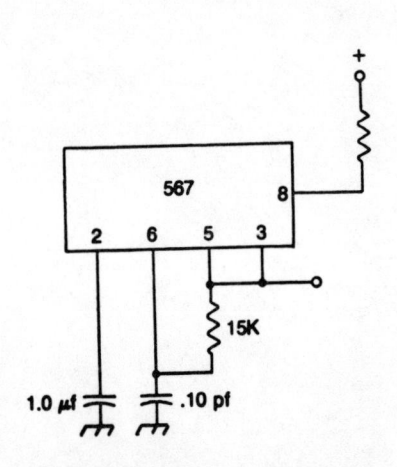

Oscillator with double frequency output

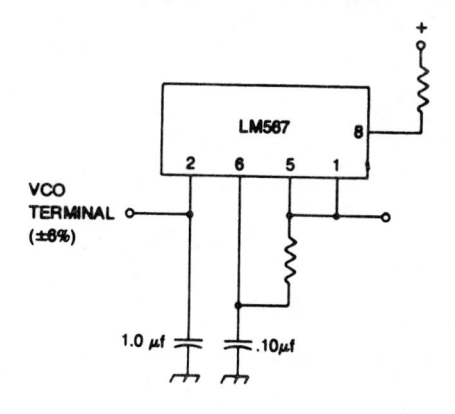

Precision oscillator drive 100 mA loads

1913 — TEMPERATURE TO FREQUENCY CONVERTER

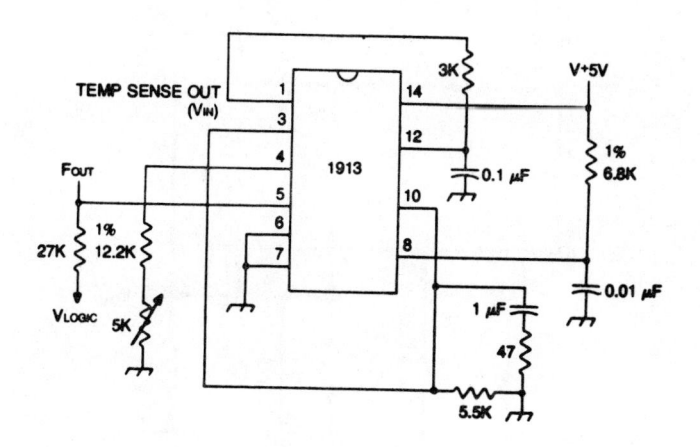

Fahrenheit scale temperature to frequency converter

2688 — RANDOM NOISE GENERATOR

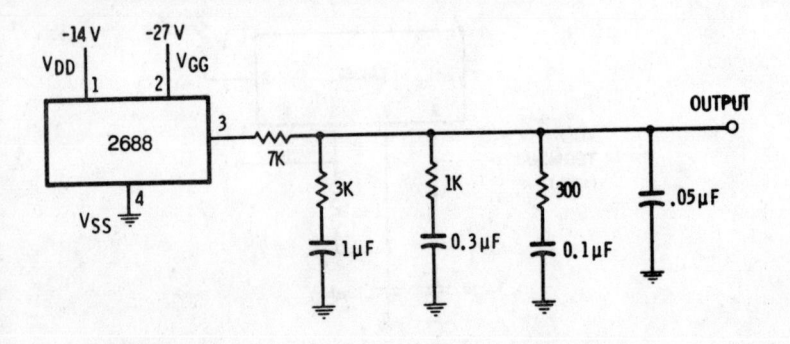

Pink noise generator

2907 — FREQUENCY TO VOLTAGE CONVERTER

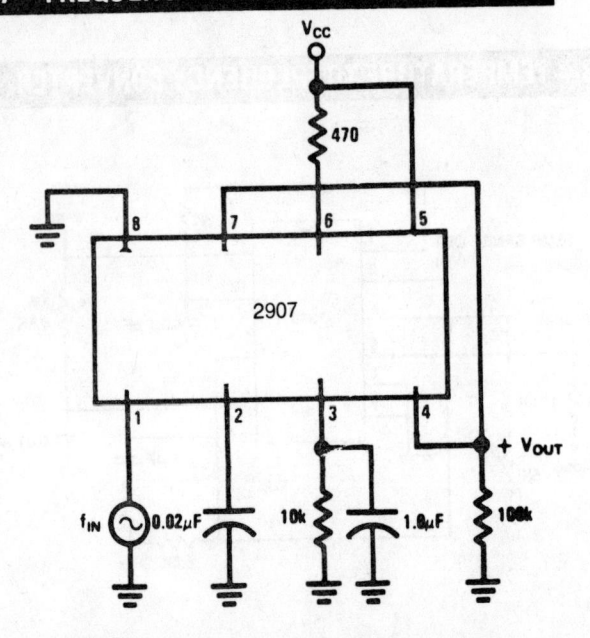

Zener regulated frequency to voltage converter

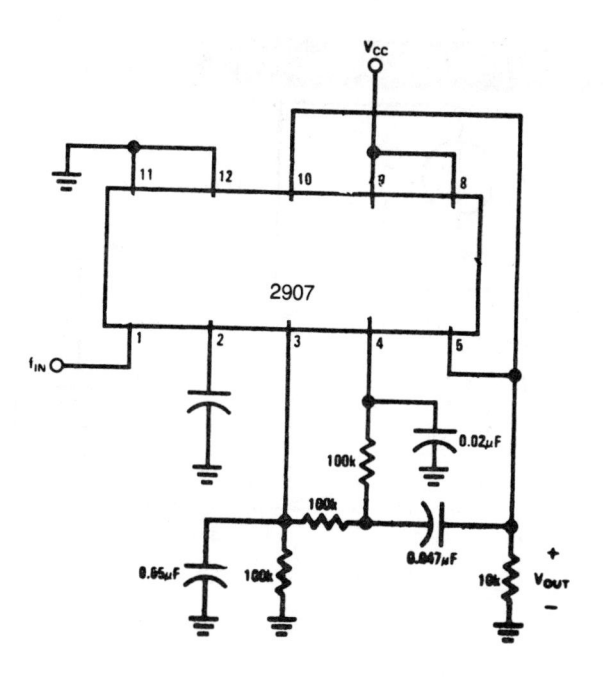

Frequency to voltage converter with 2 pole Butterworth filter to reduce ripple

3340—ELECTRONIC ATTENUATOR

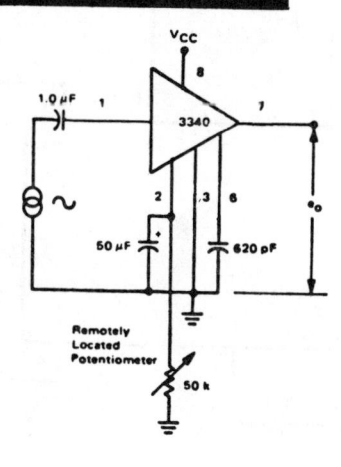

DC "remote" volume control

3909—LED FLASHER/OSCILLATOR

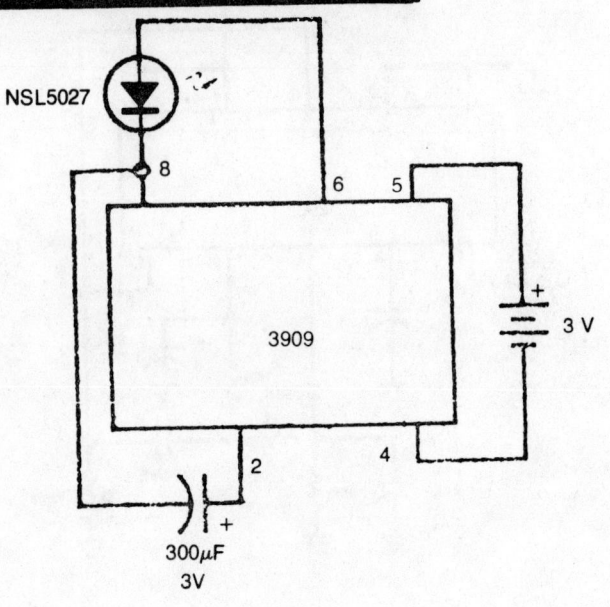

3V flasher

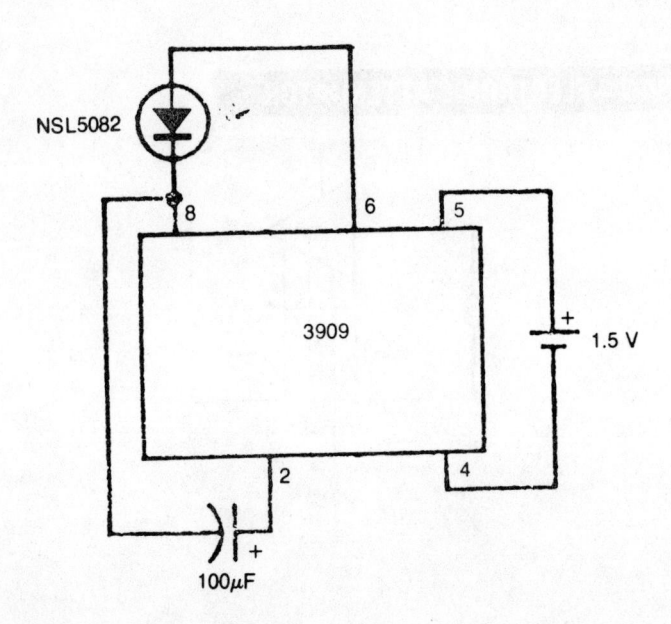

Minimum power at 1.5V

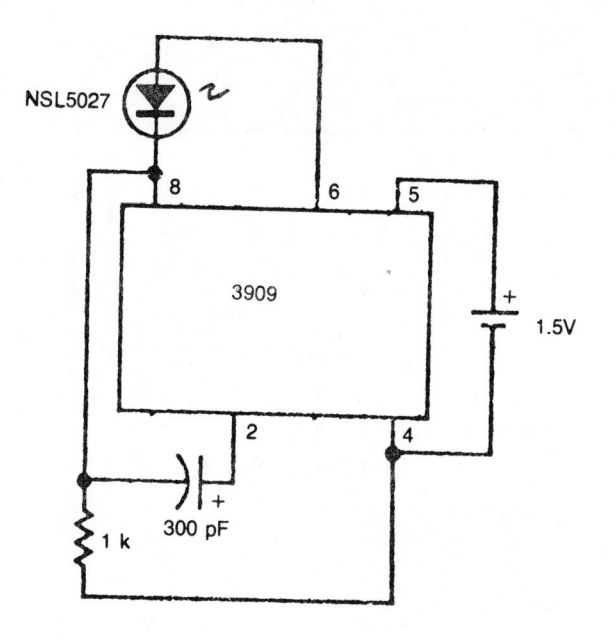

Fast blinker

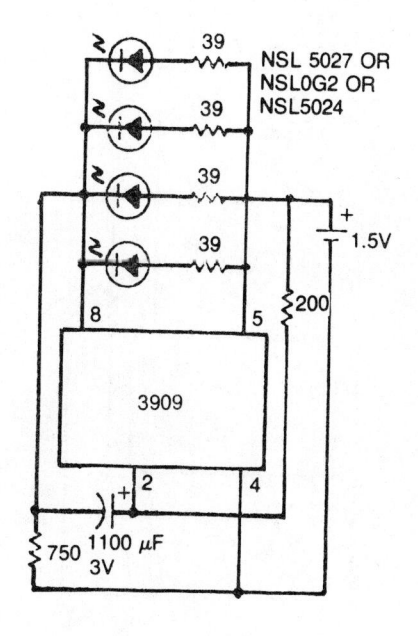

4 parallel LEDs

505

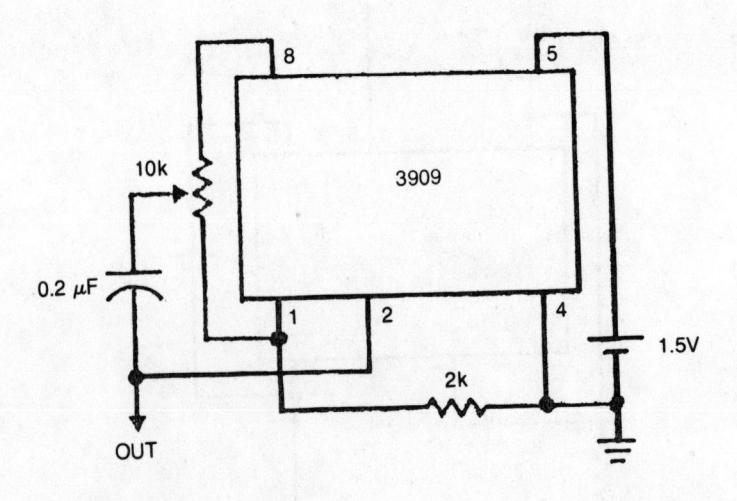

1 kHz square wave oscillator

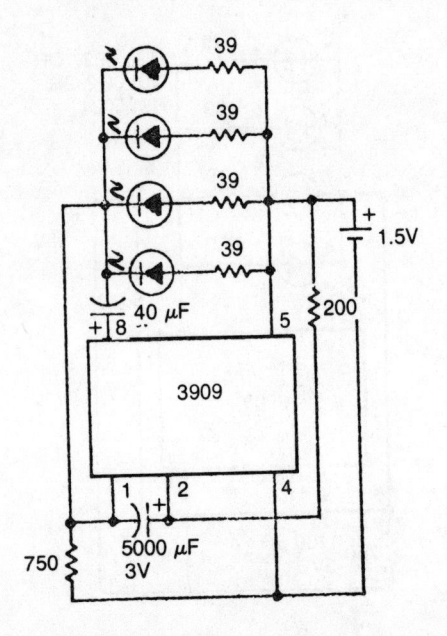

High efficiency parallel circuit

506

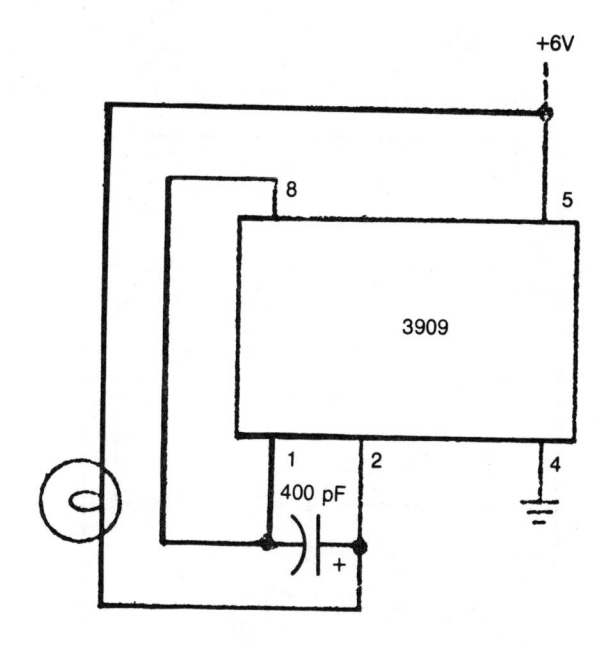

Incandescent bulb flasher

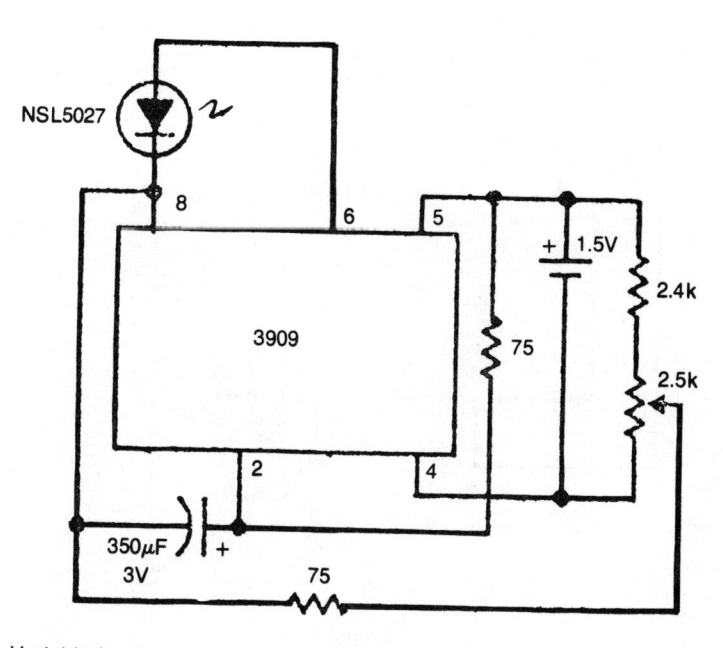

Variable flasher

507

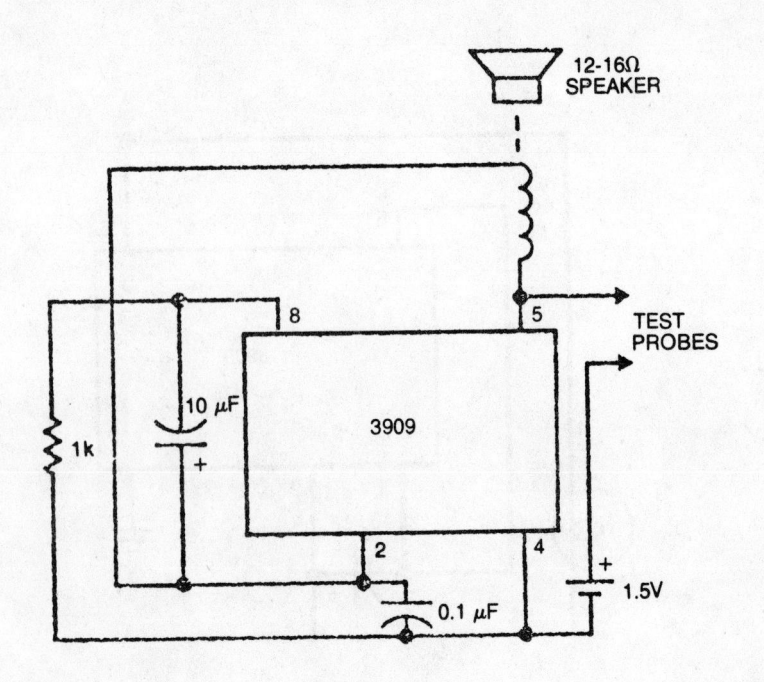

"Buzz box" continuity and coil checker

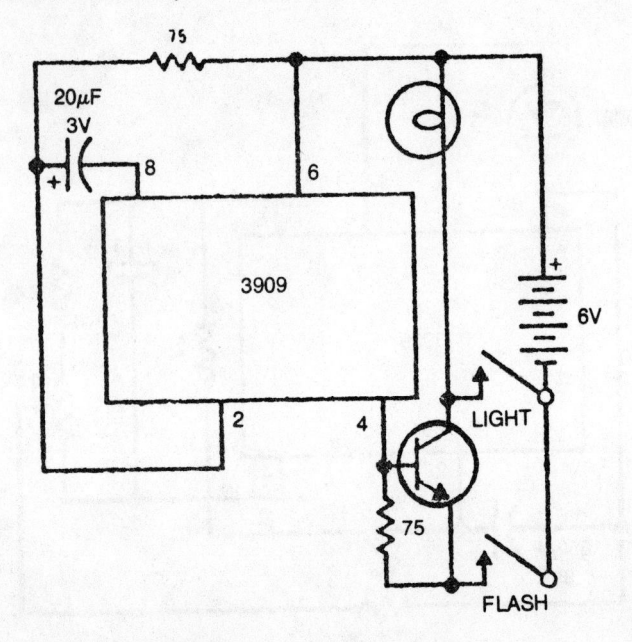

Emergency lantern/flasher

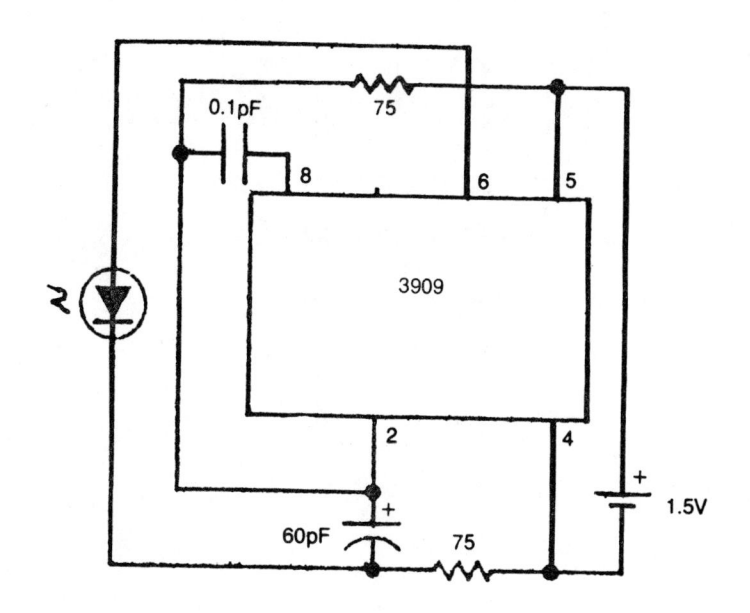

LED booster

3911—INTEGRATED TEMPERATURE CONTROLLER

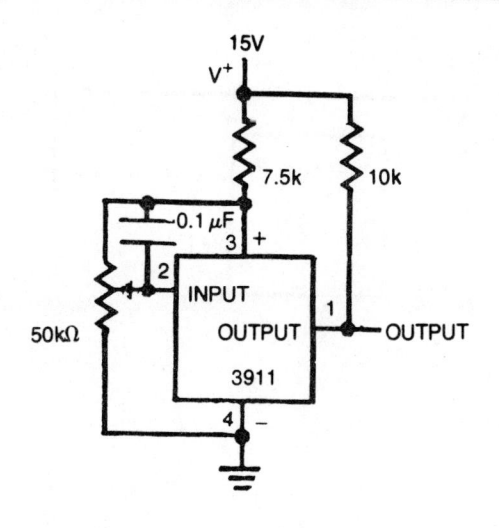

Basic temperature controller

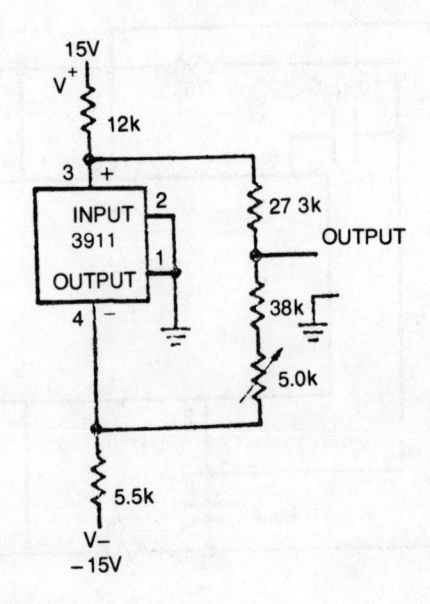

Ground referred centigrade thermometer

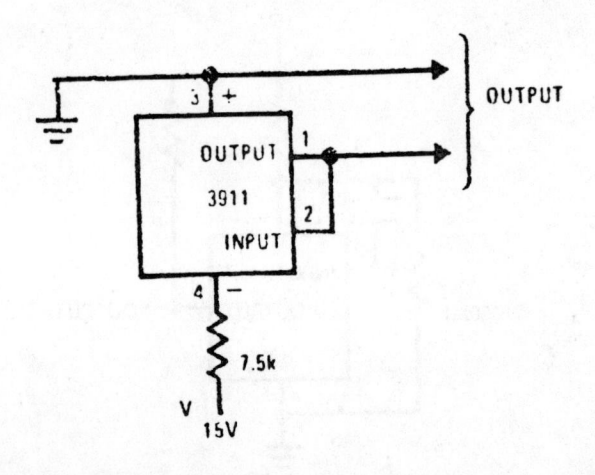

Basic thermometer for negative supply

510

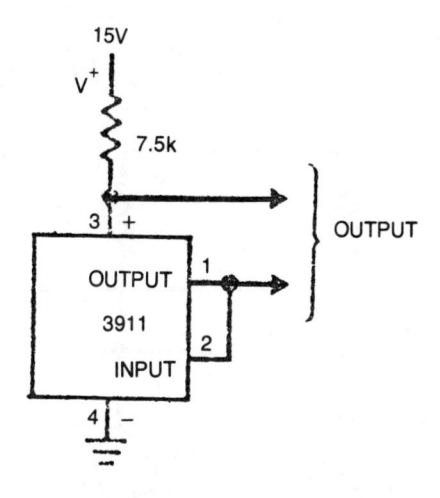

Basic thermometer for positive supply

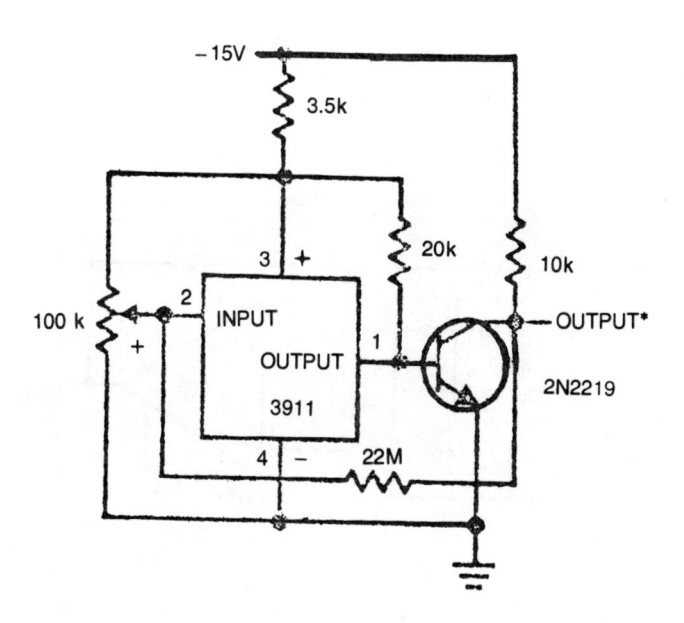

Temperature controller with hysteresis

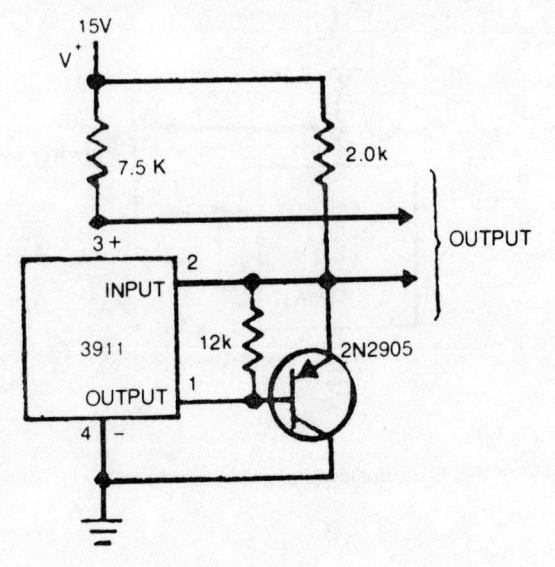

Increasing gain and output drive

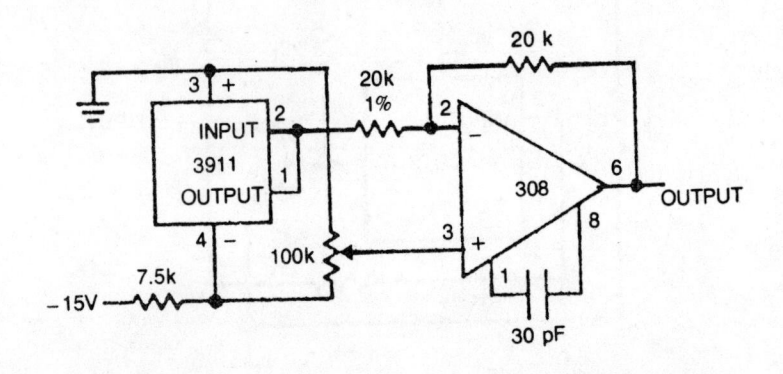

Ground referred centigrade thermometer

3914 — DOT/BAR DISPLAY DRIVER

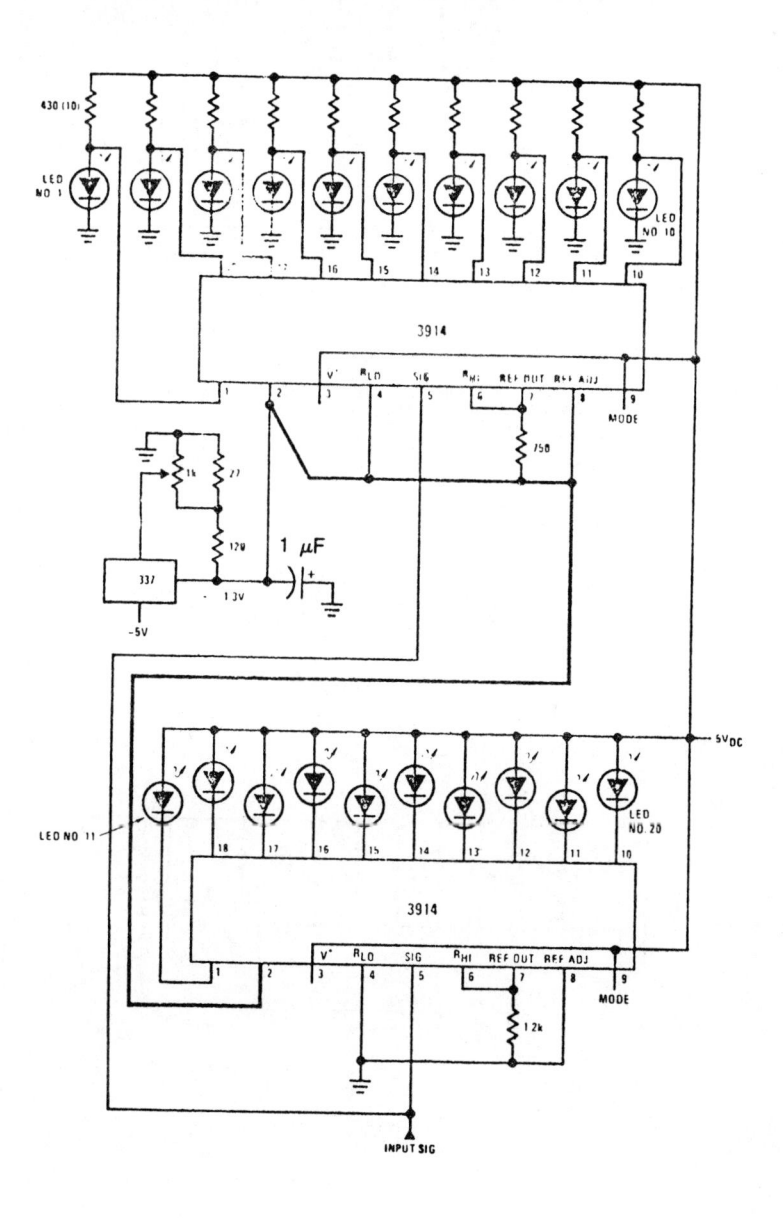

Zero-center meter, 20-segment

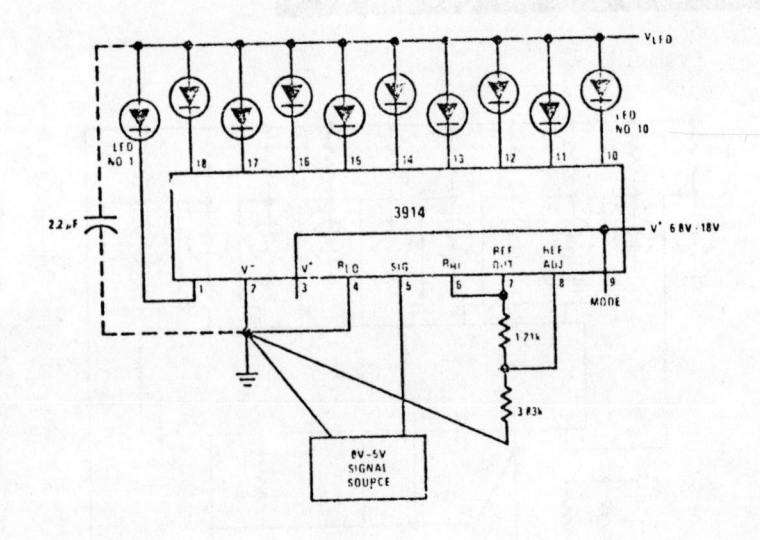

0V to 5V bar graph meter

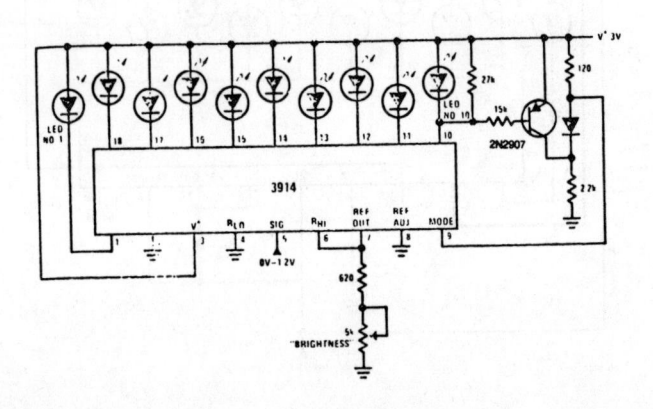

Indicator and alarm, full-scale changes display from dot to bar

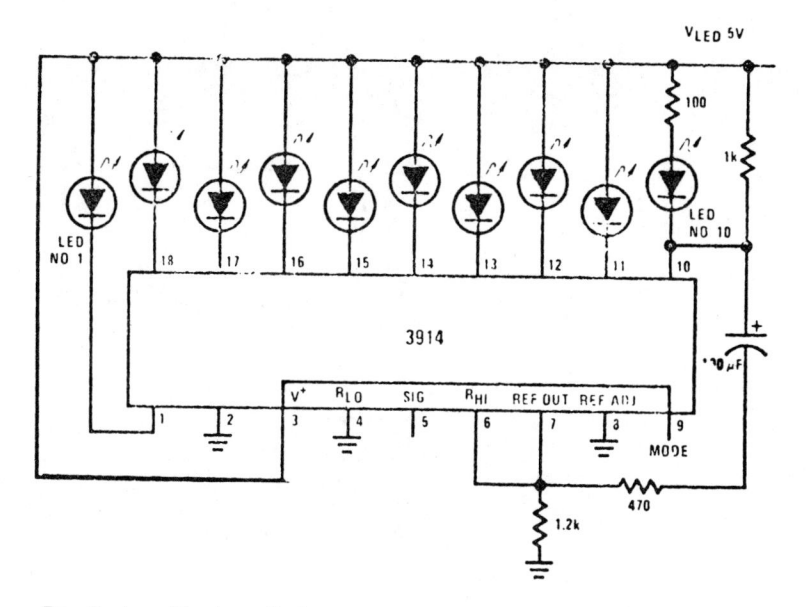

Bar display with alarm flasher

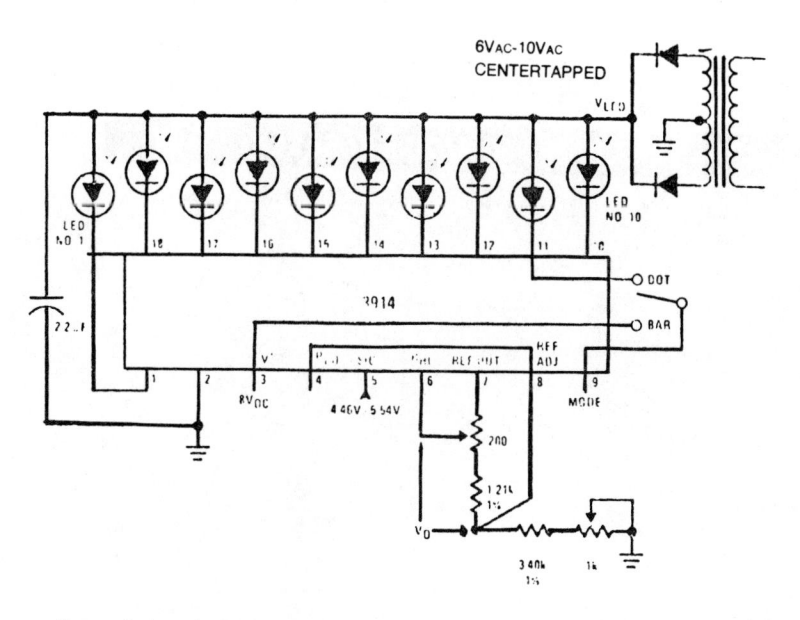

Expanded scale meter, dot or bar

515

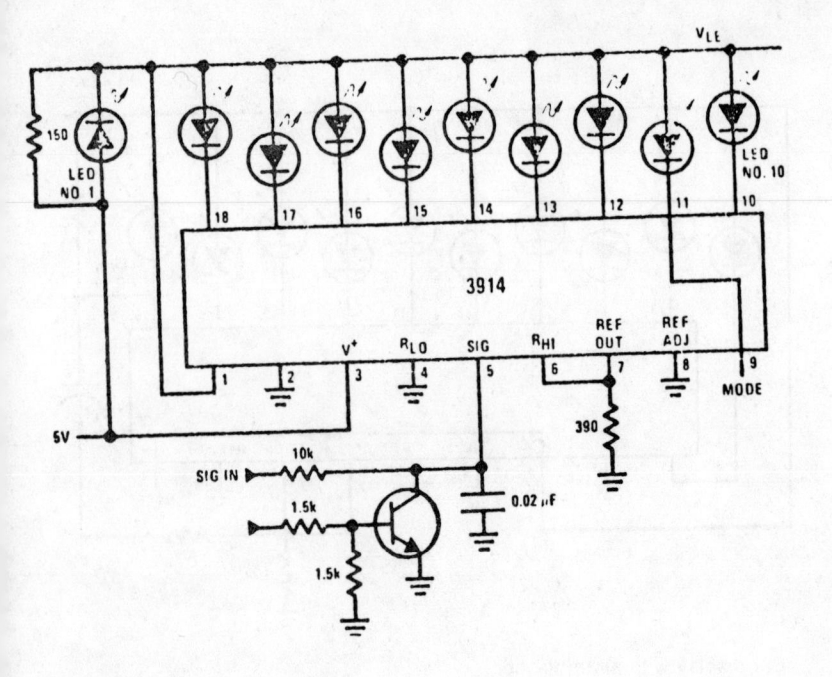

"Exclamation point" display

5369—INTEGRATED 60 HZ TIMEBASE GENERATOR

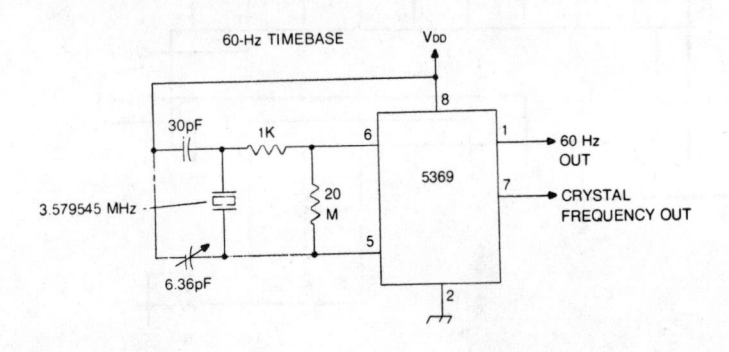

60-Hz timebase

5600—MONOLITHIC TEMPERATURE TRANSDUCERS

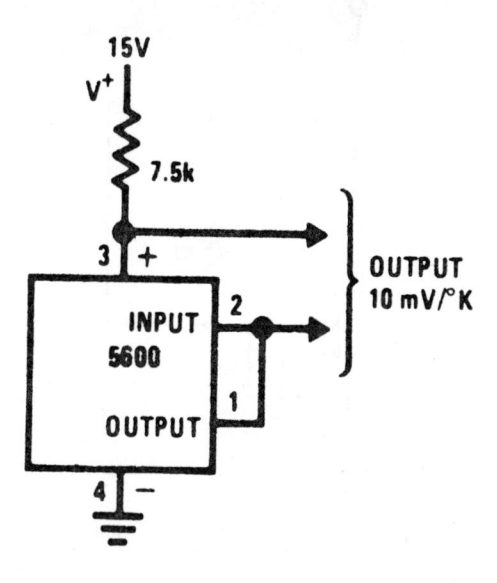

Basic thermometer for positive supply

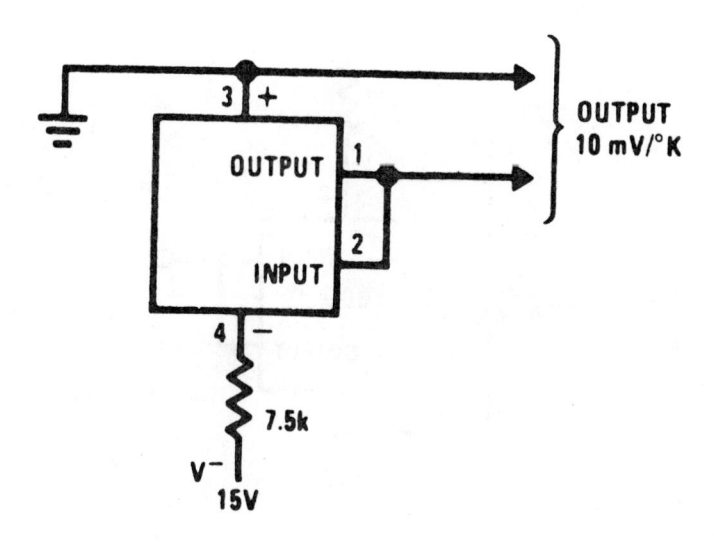

Basic thermometer for negative supply

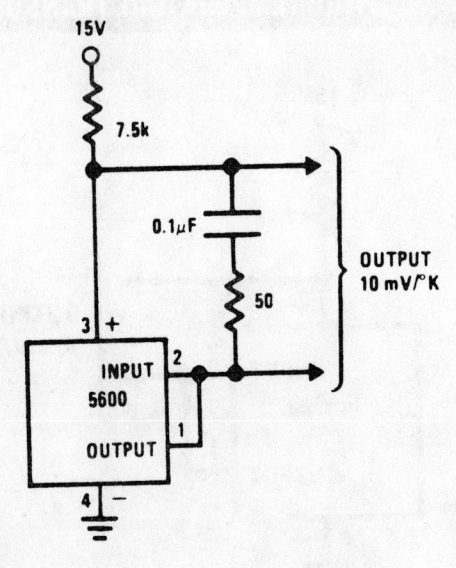

External frequency compensation for greater stability

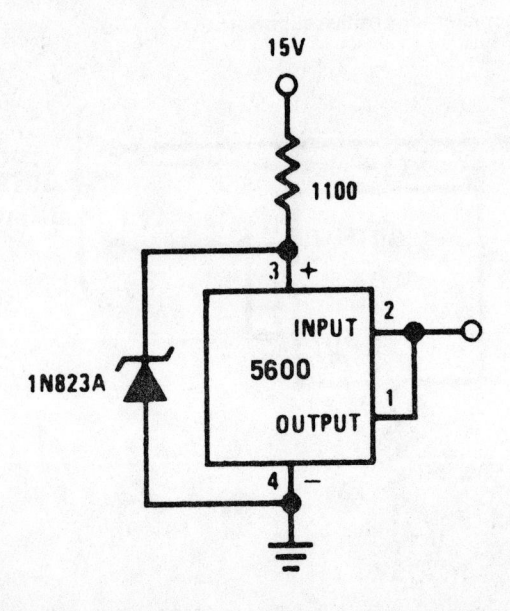

Operating with external zener for lower power dissipation and better reference stability

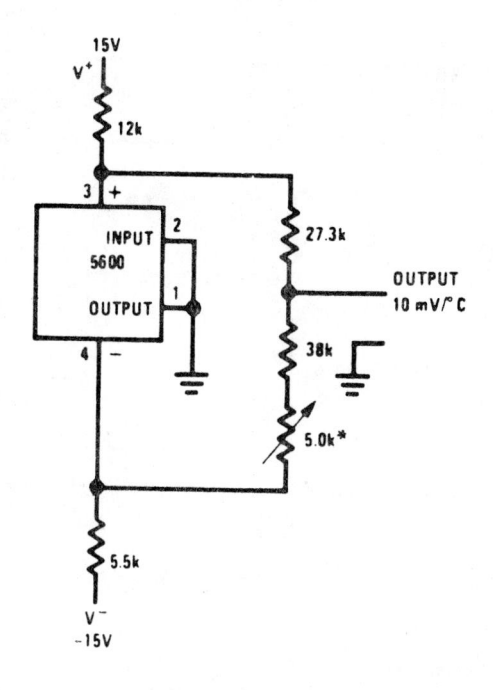

Ground referred centigrade thermometer

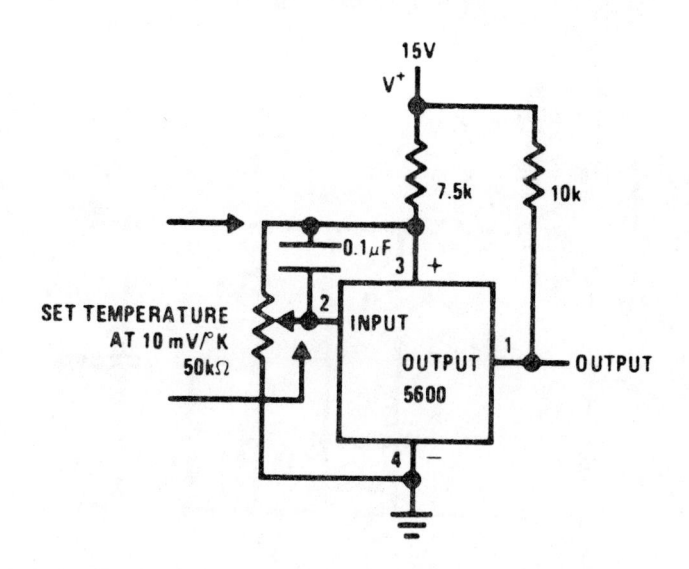

Basic temperature controller

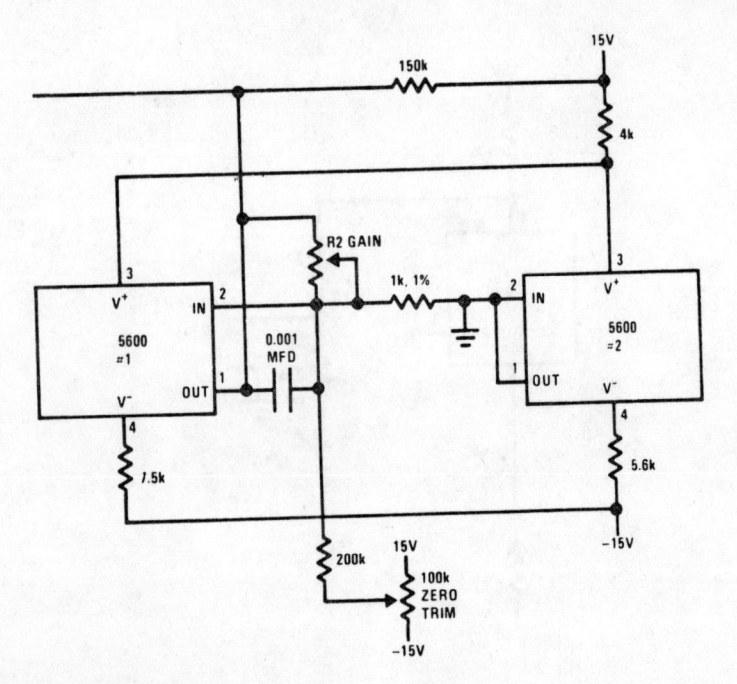

Differential thermometer

5837—DIGITAL NOISE SOURCE INTEGRATED CIRCUIT

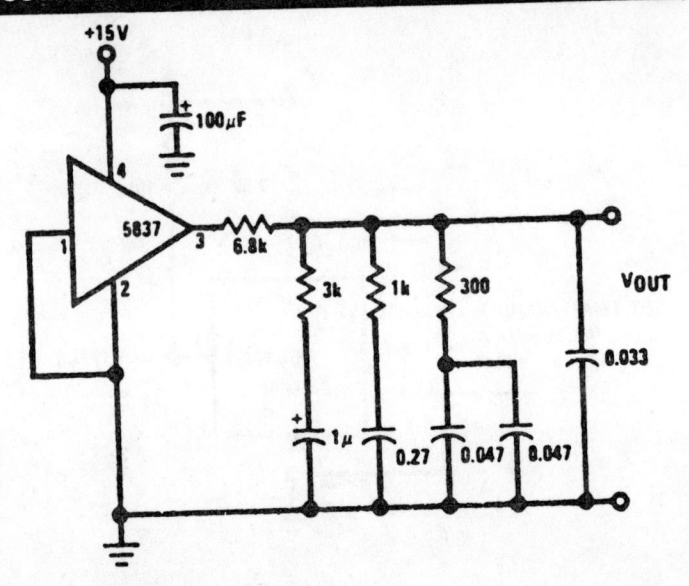

Pink noise generator

8281 — INTEGRATED DIVIDER NETWORK

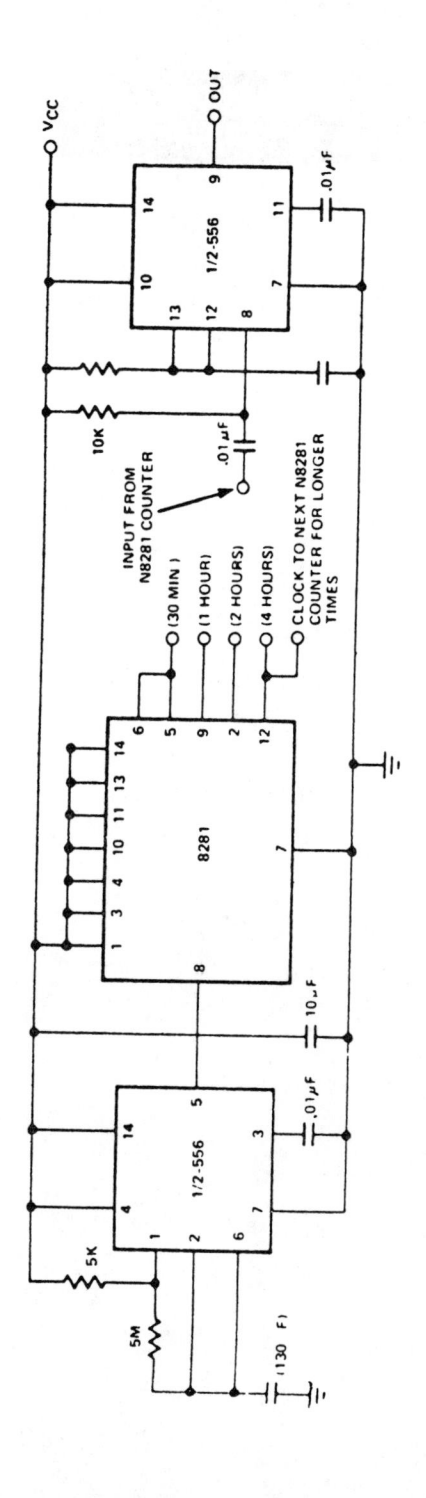

Four hour sequential timer

9400—VOLTAGE TO FREQUENCY /FREQUENCY TO VOLTAGE CONVERTER

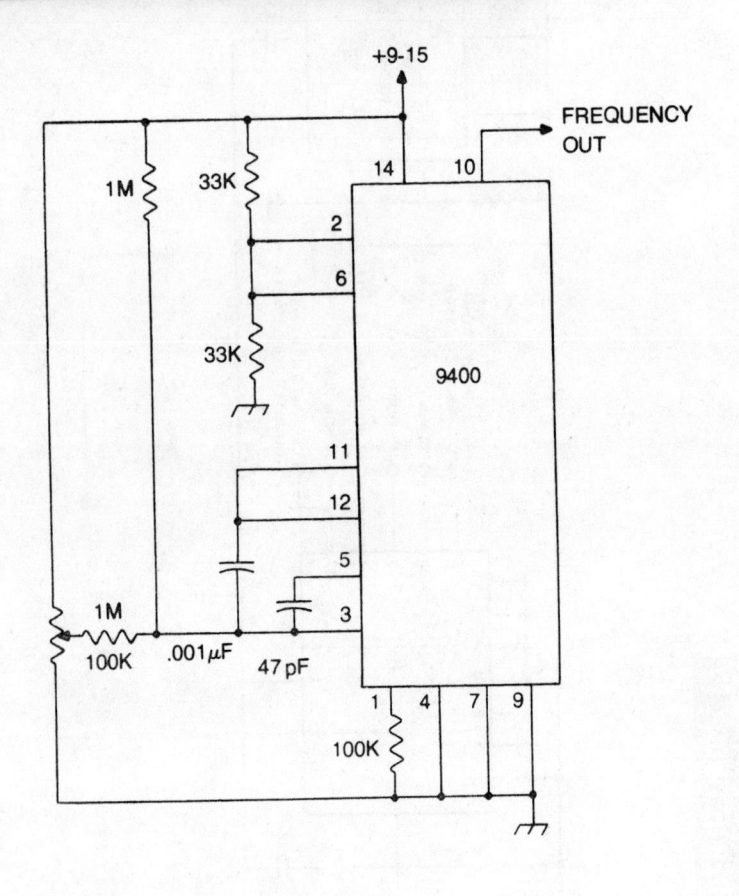

Basic voltage to frequency converter

76477 — COMPLEX SOUND GENERATOR

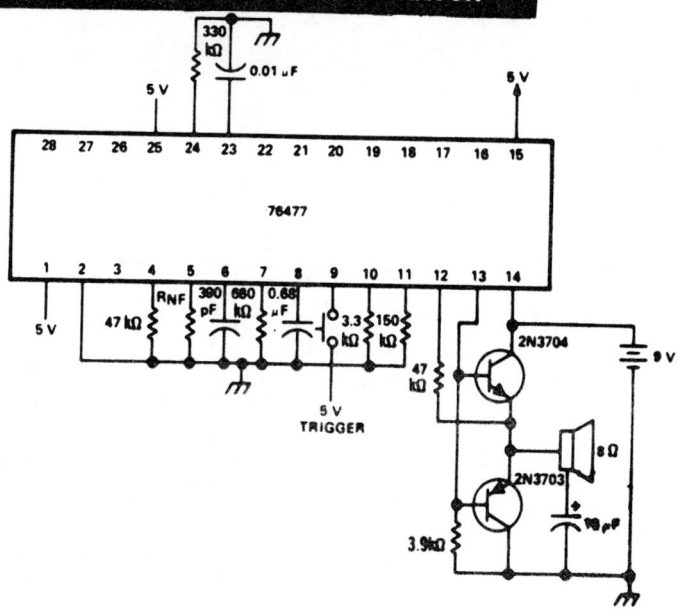

Gunshot/explosion

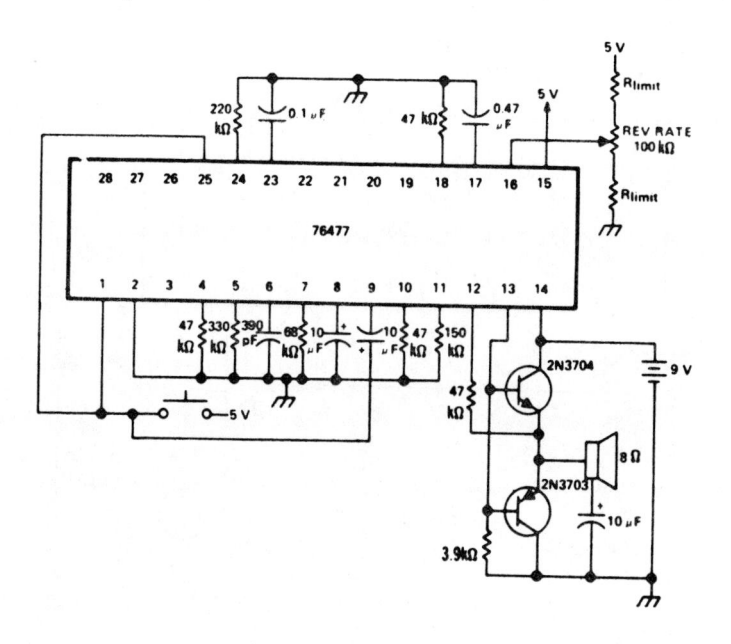

Race car motor/crash

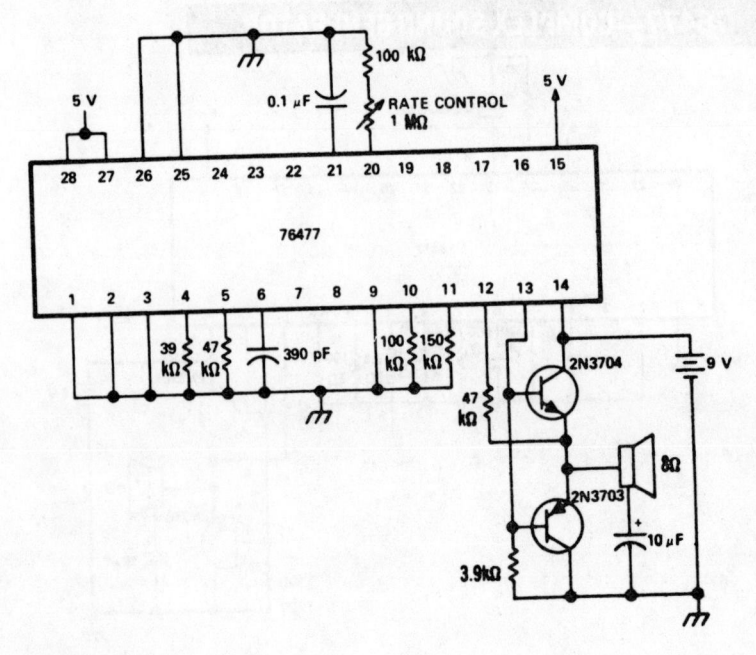

Steam train/prop plane

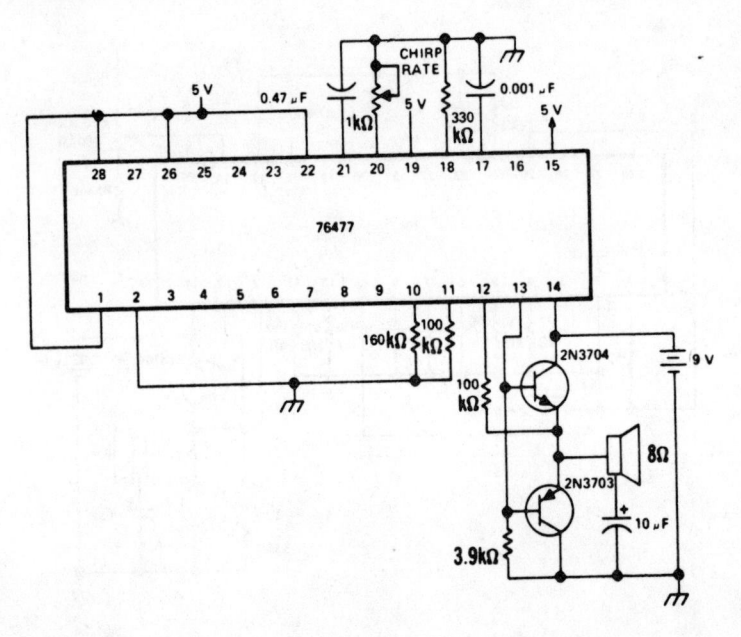

Bird chirp

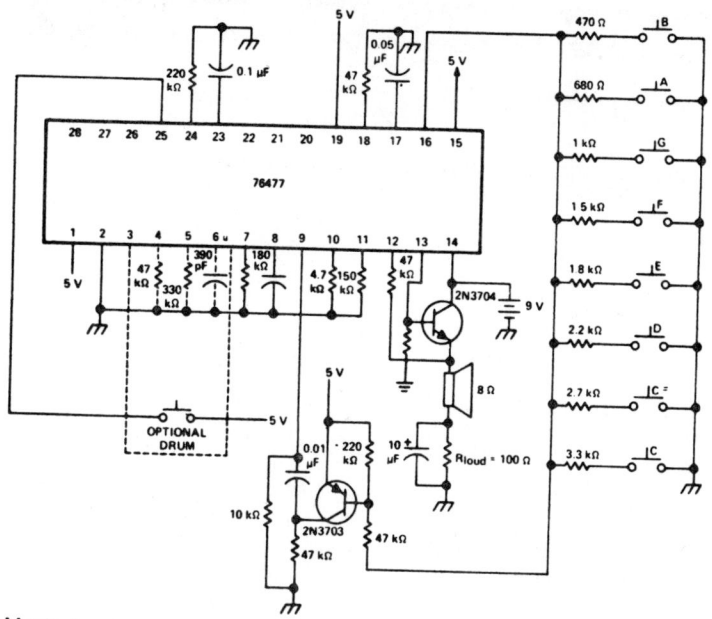

Musical organ

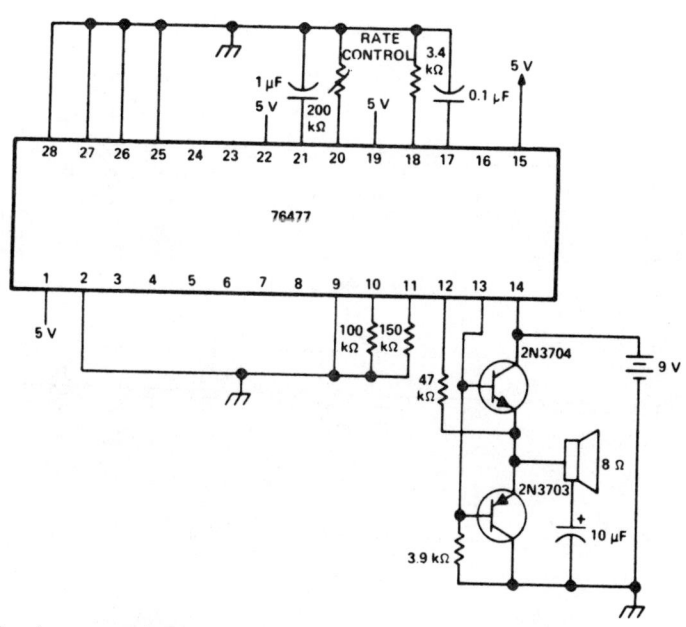

Siren/space war/phasor gun

76488 — SOUND GENERATOR WITH AUDIO AMPLIFIER

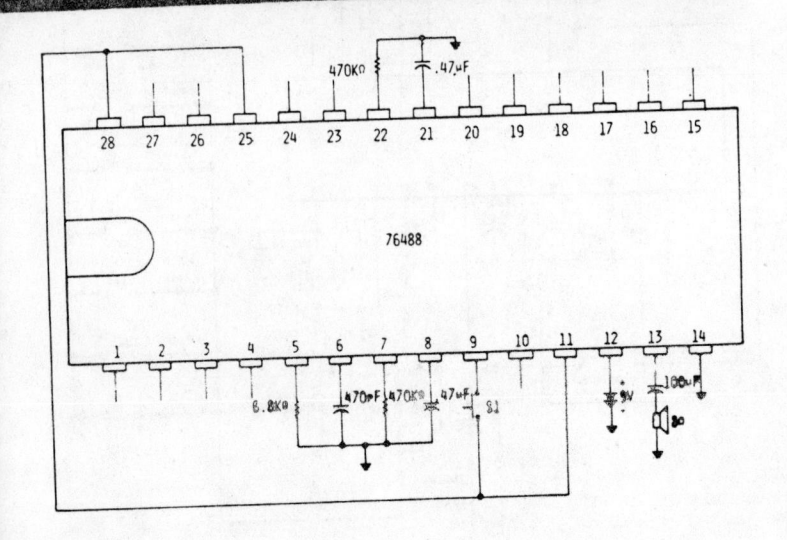

Gunshot

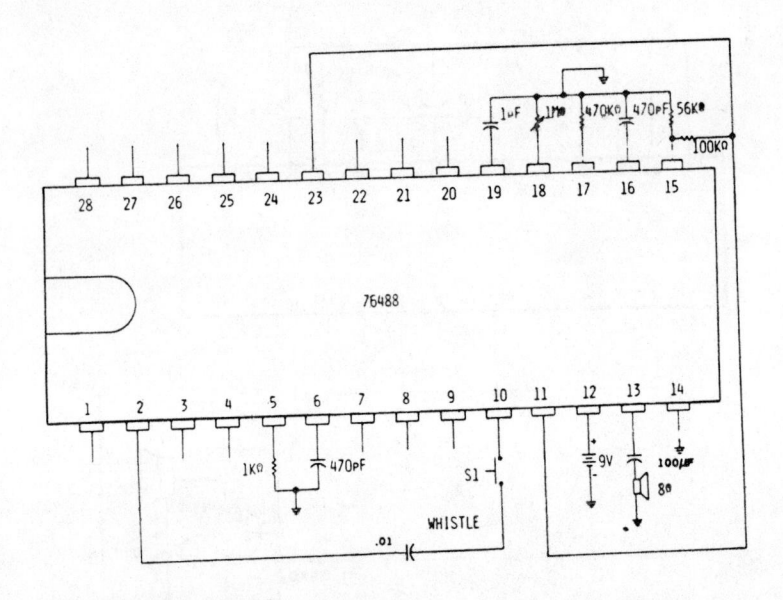

Toy steam engine and whistle

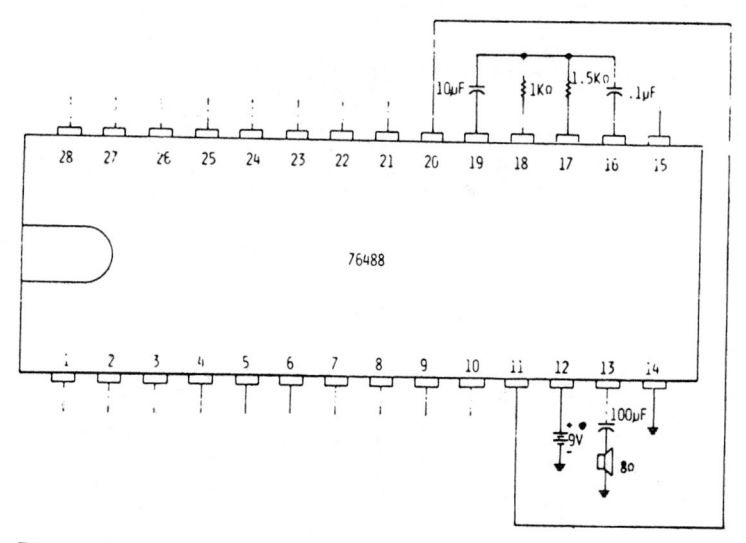

Phasor

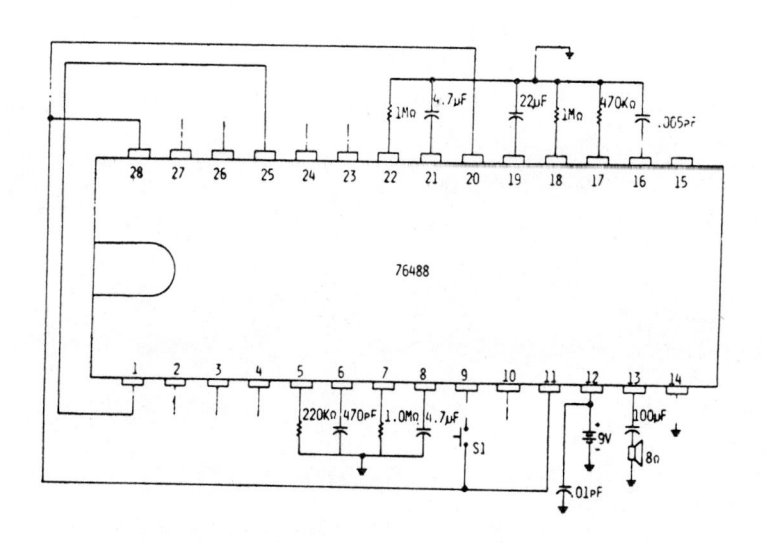

Bomb drop and explosion

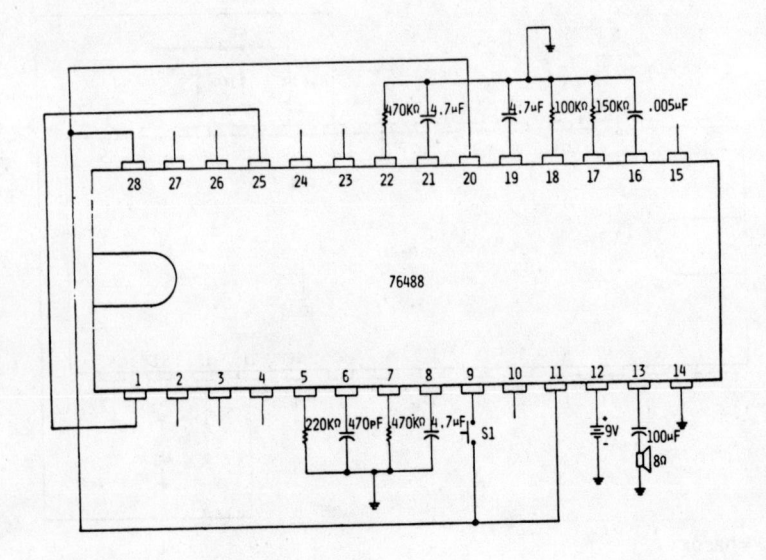

Phasor and explosion

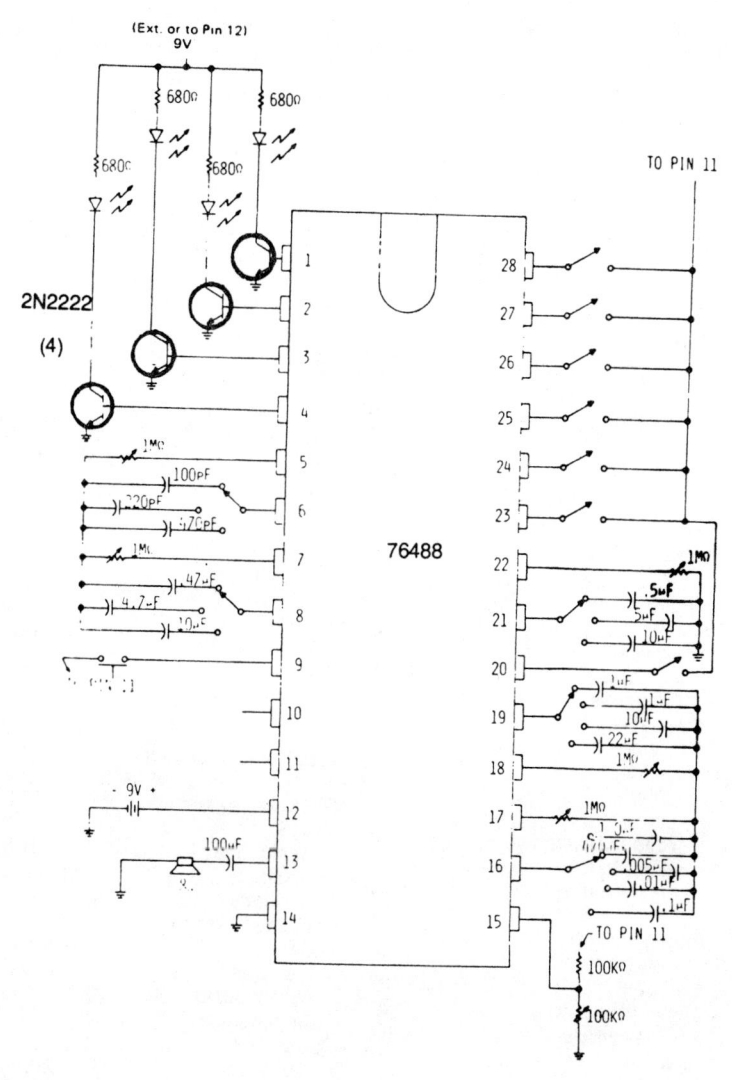

Multiple sound generator

Index

A

Ac millivoltmeter	355
test circuit	494
voltmeter	391
Active bandpass filter	109, 112
Active filter	147
high pass	40, 61
low pass	40, 61, 87, 112
AM modulator	483
Ammeter	344
AM radio	468
Amplifier	131, 178, 179, 180
ac	39, 58
general purpose	76
isolation	42
noninverting	48, 89
photo voltaic-cell	83
Analog multiplier	221
switch	45
AND gate	136, 298, 409, 427
Anti-log generator	31
Astable multivibrator	270, 306
Attenuator	139
Audio amplifier	207, 358
line driver	95
mixer	168
power amp	96

B

Bandpass active filter	300
filter	60, 86, 101, 149
high Q	104
Basic comparator	227
regulator	367
Battery charger	379
Bomb drop and explosion	527
Bridge amp	189
Bridge amplifier	154
Buffer amplifier	293
Butterworth filter	97, 144

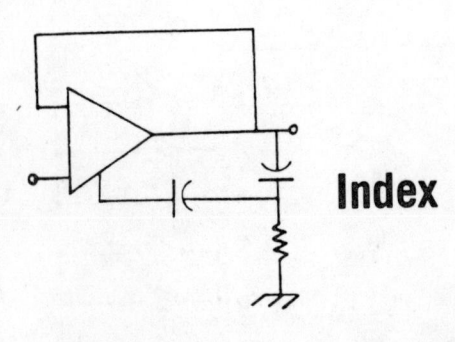

C

Capacitance multiplier	44, 52, 232, 332
Clock pulse generator	412
CMOS	406
Code practice oscillator	348
Comparator	45, 81, 110
Control gate	426
Converter, ac to dc	35
Counters	416, 440
cascaded	447
Crystal oscillator	120, 130, 347, 466
Cube generator	32
Current limiter	375
bilateral	26, 41
monitor	38, 80
sink	16, 49, 255
source	50, 94, 385, 402

D

D/A converter	73, 152
Data selector	423
Detector	64, 66

530

Differential amp	113, 148, 259
amplifier	13, 19, 73, 87
thermometer	520
Driving CMOS	118, 135
TTL	106, 114, 118, 136

E
Enable input	413

F
Fast blinker	505
integrator	29
Filter, active bandpass	84
Fixed regulator	382
Frequency divider	205
doubler	65, 283
Full-wave rectifier	199
Function generator	33, 242, 312

G
Gated amplifier	290
oscillator	449
Gunshot	526
Gunshot/explosion	523

H
Half-wave rectifier	198
High/low limit alarm	323
High current regulator	368
output regulator	390
speed integrator	210
speed inverter	197

I
Instrumentation amp	75, 88, 94, 113
amplifier	33, 43, 54, 59
Integrated voltage comparator	64
Integrator	30, 52, 62
Intercom	163, 178, 188
Inverting amp	76, 148
amplifier	53, 85, 304, 316
Isolation amplifier	25

L
Lamp driver	79, 335
LED booster	509
driver	79, 107
flasher	349
Level detector	46
Limit comparator	93
detector	28
Linear integrated circuits	11
Logic generator	361
Logrithmic amplifier	330

M
Motor drive	156, 160
Multiplier/divider	37
analog	13, 24
Multivibrator	69, 123, 130, 137
one shot	93
Musical organ	525

N
NAND gate	408
Nano-watt amplifier	229
Negative capacitance multiplier	58
regulator	396
Noise eliminator	453
Noninverting amp	111, 147
amplifier	12, 16, 86
power amp	268
Notch filter	149, 255, 217, 341
high Q	100

O
One-shot timer	204
Op amp	265
Operational amplifier	12, 25
OR gate	297, 407, 409, 427
Oscillators	65, 85, 500
Output buffer	20

P
Peak detector	47, 84, 110
Peak voltage detector	102
Phase detector	438
Phase-lock loop	481
Phase shift oscillator	167
Phase shifter	142
Phasor	527
Phasor and explosion	528
Phono amp	166
Photodiode amplifier	213
Piezoelectric transducer amplifier	53
Pink noise generator	502, 520
Positive voltage regulator	235
Power amp	78, 200
Power booster	56
Power op amp	57
Power amplifier	68, 337
Power op amp	161
Preamp	154, 172-175
Preamplifier	23, 116
Precision clamp	34
diode	34
Programmable gate	454
Pulse generator	82, 83, 237, 92, 134, 275, 342
Pulse width detector	77
modulator	29, 91

531

Q

Quadrature oscillator	220, 287

R

Race car motor/crash	523
Ramp generator	125, 195
Random number generator	414
Receiver, broadcast	463
Rectifier, full-wave	15
half wave	35
Reference source	63
Regulator, adjustable	371
variable output	405
Relay driver	46, 64
Root extractor	14
RS latch	430
Rumble filter	181

S

Sample and hold	39, 44, 51, 127
Sawtooth generator	302
Schmitt trigger	263, 295
Scratch filter	183
Simulated inductor	36
Sine wave oscillator	49, 55, 74
Siren	190, 329
Solenoid driver	66
Speech compressor	487
Square root circuit	276
Square wave generator	27
oscillator	92, 119, 134, 180
Staircase generator	128, 197, 299
Stereo amplifier	157, 292
Stereo tape preamplifier	225
Summing amp	109
amplifier	21, 27, 72, 78

T

Temperature controller	509
probe	333
Thermocouple amplifier	211
Thermometer	510, 517
Timer	77, 521
long interval	51
Tone burst generator	326
Triangle generator	74
Triangle wave source	421
Triangular-wave generator	23
TTL/LS	425
Tunable oscillator	420
Tuned circuit	36, 43
Typical afc	480

U

Unity gain follower	245

V

Variable filter	141
flasher	507
VCO	11, 120, 151
Video amp	251, 467
amplifier	489
i-f amplifier	472
Voltage comparator	20, 26
follower	17, 25, 72
reference	100
regulators	362
Voltmeter	260, 344
Vtvm	257

W

Wave shaper	437
Wien bridge oscillator	31, 115, 240
Wien bridge power oscillator	155

Z

Zero crossing detector	309

Edited by Roland Phelps